THE PLANTS *of* MOUNT KINABALU

3. GYMNOSPERMS AND NON-ORCHID MONOCOTYLEDONS

THE PLANTS *of* MOUNT KINABALU

3. GYMNOSPERMS AND NON-ORCHID MONOCOTYLEDONS

John H. Beaman and Reed S. Beaman

with photographs by

Au Kam Wah, J. H. Beaman, R. S. Beaman,
P. C. Boyce, C. L. Chan, J. Dransfield, A. Lamb, S. P. Lim,
A. Phillipps, W. M. Poon, D. A. Simpson and K. M. Wong

Natural History Publications (Borneo)
Kota Kinabalu

in association with

1998

Published by

Natural History Publications (Borneo) Sdn. Bhd.
A913, 9th Floor, Wisma Merdeka
P. O. Box 13908
88846 Kota Kinabalu, Sabah, Malaysia
Tel: 6088-233098; Fax: 6088-240768; e-mail: chewlun@tm.net.my

in association with

The Royal Botanic Gardens, Kew
Richmond, Surrey TW9 3AB
United Kingdom
Tel: 44-181-332-5000; Fax: 44-181-3325197

The Plants of Mount Kinabalu, 3: Gymnosperms and Non-orchid Monocotyledons
by John H. Beaman and Reed S. Beaman

General Editor of Series: Suzy Dickerson

Date of publication: 20 July 1998

Frontispiece: Mount Kinabalu from Melangkap Kapa. (Photo: W. M. Poon)
Dedication page: *Hornstedtia incana*. (Photo: C. L. Chan)

Addresses of authors:

John H. Beaman, The Herbarium, Royal Botanic Gardens, Kew, Richmond, Surrey, TW9 3AB, England and Department of Botany and Plant Pathology, Michigan State University, East Lansing, Michigan 48824, U. S. A.

Reed S. Beaman, Department of Botany, University of Florida Herbarium, 379 Dickinson Hall, Gainesville, Florida 32611, U. S. A.

Cover design by C. L. Chan

Perpustakaan Negara Malaysia Cataloguing-in-Publication Data

Beaman, John H.
    The Plants of Mount Kinabalu, 3: Gymnosperms and non-orchid monocotyledons /
    by John H. Beaman & Reed S. Beaman.
    Includes index
    ISBN 983-812-026-X
    I. Gymnosperms. 2. Monocotyledons. I. Beaman, Reed S.
    II. Title.
    585

Printed in Malaysia

In memory of

Otto Stapf (1857–1933)

and

Lilian Gibbs (1870–1925)

*who published the first definitive works
on the flora of Mount Kinabalu,
and whose contributions a century later
are still essential components to
the knowledge of this flora.*

CONTENTS

LIST OF FIGURES AND TABLES

Fig. 1. Otto Stapf. From the portrait files of the Library of The Royal Botanic Gardens, Kew.

Fig. 2. Lilian Gibbs. From Bol. Soc. Brot. ser. 2, 3: Pl. V (1925).

PREFACE

The present work is the third in the series enumerating the flora of Mount Kinabalu. Previous volumes (Parris *et al*. 1992; Wood *et al*. 1993) have enumerated the pteridophytes and orchids, respectively. A preliminary account also has been published of the gymnosperms (Beaman & Beaman 1993). The general plan of the project was described by Beaman and Regalado (1989). Research on the dicotyledons is continuing, which, when completed, is expected to include more species than are contained in all three preceding works. Because the gymnosperms and non-orchid monocotyledons are distinct and unrelated groups of seed plants they are treated separately here, but together they constitute a convenient size for this volume. This Preface includes sections on methodology and vegetation that relate to both groups.

Mount Kinabalu, in the Malaysian state of Sabah in northern Borneo, is the highest mountain between the Himalayas and New Guinea. It encompasses an area of about 700 km², and ranges in elevation from slightly above sea-level to a height of 4,094 m on Low's Peak. Geologically, Mount Kinabalu is one of the most recent major massifs in the world. It is an adamellite (granitic) pluton that has been uplifted diapirically in the last 1.5 million years, and may still be rising at a rate of about 0.3 cm/year. During the Pleistocene the summit supported an ice cap 5 km² in extent. Deglaciation of the summit occurred about 9,200 years ago. Moraines have been observed as low as about 3,000 m (Jacobson 1978; Myers 1978).

In terms of species per unit area, Mount Kinabalu must have one of the most diverse floras on earth. In an analysis of the global distribution of species diversity in vascular plants Barthlott *et al*. (1996) ranked a northern Borneo centre (principally Mount Kinabalu) as one of the six highest global diversity centres in the world, with more than 5,000 species per 10,000 km². Earlier we suggested that the Kinabalu flora might include more than 4,000 species of vascular plants (Beaman & Beaman 1990). Now we think the total may come to 5,000–6,000 species in an area of about 700 km². Over 200 families and 1,000 genera occur in the flora. This high taxon diversity apparently results from a combination of factors, among which the most important are: 1) great altitudinal and climatic range from tropical rain forests near sea level to freezing alpine conditions at the summit; 2) precipitous topography causing effective geographic and reproductive isolation of species over short distances; 3) geological history of the Malay Archipelago; 4) a diverse geology with many localized edaphic conditions, particularly the serpentine or ultramafic substrates; 5) frequent climatic oscillations influenced by El Niño events; and 6) enviromental instability, resulting from such causes as landslides, droughts, river flooding and glaciation.

1

METHODS

The overall concept of a floristic enumeration of the type provided in this book was outlined by Beaman and Regalado (1989). An integrated system of computer programs used for data editing and printing enumerations (*e.g.*, Parris *et al.* 1992, Wood *et al.* 1993) was written in the dBASE IV© programming language by Reed Beaman. Some of the programs have been converted to Visual FoxPro©. The programs allow access to any aspect of the database through a menu system. Six principal relational data files were employed. Two of these contain data on specimens, including types. Taxonomic, nomenclatural and bibliographic information is linked from other files. Menus facilitate entering and editing specimen and taxon data, globally replacing various expressions, such as changing an author's name or abbreviation, indexing and querying the database, computing a summary of elevation ranges for taxa, numbering taxa, making an index to numbered collections, and printing enumerations of all taxa in the database or of selected families or genera.

The monocot enumeration includes a rather high percentage of incompletely determined specimens and taxa named with varying levels of uncertainty. When the expression 'cf.' separates a generic name and specific epithet, this means that we think the specimens so identified might be a particular species but are uncertain of the identification. This is equivalent to putting a query (?) after the name. When the expression 'aff.' separates a generic name and specific epithet, this means that we are fairly sure that the taxon so identified is not the one named, but is allied to it. We have used the expression 'sp.' (or sp. 1, sp. 2) when we are unable to identify a specimen(s) but believe it (them) to be different from other taxa in the list. The expression 'indet.' has been used when certain material of a particular genus is undetermined, sometimes because of insufficient study but often because the material is inadequate to confirm an identification.

In the Enumerations are several instances where the herbarium symbols 'BM', 'K' and 'L' are repeated for unnumbered (*s.n.*) collections. This indicates that two (or three) unnumbered specimens were collected from a particular locality at different times. This circumstance relates mostly to Clemens collections; they generally did not assign a number to collections for which duplicates were not obtained.

For purposes of brevity, and especially to facilitate effective queries of the database, we have used standardized locality data in the enumerations. A list of most of the the standardized locality names, with their geographical coordinates, has been published by Beaman *et al.* (1996). It may be noted that some locality names are rather different from those on specimens and in the literature, because we have attempted to use spellings established in accordance with the modern Dusun language. Thus, the Dahobang River becomes the Tahubang, the Pinokok River becomes the Tinekek, the Columbon (or Colombon) River is the Kilembun, Bukit Hampuan is Hempuen Hill, and Kamborangah (and numerous other spellings) is standardized as Kemburongoh. For the gymnosperms we have recorded 93 standardized localities and for the non-orchid monocotyledons 247. Collections of the *Projek Etnobotani Kinabalu* (*PEK*) add a level of complexity to the collecting localities, and are discussed in a separate section on *PEK*.

In the specimen citations elevation data from labels stated in feet have been converted to metres, and all elevations are rounded to the nearest 100 m. The stated elevations of 6,000–13,500 ft for Clemens collections from 'Upper Kinabalu' have been ignored. On the field labels for these collections, usually found in BM and sometimes also in other herbaria, a specific locality such as 'Lumu' or 'Kamboranga' is often indicated. When we have been able to associate specific localities with the numbers accompanying 'Upper Kinabalu' specimens, we have used the specific localities even though specimens with printed secondary labels in different herbaria may lack those data.

Elevation data are summarized for all taxa for which these data were available on specimen labels. The elevation range indicated for each taxon is based on the lowest and highest elevations recorded (whether in feet or metres) for specimens of that taxon and rounded to the nearest 100 m. The elevation ranges pertain only to the particular species on Mount Kinabalu, not to any other part of the distribution of the species. In various taxa some specimens have no elevation data while others do. It may be apparent from the locality data that a particular taxon must occur at lower or higher elevations than indicated by the elevation printed. However, we have resisted the temptation to provide elevation ranges for taxa when the specimen labels do not provide this information, with the result that the elevations stated are not necessarily complete for a particular taxon. The collections from the *PEK* project are particularly problematical in this respect, because the *PEK* collectors have not had access to altimeters or detailed maps. They are, however, recording precise locality data that appear on the labels as toponyms. At some future time we anticipate publishing a list of the toponyms with latitude-longitude coordinates and elevations determined by GPS and altimeters.

In order to produce an index of determined specimens, it is necessary to have the taxa numbered. In the treatment of pteridophytes for the Kinabalu inventory, 30 families were recognized and numbered in alphabetical order, with the fern allies first, followed by the ferns. Four gymnosperm families occur in the Kinabalu flora and start with family 31. They are also numbered in alphabetical order, except that the Gnetaceae are placed after the conifers as family 34. Therefore, the numbering of the gymnosperms begins with the Araucariaceae as family 31 and goes on to the other families. Within each family the genera are alphabetically numbered and likewise species within genera. Nominate subspecies or varieties always precede other infraspecific taxa, regardless of alphabetical order. The monocots are now known to include 35 families (including Orchidaceae) on Mount Kinabalu. At the time the orchid treatment (Wood *et al.* 1993) was published, however, only 28 families were known, with the result that, in alphabetical order, the Orchidaceae were numbered family 55. The seven families added all come before Orchidaceae in alphabetical order. Therefore, the orchids have been renumbered as family 62. The families are recognized according to the list of Brummitt (1992).

Taxonomic treatments for various families and genera that are most relevant to those groups are indicated at the beginning of each family and genus. We realise that old treatments such as those in *Das Pflanzenreich* are likely to be of limited utility for the flora of Mount Kinabalu. These treatments are relatively obsolete but sometimes still provide the best taxonomic information available. Type specimens are cited only for taxa described from Mount Kinabalu. The synonymy includes names based on types from Mount Kinabalu. Names of authors of taxa in the Enumerations are abbreviated in accordance with the standardized list of author abbreviations (Brummitt & Powell 1992).

VEGETATION

A summary discussion of the vegetation of Mount Kinabalu was provided in Wood *et al.* (1993). It is repeated here with minor changes to make the present volume as complete as possible concerning information about the gymnosperms and monocotyledons of Mount Kinabalu.

A vegetation map of Mount Kinabalu Park was published by Kitayama (1991) in which 21 vegetation map units were recognized. Kinabalu Park does not include all the area covered by this study, but enough is included to make the map useful for understanding the vegetation of the entire Kinabalu massif. Kitayama indicated that diagnostic canopy tree species could be used to distinguish vegetation zones. He considered these species to be mutually exclusive in distribution and their occurrence correlated with elevation. The upper boundary of lowland rain forest, where the majority of emergent trees (mostly Dipterocarpaceae) disappear from the canopy, is at about 1200 m. The upper limit of lower montane forest is at between 2000 and 2350 m, and that of upper montane forest at between 2800 and 3000 m, the latter particularly marked by the upper limit of *Lithocarpus havilandii*. Above this level is a lower-subalpine coniferous forest dominated by *Dacrycarpus kinabaluensis* and two or three angiosperm species. The upper limit of this forest is at about 3400 m and corresponds to the closed forest line. A fragmented upper-subalpine forest extends above this level to the tree line at about 3700 m. Above this level is a summit-area zone of alpine rock-desert with scattered communities of alpine scrub. Kitayama suggested that the tree line may coincide with the lowest elevation where nocturnal ground frost is frequent. He noted that great variations in dominance type, species composition, and forest structure occur within each zone. These were attributed to altitudinal temperature effects, soil nutrient status in relation to topography (particularly ridge and valley differences) and slope aspect.

Serpentine (ultramafic) vegetation was noted by Kitayama to be strikingly different from that of surrounding forests on non-ultramafic substrates. He indicated that there are at least three altitudinally recognizable subdivisions in the woody serpentine vegetation. These are vegetation with *Tristania* (*Tristaniopsis*) *elliptica* dominance, that with *Leptospermum javanicum-Tristania* dominance, and that with *Leptospermum recurvum-Dacrydium gibbsiae* dominance (Plate 1A). In addition, there are graminoid (or cyperoid) ultramafic communities at Marai Parai and on the summit of Mt. Tembuyuken. Kitayama did not comment on casuarina-dominated forests (*i.e.*, those with *Gymnostoma sumatranum* or *Ceuthostoma terminale*), which we consider to be particularly distinctive markers of lower- and mid-elevation ultramafic vegetation.

In designating the habitats of the gymnosperms and monocotyledons included in this treatment we have used elevation as a primary basis for recognition of vegetation types. "Hill forest" is a much used term for the upper part of the dipterocarp-dominated lowland forest, and we have applied this to habitats between the levels of about 600 and 1400 m. Below 500–600 m is what Kitayama refers to as "substituted vegetation" that we have referred to as lowlands. Our designation of lower montane forest is in the elevation range of about 1200– (1400) 2200 m, and corresponds essentially to that indicated by Kitayama. Two families particularly important as dominants in the lower montane forest are the Fagaceae and Lauraceae.We have used the designation of upper montane forest for all forest vegetation above the lower montane forest. We refer to the area above the continuous upper montane forest as the summit area. Additional considerations of the vegetation of Mount Kinabalu are to be found especially in Stapf (1894), Gibbs (1914) and other references cited by Kitayama (1991, 1992, 1994). It should be noted that many specimens, especially those of the early collectors, lack details on the nature of the vegetation. Specimens made by the *Projek Etnobotani Kinabalu* briefly record in Dusun the vegetation type in which they were collected. A discussion of terms used for vegetation by the Dusun people is provided below in the section on Habitat Data in the Dusun Language (p. 60).

LITERATURE CITED

Barthlott, W., Lauer, W. & Placke, A. (1996). Global distribution of species diversity in vascular plants: towards a world map of phytodiversity. *Erdkunde* 50: 317–327 + Supplement VIII.

Beaman, J. H., & Beaman, R. S. (1990). Diversity and distribution patterns in the flora of Mount Kinabalu. Pp. 147–160 in Baas, P., Kalkman, K. & Geesink, R. eds. *The Plant Diversity of Malesia*. Kluwer Academic Publishers, Dordrecht/ Boston/London.

Beaman, J. H., & Beaman, R. S. (1993). The gymnosperms of Mount Kinabalu. *Contr. Univ. Michigan Herb.* 19: 307–340.

Beaman, J. H., Aman, R. H., Nais, J., Sinit, G. & Biun, A. (1996). Kinabalu place names in Dusun and their meaning. Pp. 489–510 in Wong, K. M., & Phillipps, A., eds. *Kinabalu: Summit of Borneo, a revised and expanded edition.* The Sabah Society in association with Sabah Parks, Kota Kinabalu, Malaysia.

Beaman, J. H., & Regalado, J. C. (1989). Development and management of a microcomputer specimen-oriented database for the flora of Mount Kinabalu. *Taxon* 38: 27–42.

Brummitt, R. K. (1992). *Vascular Plant Families and Genera.* Royal Botanic Gardens, Kew.

Brummitt, R. K. & Powell, C. E. eds. (1992). *Authors of Plant Names.* Royal Botanic Gardens, Kew.

Jacobson, G. (1978). Geology, Ch. 5, pp. 101–110 in Luping, M., Chin Wen & Dingley, E. R. eds. *Kinabalu: Summit of Borneo.* Sabah Society Monograph, Sabah Society, Kota Kinabalu, Sabah, Malaysia.

Kitayama, K. (1991). *Vegetation of Mount Kinabalu Park, Sabah, Malaysia. Map of physiognomically classified vegetation.* East-West Center, Honolulu, Hawaii.

Kitayama, K. (1992). An altitudinal transect study of the vegetation on Mount Kinabalu. *Vegetatio* 102: 149–171.

Kitayama, K. (1994). Biophysical conditions of the montane cloud forests of Mount Kinabalu, Sabah, Malaysia. Pp. 183–197 *in* Hamilton, L. S., Juvik, J. O. & Scatena, F. N. (eds.). *Tropical Montane Cloud Forests.* Springer-Verlag, New York.

Myers, L. C. (1978). Geomorphology, Ch. 4, pp. 91–94 in Luping, M., Chin Wen & Dingley, E. R., eds. *Kinabalu: Summit of Borneo.* Sabah Society Monograph 1978. Sabah Society, Kota Kinabalu, Sabah, Malaysia.

Parris, B. S., Beaman, R. S. & Beaman, J. H. (1992). *The Plants of Mount Kinabalu. 1. Ferns and Fern Allies.* Royal Botanic Gardens, Kew.

Stapf, O. (1894). On the flora of Mount Kinabalu, in North Borneo. *Trans. Linn. Soc. London, Bot.* 4: 69–263, pl. ll–20.

Wood, J. J., Beaman, R. S. & Beaman, J. H. 1993. *The Plants of Mount Kinabalu. 2. Orchids.* Royal Botanic Gardens, Kew.

ACKNOWLEDGEMENTS

We are grateful for the opportunities provided by Datuk Lamri Ali, Director of Sabah Parks, to work in Kinabalu Park. Various members of the Park staff, particularly Jamili Nais, Director of Research and Education, former Park Ecologist Anthea Phillipps, Naturalists Alim Biun and Ansow Gunsalam, and Ranger Gabriel Sinit, were very helpful in our visits. The *Projek Etnobotani Kinabalu (PEK)* has recently added substantially to our knowledge of many Kinabalu species, notably those of lower elevations and from previously little-collected areas. Efforts of the *PEK* staff and collectors in carrying out this project have been very significant to this study. We especially thank former *PEK* coordinator, Luiza Majuakim and the present coordinator, Ludi Apin, for their efforts in making this important material available to us. Additionally, Francis Liew, Deputy Director of Sabah Parks, has greatly facilitated the work with *PEK*, and Gary Martin, ethnobotanist with the People and Plants Initiative, had a major role in developing the philosophy, procedures and funding for *PEK*.

The aesthetic and scientific aspects of this book have been substantially enhanced through participation in its publication by Chan Chew Lun of Kota Kinabalu. His own fine photographs are an important component, and his efforts in rounding up additional colour illustrations and designing the plates are much appreciated, as are the photographs by Au Kam Wah, P. C. Boyce, J. Dransfield, A. Lamb, S. P. Lim, A. Phillipps, W. M. Poon, D. A. Simpson and K. M. Wong.

A Fulbright Fellowship made it possible for the Beaman team to carry out botanical fieldwork in Sabah in 1983–84. Research on the Kinabalu inventory subsequently has been supported by U. S. National Science Foundation grants BSR-8507843, BSR-8822696 and DEB-9400888 to Michigan State University. Grants from the Worldwide Fund for Nature International and the John D. and Catherine T. MacArthur Foundation have supported the *Projek Etnobotani Kinabalu*. John Tan Jiew Hoe of Singapore helped make possible inclusion of the extensive set of colour plates and otherwise assured viability of the book. Most of the herbarium research has been done at Kew, Leiden and the Natural History Museum, London, but Michigan State University continues to provide facilities and our permanent base, for which we are grateful, particularly to Botany Department Chairman G. A. de Zoeten.

Prof. Ghazally Ismail, formerly Dean of the Faculty of Science and Natural Resources of the National University of Malaysia, Sabah Campus (UKMS), now Deputy Vice Chancellor of Universiti Malaysia Sarawak (UNIMAS), facilitated our field work in many ways while we were at UKMS in 1983–84 and at UNIMAS in 1994–96, including making funding available for the project from a MacArthur Foundation grant to UNIMAS. Dr. A. F. Clewell of A. F. Clewell, Inc. has provided continuing support for the participation of Reed Beaman in the project. Christiane Anderson, Robert Johns and Jan Salick substantially improved the manuscript with their editorial suggestions. Teofila Beaman had an important role in data recording and other aspects of the research. Sheila and Peter Collenette helped in resolving difficult locality data and in various other ways, especially with our accommodation at Kew. Jacky Chua Kok Hian of Kota Kinabalu provided assistance in formatting the tables for publication.

We greatly appreciate the facilities and resources for this project provided by the Royal Botanic Gardens, Kew, during a sabbatical leave and earlier visits, and, since 1996, providing the centre for the taxonomic research for this project. The encouragement of Professor G. Ll. Lucas, formerly Keeper of the Herbarium, Royal Botanic Gardens, Kew, was especially helpful at earlier stages of the project, and we have had help with publication from former Assistant Keeper of the Herbarium Keith Ferguson and from Alyson Prior and Suzy Dickerson of the Information Services Department. Valuable assistance with the gymnosperms was provided by Peter Edwards, Aljos Farjon and Robert Johns of the Fern and Gymnosperm Section. The participation from other Kew collaborators, specialists in various monocotyledon groups, including Peter Boyce, Martin Cheek, Jill Cowley, John Dransfield and his former student Rebecca Carrington, Soejatmi Dransfield, Brian Mathew, David Simpson, Paul Wilkin and Jeffrey Wood, adds to the quality of this work, but we take responsibility for any errors. Tom Cope of the Grass Section, although not a collaborator in the sense used here, provided help in determining a number of grasses. Collaborators with the monocotyledons outside Kew include Gary Martin of Marrakesh, Morocco, Robert Faden of the Smithsonian Institution, the late Ben Stone, Jan-Frits Veldkamp of the Rijksherbarium, Leiden, and Rosemary Smith of the Royal Botanic Garden, Edinburgh. Simon Owens, Keeper of the Herbarium, Royal Botanic Gardens, Kew, and Phillip Cribb, Assistant Keeper, have been constantly supportive of the project. Richard Brummitt helped with nomenclatural matters. Anne Morley-Smith, Secretary of the Kew Herbarium, has been a great help in many ways.

We have been fortunate to have been able to examine the relevant specimens in the Natural History Museum, London, where Roy Vickery has been most helpful. Our work in the Rijksherbarium, Leiden, was variously aided by Stans Kofman, and we greatly appreciate the courtesies that have been extended to us by the former Director, the late Prof. Cornelis Kalkman, and the present Director, Prof. Pieter Baas. Wilbert Hetterscheid of the Rijksherbarium and Hortus Botanicus, Leiden, provided information for the genus *Amorphophallus*, Leng Guan Saw of the Forest Research Institute, Malaysia, identified the *Licuala* specimens and George Argent of the Edinburgh Botanical Garden helped with *Musa*. The Singapore Botanic Gardens made their specimens available to us as did Herbarium Bogoriense. We have had the opportunity to study a considerable range of Araceae as a result of loans to Peter Boyce at Kew from a number of herbaria. To the curators of all these institutions we extend a sincere thanks.

GYMNOSPERMS

INTRODUCTION

The enumeration that forms the body of this account includes three families of Coniferales, namely Araucariaceae, Phyllocladaceae and Podocarpaceae and the Gnetaceae (Gnetales). Nine genera are recognized, with the number of species and additional infraspecific taxa indicated in parentheses as follows. Araucariaceae: *Agathis* (3); Phyllocladaceae: *Phyllocladus* (1); Podocarpaceae: *Dacrycarpus* (2); *Dacrydium* (5); *Falcatifolium* (1); *Nageia* (1); *Podocarpus* (6); *Sundacarpus* (1); Gnetaceae: *Gnetum* (6; 2). Thus, 26 species of gymnosperms and two additional varieties are currently recognized in the Kinabalu flora. The occurrence of 28 gymnosperm taxa in the limited area of Mount Kinabalu would seem to indicate that they represent a rather diverse component in terms of world-wide gymnosperm diversity, but we lack data for comparable areas.

HISTORICAL ASPECTS

The first account of Kinabalu gymnosperms was by Stapf (1894). He listed five species, only four of which were fully determined. Two of the species Stapf treated had names in approximate agreement with those in the present account. Twenty years later Stapf (1914) reported seven gymnosperms, based on the materials in his previous study with the addition of the 1910 collections by Lilian Gibbs.

Subsequent to the reports by Stapf, Meijer (1965a) mentioned nine taxa, including the genera *Gnetum*, *Phyllocladus*, *Podocarpus*, *Dacrydium*, and *Agathis*. Most of these were also illustrated by him (Meijer 1965b). Corner (1978) listed 13 species of conifers, but one of these (*Podocarpus glaucus*) is not presently considered to occur in Borneo and another (*Podocarpus polystachyus*) is a strictly coastal species. Cockburn (1980), although not attempting to list the gymnosperms of Kinabalu, included 12 species that occur on the mountain and illustrated most of them. Beaman and Beaman (1993) provided diagnostic herbarium-specimen photographs of all gymnosperm species known from Mount Kinabalu and citations of the specimens available at that time.

The two most important publications relating to the taxonomy of the gymnosperms of Mount Kinabalu are the Flora Malesiana treatments of *Gnetum* (Markgraf 1951) and of the Coniferales (de Laubenfels 1988). The present account follows those treatments in generic and species concepts, but some varieties of *Gnetum* proposed by Markgraf are not recognized. Not all gymnosperm specialists accept the generic concepts of de Laubenfels for segregates of *Podocarpus* and *Dacrydium*, nor his species distinctions in *Agathis*, but in the course of this study they appeared to be useful.

GYMNOSPERM COLLECTIONS

Up to 1931 only about 26 gymnosperm specimens had been collected on the mountain, four by H. Low in 1851 or 1858, five by G. D. Haviland in 1892, 11 by L. S. Gibbs in 1910, and 6 by M. S. Clemens in 1915. In 1931–33 J. and M. S. Clemens collected intensively on the south and west sides of the mountain, probably obtaining over 9000 numbered collections plus many others that they did not number. Our records include 106 of their gymnosperm collections from that period. Other significant collections of gymnosperms have been obtained by the Royal Society expeditions in 1961 (Chew, Corner & Stainton) and 1964 (Chew & Corner) (32 numbers), by W. Meijer (23 numbers) in 1959–1963, by S. Kokawa and M. Hotta (21 numbers) in January–February, 1969, by P. F. Cockburn and Aban Gibot (17 numbers) in 1976, by D. J. de Laubenfels (the only gymnosperm specialist who has collected on the mountain; 43 numbers) in August, 1978, and January, 1979, and by the Beaman team (29 numbers) in 1983–84 and 1992. Even now the Kinabalu gymnosperms could hardly be said to be well collected, but the present study is based on about 462 collections and 716 specimens obtained by 80 collectors or collecting teams.

Subsequent to the preliminary account on the gymnosperms of Mount Kinabalu (Beaman & Beaman 1993) the only new collections we have seen are those made by *Projek Etnobotani Kinabalu* (*PEK*) collectors. A detailed discussion of that project is included with the treatment of the monocotyledons. Because most of the gymnosperms occur at mid to high elevations, not many of them have accumulated from *PEK* activities. Sixteen *PEK* collections are listed, of which 11 belong to *Gnetum*, a gymnosperm genus that is common at lower elevations where most of the *PEK* collectors have worked.

One objective of the project has been to examine critically all specimens upon which the Enumeration is based. For the gymnosperms, these include specimens located in six herbaria: BM, K, L, MSC, SING and SNP (Sabah Parks Herbarium on Mount Kinabalu). From an examination of specimen citations it can be seen that certain species have been very much collected and others very little. In the case of the rather conspicuous gymnosperms, the number of specimens cited for a species probably is a fairly good indicator of the frequency of that taxon on the mountain.

GYMNOSPERM ECOLOGY

Gymnosperms are scattered in the Kinabalu vegetation from the lowest to highest elevations. Just below the summit area they are ecological dominants. Kitayama (1991) recognized "Tropical lower-subalpine coniferous forest" and "Tropical upper-subalpine forest," the dominant species of which include the gymnosperms *Phyllocladus hypophyllus* and *Dacrycarpus kinabaluensis*, along

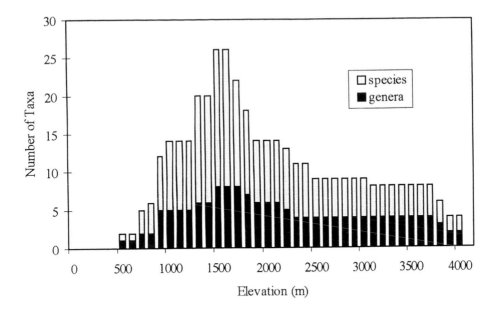

Fig. 3. Elevational distribution of gymnosperm genera and species on Mount Kinabalu.

with the angiosperm trees (or shrubs) *Leptospermum recurvum, Eugenia (Syzygium) kinabaluensis, Rhododendron buxifolium* and *Schima brevifolia*. In addition to these two important gymnosperms in the summit flora, Fig. 3 shows that two other species (*Podocarpus brevifolius* and *Dacrydium gibbsiae*) also extend to high elevations.

In spite of the relative importance of gymnosperms in the high-elevation vegetation of Kinabalu, maximum gymnosperm generic and specific diversity occurs at about 1500 m (Fig. 3), a circumstance that also pertains with the pteridophytes (Parris *et al.* 1992) and orchids (Wood *et al.* 1993). Corner (1978) noted that *Agathis* was a common big tree from the Mesilau River across the Pinosuk Plateau. Much of that area is at an elevation of about 1500 m. The natural vegetation of the Pinosuk Plateau, formerly part of Kinabalu Park, has been virtually destroyed by various development projects in the past 15 years. Ironically, some spindly pines (*Pinus*) have been planted on the Pinosuk Plateau golf course, probably in the same places where magnificent *Agathis* trees once stood. *Agathis* also was common in the lower elevation hill forest on the ultramafic Hempuen Hill, but after that area was degazetted from the Park in 1984, it became a virtual free-for-all for various land grabs, logging and unsuccessful slash and burn agriculture. Most of that forest spared from such activities was consumed by forest fires in 1990.

The *Gnetum* species are all lianas, often of large size and extending high into tree crowns in the lowlands, hill forest and lower montane forest. *Gnetum latifolium* var. *minus* extends up to 1800 m at Tinekek Falls and *G. leptostachyum* var. *abbreviatum* to the same elevation in the Kilembun basin and at Marai Parai, if Clemens elevation data are to be believed. The relatively rare species *G. gnemonoides*, with huge seeds, and *G. klossii*, with an unusual seed-surface texture, are known only from lowlands and hill forest, respectively. *Gnetum cuspidatum* and *G. neglectum* are not readily distinguishable from the relatively common *G. leptostachyum* var. *leptostachyum*, which occurs in hill forest and lower montane forest at a number of localities.

One of the most abundantly collected of all species in the Kinabalu flora is *Phyllocladus hypophyllus*, now known from 62 collections and 93 specimens. As noted by Corner (1996), it is an Australasian element and extends from Borneo to Sulawesi, Maluku, the Philippines and New Guinea, with other species known from New Zealand and Tasmania. We have seen large trees of it on the Pinosuk Plateau, but at the highest elevations it is reduced to a low shrub or treelet. *Dacrycarpus imbricatus* var. *patulus* and *D. kinabaluensis* are related taxa that were originally treated as the same species. However, they seem readily distinguishable and occur exclusively at different elevations. *Dacrycarpus imbricatus* var. *patulus* is a species of lower montane forest, hardly occurring above 1700–1800 m. *Dacrycarpus kinabaluensis*, on the other hand, is a strictly high-elevation species of upper montane forest and summit scrub rarely occurring below 2500 m. De Laubenfels (1988) noted that on Kinabalu *D. imbricatus* does not occur above 2000 m; indeed the high-elevation figure we have is based on the specimen *Clemens 33618* from the Penataran basin and may be erroneous. *Dacrycarpus imbricatus* is widely distributed in Borneo, Southeast Asia, the Malay Peninsula, Sumatra, Mindanao, Sulawesi, New Guinea, Vanuatu and Fiji, whereas *D. kinabaluensis* is endemic to Mount Kinabalu.

One of the most distinctive trees in the upper montane forest is *Dacrydium gibbsiae*, conspicuous along the summit trail from about 2500 m to 3000 m. The species is a very graceful plant of 'Christmas tree' aspect with drooping lower branches. It appears to be restricted to ultramafic substrates, such as those in the area of mossy forest on ridges between Layang-layang and Paka-paka Cave. Found just below this area, in a zone from about 2000 to 2500 m, is *Dacrydium xanthandrum*, readily distinguishable because its leaves are slightly longer than those of *D. gibbsiae*, and extend perpendicular to the stem rather than curving stiffly upward as in *D. gibbsiae*.

Along the road from Park Headquarters to the Power Station, in the elevation range of 1500 to 1800 m, several gymnosperms can be seen, including *Agathis kinabaluensis*, *A. lenticula*, *Dacrycarpus imbricatus* var. *patulus*, *Dacrydium gracilis*, *Falcatifolium falciforme*, *Podocarpus laubenfelsii* and *Sundacarpus amara*. In this area Cockburn performed a useful service by collecting lower, middle and upper branches from trees of several different species. Through these collections one can better understand the extent of foliage variation on individual trees, and thus better interpret specimens for which the collector did not indicate the part of the tree from which they were obtained.

Eight of the Kinabalu gymnosperms (*i.e.*, about one-third of the taxa) are found predominantly or entirely on ultramafic substrates. These are: *Agathis dammara, Gnetum leptostachyum* var. *abbreviatum, Dacrydium gibbsiae, D. pectinatum, Podocarpus brevifolius, P. confertus, P. gibbsii* and *P. globulus*. Ultramafic substrates are extremely important in the occurrence of orchids on Kinabalu (Wood *et al.* 1993), but proportionally fewer of the gymnosperms seem restricted to ultramafic conditions.

The following four taxa are thought to be endemic to Mount Kinabalu: *Gnetum leptostachyum* var. *abbreviatum, Dacrycarpus kinabaluensis, Podocarpus brevifolius* and *P. gibbsii*. Seven others, *i.e.*, *Agathis kinabaluensis, A. lenticula, Dacrydium gibbsiae, D. gracilis, Podocarpus confertus, P. globulus* and *P. laubenfelsii,* are endemic to Borneo. The other species are largely centred in the Malesian region, particularly Borneo, with *Phyllocladus hypophyllus* being the conspicuous Australasian element.

TAXONOMIC PROBLEMS

The enumeration of *Gnetum* presented here may not be very sound. Markgraf's treatment in Flora Malesiana was based largely on his earlier (1930) monograph. At the time of that work hardly any specimens of *Gnetum* from Kinabalu had been collected. Relatively few of the specimens we examined have his annotations. *Gnetum* is particularly difficult because some of the best characters are in the seeds and the branches on which they are borne. Since the plants are dioecious, a rather high percentage of specimens lack these reproductive structures. The characters of leaf venation that Markgraf used are difficult to recognize and probably do not hold up very well. A modern taxonomic account of the genus is much needed.

Agathis is also a difficult genus, and the two higher-elevation species recognized in this account (*A. kinabaluensis, A. lenticula*) may be local ecological variants. On Kinabalu, however, they seem reasonably distinguishable and are therefore retained. An alternative view of *Agathis* is provided by Whitmore (1980), who included material here recognized in both *A. kinabaluensis* and *A. lenticula* as *A. dammara* (Lamb.) Rich. & A. Rich. subsp. *dammara*. According to Whitmore's treatment, *A. dammara* occurs in Borneo only in some mountainous areas. Whitmore emphasized characters of the male cones and considered leaf shape and size to be highly and continuously variable.

Cockburn (1980) indicated that the splitting of *Podocarpus* into several smaller genera was difficult to condone when even the differences between *Dacrydium* and *Podocarpus* are so fine as to make these genera unworthy of separation. He regarded *Podocarpus imbricatus* (*i.e.*, *Dacrycarpus imbricatus*) as a large, widespread species with a number of ecotypes associated with altitude and exposure, but in which the proposed varieties grade imperceptibly into one another. We find, however, that *Dacrycarpus kinabaluensis* and *D. imbricatus* are almost always readily distinguishable. Page (1988) noted that the diagnoses of many of these genera often depend heavily on vegetative aspects of the plants, in a group whose reproductive aspects offer a rather limited array of features. He further indicated that most of the small genera in the Podocarpaceae are fairly natural groupings with good geographic and probably evolutionary cohesion, supported in many cases by cytological and phytochemical data.

A few *Podocarpus* specimens are tentatively determined. One of these is *Beaman 10362* from Mamut Copper Mine. This collection lacks the distinctively erect disposition of the leaves, characteristic of *Podocarpus brevifolius*, and occurs at a lower elevation than normal for the species. However, it seems to be more in accord with that taxon than any other currently recognized species.

The juvenile material, or perhaps lower branches, that collectors have obtained of certain species sometimes makes identification difficult. For example, in the mature state and from upper branches, *Dacrydium gibbsiae* and *D. gracilis* are highly distinctive, but young individuals or lower branches can have a different aspect with much longer and more similar leaves.

VEGETATIVE KEY TO THE GYMNOSPERM GENERA

1. Large woody climbers. Leaves opposite, of dicotyledonous aspect .... *Gnetum.*

1. Trees, or, at high elevations, shrubs. Leaves or phyllodes opposite or alternate, more or less scleromorphic

    2. True leaves absent, these replaced by cladodes or flattened shoots .... *Phyllocladus.*

    2. Leaves present, needle-like, scale-like, or expanded into broad blades

        3. Leaves opposite, the largest more than 2 cm wide

            4. Terminal buds globose .... *Agathis.*

            4. Terminal buds acute .... *Nageia.*

        3. Leaves spirally arranged, less than 2 cm wide

            5. Leaves, at least some of them, more than 3 cm long

6. Principal leaves bilaterally flattened, falcate in shape . *Falcatifolium.*

6. Principal leaves with distinct upper and lower surface

    7. Mid-vein on the upper surface with a longitudinal groove .... *Sundacarpus.*

    7. Mid-vein not grooved .... *Podocarpus.*

5. Leaves all relatively small, needle-like or scale-like, less than 3 cm long

    8. Foliage dimorphic .... *Dacrycarpus.*

    8. Foliage not dimorphic, needle-like .... *Dacrydium.*

ENUMERATION OF THE GYMNOSPERMS

Kramer, K. U., & Green, P. S., eds. (1990). *The Families and Genera of Vascular Plants.* Kubitzki, K., ed. *I. Pteridophytes and Gymnosperms.* Springer-Verlag, Berlin etc.; Gymnosperms, pp. 279–391.

31. ARAUCARIACEAE

De Laubenfels, D. J. (1988). Araucariaceae. *Fl. Males.* I, 10: 419–442.

31.1. AGATHIS Salisb.

De Laubenfels, D. J. (1988). *Agathis. Fl. Males.* I, 10: 429–442. De Laubenfels, D. J. (1979). The species of *Agathis* (Araucariaceae) of Borneo. *Blumea* 25: 531–541.

31.1.1. Agathis dammara (Lamb.) Rich. & A. Rich. in A. Rich., *Comm. Bot. Conif. Cycad.*: 83 (1826). Plate 1B.

Large tree. Hill forest, sometimes on ultramafic substrate. Elevation: 800–1400 m.

Material examined. DALLAS: 900 m, *Clemens 27302* (BM); HEMPUEN HILL: 800–1000 m, *Beaman 7425* (MSC), 800–1200 m, *7694* (K), 1400 m, *Beaman et al. SNP 5068* (SNP); HIMBAAN: *Doinis Soibeh 715* (K); MT. NUNGKEK: 900–1200 m, *Clemens 32821* (BM, L).

31.1.2. Agathis kinabaluensis de Laub., *Blumea* 25: 535 (1979). Plate 2A, 2B. Type: SUMMIT TRAIL, 2000 m, *de Laubenfels P 625* (holotype L!).

Small to large tree. Lower montane forest, sometimes in mossy forest on ridges. Elevation: 1500–2200 m. Also known from Mt. Murud, Sarawak.

Additional material examined. BAMBANGAN RIVER: 1500 m, *RSNB 4457* (K, SING); EASTERN SHOULDER: *Phillipps SNP 2165* (SNP); KIAU: *Clemens 10004* (BM, K); KIAU VIEW TRAIL: 1600 m, *Beaman et al. SNP 5077* (SNP), 1600 m, *Justine SNP 675* (SNP), 1600 m, *SNP 767* (SNP); KILEMBUN BASIN: 1700 m, *Clemens 34496* (BM, K, L); MEMPENING TRAIL: 1700 m, *de Laubenfels P 646* (L); MESILAU CAVE: 1900–2200 m, *Beaman 9556* (K), 1800 m, *RSNB 4778* (K, SING); MT. TEMBUYUKEN: *Nais et al. SNP 4826* (SNP); PARK HEADQUARTERS: 1700 m, *Gimpiton et al. SNP 937* (SNP), 1700 m, *de Laubenfels P 644* (L), 1700 m, *P 720* (L).

31.1.3. Agathis lenticula de Laub., *Blumea* 25: 537 (1979). Plates 2C, 3A.

Large tree. Lower montane forest. Elevation: 900–1800 m.

Material examined. KILEMBUN BASIN: 1700 m, *Clemens 34496* (BM); LIWAGU RIVER TRAIL: 1500 m, *Lajangah SAN 44400* (K, L), 1500 m, *Sadau SAN 42812* (K, L), 1500 m, *de Laubenfels P 637* (L); MAMUT HILL: 1500–1800 m, *Kokawa & Hotta 5676* (L); MESILAU RIVER: 1500 m, *RSNB 4249* (K, L, SING), 1500 m, *RSNB 4330* (K, L, SING), 1500 m, *de Laubenfels P 621* (L); MESILAU/BAMBANGAN RIVERS: 1600–1700 m, *Kokawa & Hotta 4280* (L); MOUNT KINABALU: *Aban Gibot SAN 56636* (SNP), 1700 m, *Binideh SAN 65171* (SING), 1600 m, *Dolois & Ansow SNP 1941* (SNP); PARK HEADQUARTERS: 1400 m, *Abbe et al. 9973* (L), 1600 m, *Beaman & Ansow SNP 5060* (SNP), 1600 m, *Binideh SAN 65143* (K), *SAN 65144* (SNP), 1600 m, *Kokawa & Hotta 6128* (L), 1500 m, *Lowry 649* (L), 1200 m, *Meijer SAN 22111* (K), 1600 m, *Phillipps SNP 1516* (SNP), 1500 m, *de Laubenfels P 620* (L); PENIBUKAN: 1200 m, *Clemens 40732* (BM); TENOMPOK: 1500 m, *Clemens 28145* (BM), 1500 m, *28390* (BM, K, L), 1500 m, *28729* (BM, K), 1400 m, *Melegrito A 473* (K, L, SING), 1400 m, *Smythies S 10602* (K, L, SING); TENOMPOK/RANAU: 1500 m, *Carr SFN 27005* (SING).

32. PHYLLOCLADACEAE

32.1. PHYLLOCLADUS Rich. ex Mirb.

De Laubenfels, D. J. (1988). *Phyllocladus. Fl. Males.* I, 10: 355–360. De Laubenfels, D. J. (1969). A revision of the Malesian and Pacific rainforest conifers, I. Podocarpaceae, in part. *J. Arnold Arbor.* 50: 277–282.

32.1.1. Phyllocladus hypophyllus Hook. f., *Icon. Pl.* n.s. 5: t. 889 (1852). Plate 3B. Type: MOUNT KINABALU, 2400 m, *Low s.n.* (holotype K!).

Large to small tree, shrubby and gnarled at high elevations. Lower montane forest, upper montane forest, mossy forest on ridges, low mossy and xerophyllous scrub. Elevation: 1200–4000 m.

Additional material examined. EASTERN SHOULDER: 2300 m, *RSNB 710* (K, L, SING); GURULAU SPUR: 2400–2700 m, *Clemens 50626* (BM), *50784* (BM, L), 3000 m, *50797* (BM, L), 3400–3700 m, *51220* (BM); KEMBURONGOH: 2100 m, *Mikil SAN 56277* (K, L), 2100 m, *Price 183* (K), 2100 m, *Sinclair et al. 9053* (SING), 2200 m, *Smith 453* (L); KINATEKI RIVER HEAD: 2100 m, *Clemens 31838* (L); KUNDASANG: *Burgess SAN 25167* (K), 1500–1800 m, *Meijer SAN 21968* (K); LAYANG-LAYANG: 2200 m, *Andrews 883* (K); LUBANG: 1800 m, *Gibbs 4152* (BM, K); LUGAS HILL: *Doinis Soibeh 736* (K); MAMUT COPPER MINE: *Aban Gibot SAN 66823* (SING); MARAI PARAI: 1500 m, *Clemens 31927* (K), 1500 m, *32459* (BM, L); MARAI PARAI SPUR: *Clemens 10957* (BM), 2100 m, *Gibbs 4088* (BM, K); MESILAU BASIN: 2400–2700 m, *Clemens 29743* (BM, K, L); MESILAU CAVE: 1800 m, *RSNB 4824* (K, SING); MESILAU CAVE TRAIL: 1700–1900 m, *Beaman 8000* (MSC); MESILAU RIVER: 1500 m, *RSNB 4172* (K, L, SING); MESILAU TRAIL: *Chow & Leopold SAN 74513* (K, L); MINETUHAN SPUR: 1800–2100 m, *Clemens 33864* (BM); MOUNT KINABALU: 3400 m, *Haviland 1092* (BM, K, L, SING), 3000 m, *Low s.n.* (K), 3400 m, *Native Collector 39* (K), 2700 m, *Nicholson SAN 17823* (K, SING), *Rao et al. 76* (SING), 3100 m, *Rickards 161* (K); MT. TEMBUYUKEN: *Nais et al. SNP 4830* (SNP); PAKA-PAKA CAVE: 3100 m, *Carr SFN 27632* (SING), *Clemens 10565* (K), 3400 m, *27930* (BM, L), 3000 m, *29328* (BM, K, L, SING), 3200 m, *30030* (K), 3000 m, *Gibbs 4238* (BM), 3100 m, *Holttum s.n.* (SING), 3000 m, *Meijer SAN 29271* (K, L), 2700 m, *Smythies S 10622* (K, L, SING), *Wyatt-Smith 80371* (K, L, SING); PAKA-PAKA CAVE/PANAR LABAN: 3200–3400 m, *Kokawa & Hotta 3451* (L), 3200–3400 m, *3497* (L); PAKA-PAKA CAVE/SUMMIT AREA: 2700–3700 m, *Gibbs 4273* (BM, K); PANAR LABAN: 3500 m, *Beaman 8297* (MSC), 3400 m, *Nais & Dolois SNP 3277* (SNP), 3300 m, *Smith 474* (L), 3400 m, *Stone 11348* (L), 3400 m, *11373* (L); PARK HEADQUARTERS: 1200 m, *Meijer SAN 22114* (K), 1700 m, *de Laubenfels P 645* (L); PARK HEADQUARTERS/POWER STATION: 1600 m, *Cockburn & Aban SAN 82973* (K, L); PIG HILL: 2000–2300 m, *Beaman 9843* (MSC); SAYAT-SAYAT: 3500 m, *Carr SFN 27617* (BM, SING); SHEILA'S PLATEAU: 3400 m, *Fuchs & Collenette 21430* (K); SUMMIT AREA: 3600 m, *Anderson S 27089* (K), 3600 m, *S 27090* (L), 3400–4000 m, *Kokawa & Hotta 3563* (L); SUMMIT TRAIL: 3100 m, *de Laubenfels P 636* (L).

33. PODOCARPACEAE

De Laubenfels, D. J. (1988). Podocarpaceae. *Fl. Males.* I, 10: 351–419.

33.1. DACRYCARPUS (Endl.) de Laub.

De Laubenfels, D. J. (1988). *Dacrycarpus. Fl. Males.* I, 10: 374–384. De Laubenfels, D. J. (1969). A revision of the Malesian and Pacific rainforest conifers, I. Podocarpaceae, in part. *J. Arnold Arbor.* 50: 315–337 (1969).

33.1.1. Dacrycarpus imbricatus (Blume) de Laub.

a. var. **patulus** de Laub., *J. Arnold Arbor.* 50: 320 (1969). Plate 3C.

Small to large tree. Lower montane forest. Elevation: 1400–2400 m.

Material examined. GURULAU SPUR: 1700 m, *Clemens 50696* (BM), 2300 m, *51024* (BM); KADAMAIAN RIVER: 2000 m, *Carr SFN 27735* (SING); KEMBURONGOH: 2000 m, *Carr SFN 27553* (BM, SING), 1500 m, *Clemens 28954* (K), 1500 m, *Fosberg 44128* (K, L); KIAU VIEW TRAIL: *Aban Gibot SAN 56305* (SING); KINATEKI RIVER HEAD: 2400 m, *Clemens 35011* (BM); KUNDASANG: *Burgess SAN 25162* (SING); LIWAGU RIVER TRAIL: 1500 m, *Sadau SAN 42811* (SNP); MAMUT HILL: 1400–1700 m, *Kokawa & Hotta 5384* (L); MAMUT/BAMBANGAN RIVER: 1400–1700 m, *Kokawa & Hotta 5514* (L); MESILAU CAVE TRAIL: 1700–1900 m, *Beaman 8008* (MSC); MESILAU RIVER: 1500 m, *RSNB 4084* (K, L, SING), *Clemens 51635* (BM, K, L); MOUNT KINABALU: *Lajangah SAN 33085* (K, L, SING); PARK HEADQUARTERS: *Abbe et al. 9994* (SING), *Tan & Gimpiton SNP 507* (SNP), 1600 m, *Thomas & Patrick SNP 235* (SNP); PARK HEADQUARTERS/POWER STATION: 1600 m, *Cockburn & Aban SAN 82961* (K, L), 1700–1900 m, *Kokawa & Hotta 3217* (L), 1600 m, *Mikil SAN 33930* (SNP); PENATARAN BASIN: 2400 m, *Clemens 33618* (BM, K, L); PINOSUK PLATEAU: *Chow & Leopold SAN 74521* (K, L, SING), 1500 m, *Sadau SAN 42890* (K, L); SOSOPODON: 1500 m, *Sario SAN 32246* (K); TENOMPOK: 1500 m, *Clemens 28631* (BM, K, L, SING), 1500 m, *29779* (BM, K, L), 1400 m, *Melegrito A 471* (K, L, SING), 1400 m, *Smythies S 10601* (K, L, SING); TENOMPOK/RANAU: 1500 m, *Carr SFN 27010* (SING).

33.1.2. Dacrycarpus kinabaluensis (Wasscher) de Laub., *J. Arnold Arbor.* 50: 330 (1969). Plate 4A.

Podocarpus cupressina Stapf non R. Br. ex Mirbel, *Trans. Linn. Soc., Bot.* 4: 249 (1894).
Podocarpus imbricatus Blume var. *kinabaluensis* Wasscher, *Blumea* 4: 400 (1941). Type: ABOVE PAKA-PAKA CAVE, 3900 m, *Clemens 27854* (holotype B†; isotype BM!).

Shrub, treelet or small tree, frequently gnarled. Upper montane forest, especially on ultramafic substrate, extending to upper limit of scrub vegetation. Elevation: 2100–4000 m. Endemic to Mount Kinabalu.

Additional material examined. EASTERN SHOULDER: 3200 m, *RSNB 868* (K, L, SING), *Phillipps et al. SNP 2388* (SNP); GURULAU SPUR: 3000 m, *Clemens 51066* (BM), 3400 m, *51201* (BM, L); JANET'S HALT: 2400 m, *Collenette 579* (K); LAYANG-LAYANG/PANAR LABAN: 2700–3400 m, *Kokawa & Hotta 3380* (L); LUBANG/GRANITE CAP: *Gibbs 4216* (K); MARAI PARAI: 3000–3400 m, *Clemens 32316* (K, L), 3000–3400 m, *32316A* (BM), 3400 m, *32317* (BM, L), 3400 m, *32318* (BM, L); MARAI PARAI SPUR: *Gibbs 4216* (BM); MESILAU FRONT: 2700–3400 m, *RSNB 5887* (K, L); MOUNT KINABALU: *Binideh SAN 65173* (K), 3100 m, *Cockburn SAN 82988* (K), 3200 m, *Cockburn & Aban SAN 82972* (K), 3200 m, *SAN 82978* (K, L), 3200 m, *SAN 82981* (K), 3400 m, *Haviland 1094* (K, SING), 3400 m, *1095* (K, SING), 3000 m, *Nicholson SAN 17825* (K, SING), 2700 m, *SAN 39766* (K, L, SING), 2100 m, *Rao et al. 83* (SING); PAKA-

19

PAKA CAVE: 3100 m, *Carr SFN 28052* (SING), 3400 m, *Clemens 28910* (K, L), 3100 m, *Holttum s.n.* (SING), 3000 m, *Meijer SAN 21988* (K), 3000 m, *SAN 29265* (K, L), 3300 m, *Sinclair et al. 9146* (K, SING), 3200 m, *Wood & Wyatt-Smith SAN 4493* (SING); PAKA-PAKA CAVE/PANAR LABAN: 3200 m, *Wood & Wyatt-Smith SAN 4493* (L); PANAR LABAN: 3500 m, *Anonymous SNP 2337* (SNP), 3400 m, *Boeriaatmadja 89* (L), *Justine SNP 676* (SNP), 3400 m, *Phillipps & Tan SNP 1594* (SNP), 3300 m, *Smith 471* (L), 3400 m, *Stone 11368* (L); PANAR LABAN/SAYAT-SAYAT: 3400–3700 m, *Sato UKMS 764* (SNP); SAYAT-SAYAT: 3800 m, *Cockburn & Aban SAN 82771* (K, L); SUMMIT AREA: 3400 m, *Anderson S 27079* (K), 4000 m, *Clemens 27092* (BM, K, L), 3800 m, *29914* (K, L), 3700 m, *Kokawa & Hotta 3640* (L), 3600 m, *3642* (L); SUMMIT TRAIL: 2800–3000 m, *Beaman SNP 5062* (SNP), 3000 m, *8305* (K), *Clemens 10636* (K), 3600 m, *Jacobs 5755* (K, L), 2800 m, *de Laubenfels P 631* (L), 2800 m, *P 632* (L), 3100 m, *P 635* (L).

33.2. DACRYDIUM Lamb.

De Laubenfels, D. J. (1988). *Dacrydium. Fl. Males.* I, 10: 360–371. De Laubenfels, D. J. (1969). A revision of the Malesian and Pacific rainforest conifers, I. Podocarpaceae, in part. *J. Arnold Arbor.* 50: 282–308.

33.2.1. Dacrydium beccarii Parl. in DC., *Prodr.* 16, 2: 494 (1868).

Shrub or small tree. Lower montane forest, probably on ultramafic substrate. Elevation: 1500–1600 m. This identification may be questionable, but the specimen cited was determined by de Laubenfels.

Material examined. MAMUT HILL: 1500–1600 m, *Kokawa & Hotta 6051* (L).

33.2.2. Dacrydium gibbsiae Stapf in Gibbs, *J. Linn. Soc., Bot.* 42: 192 (1914). Plates 1A, 4B. Type: LUBANG/GRANITE CORE, 1800–3700 m, *Gibbs 4162* (holotype BM!; isotype K!).

Dacrydium beccarii Parl. in DC. var. *kinabaluense* Corner, *Gard. Bull. Straits Settlem.* 10: 244, t. 9 (1939). Type: PENIBUKAN, 1400 m, *Carr SFN 26437* (holotype SING!).

Shrub or small tree with pendulous branches. Upper montane forest on ridges, rarely lower montane forest, often on ultramafic substrate. Elevation: 1200–3700 m. Also known from Mt. Murud, Sarawak.

Additional material examined. KEMBURONGOH/PAKA-PAKA CAVE: 2700–3400 m, *Clemens s.n.* (BM); LAYANG-LAYANG: 2700–2900 m, *Hotta 3897* (L), 2700–2900 m, *3900* (L); LAYANG-LAYANG/ PAKA-PAKA CAVE: 2700–3200 m, *Kokawa & Hotta 3401* (L); LETENG TRAIL: 1500 m, *Meijer SAN 21100* (K); LUBANG/PAKA-PAKA CAVE: *Clemens 10685* (BM, K); MARAI PARAI: 1600 m, *Argent SNP 2379* (SNP), 1600 m, *Carr SFN 26588* (SING), 1500–1800 m, *Clemens 33037* (BM, L), 1500 m, *Collenette A 100* (BM); MARAI PARAI SPUR: 1500–2400 m, *Gibbs 4050* (BM), 1800 m, *Phillipps SNP 1821* (SNP); MOUNT KINABALU: 2400 m, *Low s.n.* (K, K), 2600 m, *Meijer SAN 22045* (K, SING), 2700 m, *Nicholson SAN 17826* (L, SING), 2900–3300 m, *Rickards 153* (K), 2400 m, *Whitehead s.n.* (BM); PAKA-PAKA CAVE: 2100–2700 m, *Enriquez SFN 18168* (SING); PENATARAN BASIN: 2000 m, *Clemens 40151* (BM); PENIBUKAN: 1500 m, *Clemens 30922* (BM, L); PIG HILL: 2100 m, *RSNB 4361* (K, L, SING); PINOSUK PLATEAU: 1600 m, *Collenette 542* (K); SUMMIT TRAIL: 2800–3000 m, *Beaman SNP 5063* (SNP), 3000 m, *8306* (K), 2800 m, *de Laubenfels P 628* (L), 2800 m, *P 629* (L), 2200 m, *P 630* (L).

33.2.3. Dacrydium gracilis de Laub., *Fl. Males.* I, 10(3): 367 (1988). Plate 4C. Type: PARK HEADQUARTERS, 1500 m, *de Laubenfels P 716* (holotype L!).

Large tree. Lower montane forest. Elevation: 1400–1600 m.

Additional material examined. HEMPUEN HILL: 1400 m, *Beaman et al. SNP 5070* (SNP); LETENG TRAIL: 1500 m, *Meijer SAN 21086* (K, SING), 1500 m, *SAN 21098* (K, SING), 1500 m, *SAN 21100* (SING); LIWAGU RIVER TRAIL: 1500 m, *de Laubenfels P 638* (L), 1500 m, *P 642* (L); MELANGKAP TOMIS: *Lorence Lugas 1962* (K); MESILAU RIVER: 1500 m, *RSNB 4303* (K, SING), 1500 m, *RSNB 4305* (K, SING); PARK HEADQUARTERS: 1500 m, *de Laubenfels P 717* (L); PARK HEADQUARTERS/POWER STATION: 1600 m, *Cockburn SAN 82963* (K), 1600 m, *SAN 82965* (K), 1600 m, *Cockburn & Aban SAN 82959* (K), 1600 m, *SAN 82962* (K); TENOMPOK: 1600 m, *Mujin SAN 33774* (K, L).

33.2.4. Dacrydium pectinatum de Laub., *J. Arnold Arbor.* 50: 289 (1969). Plate 4D.

Medium-sized tree (on Mount Kinabalu). Hill forest on ultramafic substrate. Elevation: 800–1600 m.

Material examined. HEMPUEN HILL: 800 m, *Abbe et al. 9938* (L), 800 m, *9939* (SING), 1100 m, *9952* (L), 1400 m, *Beaman et al. SNP 5071* (SNP), 900–1200 m, *Meijer SAN 20951* (L), 1200 m, *SAN 20952* (L), 1300 m, *SAN 20970* (K, SING); KIAU VIEW TRAIL: 1600 m, *Justin SNP 290* (SNP).

33.2.5. Dacrydium xanthandrum Pilger, *Bot. Jahrb.* 69: 252 (1938). Plate 4E.

Small to medium-sized tree. Lower montane forest, upper montane forest, especially in mossy forest on ridges. Elevation: 1400–3000 m.

Material examined. JANET'S HALT: 2400 m, *Collenette 543* (K), 2400 m, *Nicholson SAN 39768* (K); KEMBURONGOH: 2100 m, *Anonymous SAN 62031* (K), 2100 m, *Meijer SAN 29153* (K, L), 2100 m, *Price 205* (K), 1800 m, *Smythies S 10607* (K, SING); KILEMBUN RIVER HEAD: 1800 m, *Clemens 32502* (BM, K, L); KUNDASANG: 1800 m, *Meijer SAN 23500* (K, SING); LETENG TRAIL: 1700 m, *Meijer SAN 21097* (K, SING); MAMUT/BAMBANGAN RIVERS: 1400–1700 m, *Kokawa & Hotta 5565* (L); MESILAU HILL: 2300 m, *RSNB 8024* (K, L); MOUNT KINABALU: 1800–2100 m, *Enriquez 18169* (SING), 2000 m, *Haviland 1183* (K, SING), 2700 m, *Holttum s.n.* (SING, SING), 2700 m, *Nicholson SAN 17827* (K, SING); MURU-TURA RIDGE: 1500–1800 m, *Clemens 34341* (BM, K, L); PAKA-PAKA CAVE: 2400–3000 m, *Clemens 28542* (BM); SUMMIT TRAIL: 2100 m, *Aban Gibot SAN 62031* (L), 2100–2300 m, *Beaman & Beaman SNP 5061* (SNP), 2700 m, *Carr SFN 27599* (SING), 2000 m, *Cockburn & Aban SAN 82971* (K), 1800 m, *Smythies S 10607* (L), 2700 m, *de Laubenfels P 626* (L), 2700 m, *P 627* (L).

33.3. FALCATIFOLIUM de Laub.

De Laubenfels, D. J. (1988). *Falcatifolium. Fl. Males.* I, 10: 371–374. De Laubenfels, D. J. (1969). A revision of the Malesian and Pacific rainforest conifers, I. Podocarpaceae, in part. *J. Arnold Arbor.* 50: 308–314.

33.3.1. Falcatifolium falciforme (Parl.) de Laub., *J. Arnold Arbor.* 50: 309 (1969). Plate 4F.

Podocarpus falciformis Parl. in DC., *Prodr.* 16, 2: 685 (1868).
Dacrydium falciforme (Parl.) Pilger, *Pflanzenr.* IV. 5 (Heft 18): 45 (1903).

Small tree. Lower montane forest on ridges. Elevation: 800–2100 m.

Material examined. HEMPUEN HILL: *Abbe et al. 9940* (SING), 800–1200 m, *Beaman 7668* (MSC), 1400 m, *Beaman et al. SNP 5075* (SNP), 1300 m, *Madani SAN 89369* (K), 1200–1400 m, *Meijer SAN 20279* (L, SING), 1200 m, *SAN 20953* (K), 1500 m, *de Laubenfels P 707* (L); KEMBURONGOH: 2100 m, *Clemens 27851* (BM, K); LIWAGU RIVER TRAIL: 1500 m, *de Laubenfels P 639* (L), 1500 m, *P 640* (L); LUMU-LUMU: 1600 m, *Carr SFN 27241* (SING); MARAI PARAI: *Argent & Phillipps SNP 2732* (SNP), 1500 m, *Clemens 33078* (K, L), 1500 m, *Holttum s.n* (SING), *Phillipps SNP 1145* (SNP), *SNP 2762* (SNP); MARAI PARAI SPUR: *Clemens 10962* (K), 1500 m, *Holttum s.n.* (SING); MARAI PARAI SPUR/DAPATAN & PENIBUKAN RIDGES: 1500 m, *Gibbs 4067* (K); MESILAU RIVER: 1500 m, *RSNB 4847* (K, L, SING); PARK HEADQUARTERS: 1600 m, *Patrick & Gimpiton SNP 499* (SNP), 1600 m, *Tan & Gimpiton SNP 568* (SNP), 1500 m, *de Laubenfels P 718* (L), 1500 m, *P 719* (L); PARK HEADQUARTERS/POWER STATION: 1600 m, *Cockburn SAN 82966* (K, L), 1700–1900 m, *Kokawa & Hotta 3213* (L); PENIBUKAN: 1200–1500 m, *Clemens s.n.* (BM, BM, K, K, L, L); PINOSUK PLATEAU: 1700 m, *RSNB 1863* (K, L, SING).

33.4. NAGEIA Gaertn.

De Laubenfels, D. J. (1988). *Nageia. Fl. Males.* I, 10: 389–395. De Laubenfels, D. J. (1987). Revision of the genus *Nageia* (Podocarpaceae). *Blumea* 32: 209–211. De Laubenfels, D. J. (1969). A revision of the Malesian and Pacific rainforest conifers, I. Podocarpaceae, in part [*Nageia* as *Decussocarpus*]. *J. Arnold Arbor.* 50: 340–359.

33.4.1. Nageia wallichiana (Presl) Kuntze, *Rev. Gen. Pl.* 2: 800 (1891).

Medium-sized to large tree. Lower montane forest. Elevation: 1500 m.

Material examined. MESILAU RIVER: 1500 m, *RSNB 4878* (K, L, SING), 1500 m, *de Laubenfels P 624* (L); PENIBUKAN: *Clemens s.n.* (BM); SOSOPODON: *Aban Gibot SAN 62022* (K, SING).

33.5. PODOCARPUS L'Hér. ex Pers.

De Laubenfels, D. J. (1988). *Podocarpus. Fl. Males.* I, 10: 395–419. De Laubenfels, D. J. (1985). A taxonomic revision of the genus *Podocarpus. Blumea* 30: 251–278. Wasscher, J. (1941). The genus *Podocarpus* in the Netherlands Indies. *Blumea* 4: 359–542. Gray, N. E. (1958). A taxonomic revision of *Podocarpus*, XI. The South Pacific species of section *Podocarpus*, subsection B. *J. Arnold Arbor.* 39: 424–477.

33.5.1. Podocarpus brevifolius (Stapf) Foxw., *Philipp. J. Sci., Bot.* 6: 160, t. 29 (1911). Plate 5A.

Podocarpus neriifolius D. Don in Lamb. var. *brevifolius* Stapf, *Trans. Linn. Soc., Bot.* 4: 249 (1894). Type: MOUNT KINABALU, 3400 m, *Haviland 1093* (syntype K!), 3700 m, *Low s.n.* (syntype K!).

Small tree or shrub, often gnarled. Mostly upper montane forest, rarely lower montane forest, on ultramafic substrate or in granitic crevices. Elevation: 1200–3800 m. Endemic to Mount Kinabalu.

Additional material examined. EASTERN SHOULDER: 3000 m, *RSNB 724* (K, L, SING), 3000 m, *RSNB 756* (K, L, SING), 2400 m, *Collenette 810* (K, L, SING); GURULAU SPUR: 2100–2700 m, *Clemens 50790* (BM), 3400–3700 m, *50825* (BM, L); LAYANG-LAYANG/PAKA-PAKA CAVE: 2700–3200 m, *Kokawa & Hotta 3411* (L); LAYANG-LAYANG/PANAR LABAN: 2700–3400 m, *Kokawa & Hotta 3379* (L); LUBANG/GRANITE CORE: 1800–3700 m, *Gibbs 4166* (BM); MAMUT COPPER MINE: 1400–1500 m, *Beaman 10362* (K); MARAI PARAI SPUR: 1500–2400 m, *Gibbs 4089* (BM, K); MESILAU RIVER: 3500 m, *Smith 529* (L); MOUNT KINABALU: 2700–3000 m, *Meijer SAN 22065* (K, SING), 3000 m, *Nicholson SAN 17824* (K, SING), 3400 m, *Rao et al. 77* (SING); MT. TEMBUYUKEN: *Nais et al. SNP 4834* (SNP); PAKA-PAKA CAVE: 3400 m, *Clemens 27103* (BM), 3400 m, *28901* (K, L), 3100 m, *Holttum s.n.* (SING), 3400 m, *Lampangi SAN 29290* (K, L, SING); PANAR LABAN: 3500 m, *Anonymous SNP 2335* (SNP), *Cockburn SAN 82780* (K, L), 3400 m, *Cockburn & Aban SAN 82276* (K), 3400 m, *SAN 82776* (L), 3400 m, *Phillipps & Tan SNP 1586* (SNP), 3400 m, *Stone 11351* (L), 3400–3700 m, *Wong 21* (SING); PENIBUKAN: 1200 m, *Clemens s.n.* (BM, K); SUMMIT AREA: 3400 m, *Anderson S 27094* (K), 3800 m, *Clemens 27826* (BM); SUMMIT TRAIL: 2800–3000 m, *Beaman SNP 5064* (SNP), *Clemens 10657* (K), 3600 m, *Lee et al. SAN 69959* (SNP), 2700 m, *Meijer SAN 21975* (K), 3000 m, *de Laubenfels P 633* (L), 3100 m, *P 634* (L); UPPER KINABALU: 3800 m, *Clemens 27825* (K, L).

33.5.2. Podocarpus confertus de Laub., *Blumea* 30: 271 (1985).

Small to large tree. Hill forest on ultramafic substrate. Elevation: 600–1200 m.

Material examined. HEMPUEN HILL: *Aban Gibot SAN 90606* (K, L), 600 m, *Madani SAN 89400* (K, L), 1200 m, *de Laubenfels P 712* (L), 1200 m, *P 713* (L), 1200 m, *P 714* (L).

33.5.3. Podocarpus gibbsii N. E. Gray, *J. Arnold Arbor.* 39: 429 (1958). Type: MARAI PARAI, 1500 m, *Clemens 32021* (holotype A n.v.; isotypes BM!, K!, L!).

Small to medium-sized tree. Lower montane forest, upper montane forest, mossy ridge forest, on ultramafic substrate. Elevation: 1200–2400 m. Endemic to Mount Kinabalu.

Additional material examined. HEMPUEN HILL: 1500 m, *de Laubenfels P 709* (L), 1500 m, *P 710* (L), 1500 m, *P 711* (L); KILEMBUN BASIN: 1500 m, *Clemens 40001* (BM, K, L); MARAI PARAI: *Phillipps SNP 2761* (SNP); MARAI PARAI SPUR/PENIBUKAN: 1500–2400 m, *Gibbs 4092* (BM, K); PENIBUKAN: 1400 m, *Carr SFN 26450* (SING), 1200–1500 m, *Clemens s.n.* (L); PIG HILL: 2100 m, *RSNB 4369* (K, L, SING).

33.5.4. Podocarpus globulus de Laub., *Blumea* 30: 269 (1985).

Small to medium-sized tree. Lower montane forest on ultramafic substrate. Elevation: 1500 m.

Material examined. HEMPUEN HILL: 1500 m, *de Laubenfels P 704* (L), 1500 m, *P 705* (L), 1500 m, *P 706* (L).

33.5.5. Podocarpus laubenfelsii Tiong, *Blumea* 29: 523 (1984). Type: PARK HEADQUARTERS, 1500 m, *de Laubenfels P 715* (holotype L!).

Small to medium-sized tree. Lower montane forest. Elevation: 1400–1600 m.

Additional material examined. LIWAGU/MESILAU RIVERS: 1400 m, *RSNB 2657* (K, L, SING); MESILAU RIVER: 1500 m, *RSNB 4350* (K, L, SING); NALUMAD: *Daim Andau 838* (K); PARK HEADQUARTERS: 1600 m, *Aban Gibot SAN 49409* (K), 1500 m, *de Laubenfels P 643* (L); SOSOPODON: 1500 m, *Gintus SAN 56374* (K, L).

33.5.6. Podocarpus neriifolius D. Don in Lamb., *Gen. Pinus* ed. 1: 21 (1824). Plate 5B.

Medium-sized to large tree. Lower montane forest. Elevation: 1100–1700 m.

Material examined. GURULAU SPUR: 1700 m, *Clemens 50691* (BM, K, L); HEMPUEN HILL: 1400 m, *Beaman et al. SNP 5069* (SNP); HIMBAAN: *Doinis Soibeh 714* (K); MESILAU RIVER: 1500 m, *RSNB 4255* (K, L, SING); PENIBUKAN: 1100 m, *Clemens 50051* (BM, K, L).

33.6. SUNDACARPUS (J. Buchholz & N. E. Gray) C. N. Page

De Laubenfels, D. J. (1988). *Prumnopitys. Fl. Males.* I, 10: 384–389.

33.6.1. Sundacarpus amara (Blume) C. N. Page, *Notes Roy. Bot. Gard. Edinburgh* 45: 378 (1988).

Prumnopitys amara (Blume) de Laub., *Blumea* 24: 190 (1978).

Large tree. Lower montane forest. Elevation: 1400–1700 m.

Material examined. KINASARABAN HILL: 1400 m, *Badak SAN 32333* (L); MAMUT/BAMBANGAN RIVERS: 1400–1700 m, *Kokawa & Hotta 5541* (L); MESILAU CAMP: *RSNB 5858* (K, L); MESILAU RIVER: 1500 m, *RSNB 4211* (K, L, SING), 1500 m, *RSNB 7031* (K, L, SING), 1500 m, *RSNB 7102* (K, L, SING), 1500 m, *de Laubenfels P 622* (L), *P 623* (L); PARK HEADQUARTERS/POWER STATION: 1600 m, *Cockburn & Aban SAN 82967* (K, L); PINOSUK PLATEAU: 1500 m, *Sadau SAN 49689* (L); TENOMPOK: 1400 m, *Smythies S 10614* (K, L, SING), 1500 m, *Wood & Wyatt-Smith SAN 4500* (L, SING); TENOMPOK/KUNDASANG: 1500 m, *Meijer SAN 20411* (L).

34. GNETACEAE

Markgraf, F. (1930). Monographie der Gattung *Gnetum. Bull. Jard. Bot. Buitenzorg*, ser. 3, 10: 407–511. Markgraf, F. (1951). Gnetaceae. *Fl. Males.* I, 4: 336–347. Markgraf, F. (1972). Gnetaceae. *Fl. Males.* I, 6: 944–949 (Addenda, corrigenda et emendanda).

34.1. GNETUM L.

34.1.1. Gnetum cuspidatum Blume, *Rumphia* 4: 5 (1848).

Woody climber. Lower montane forest. Elevation: 400–1700 m.

Material examined. KAUNG: 400 m, *Carr SFN 27293* (SING); KILEMBUN BASIN: 1700 m, *Clemens 33671* (BM, L); TENOMPOK: 1400 m, *Clemens 26203* (BM, K, L), 1500 m, *26203b* (K, L).

34.1.2. Gnetum gnemonoides Brongn. in Duperrey, *Voy. Coquille,* 12 (1829).

Woody climber with enormous seeds. Hill forest. Elevation: 600 m.

Material examined. HIMBAAN: *Doinis Soibeh 289* (K); MELANGKAP TOMIS: *Lorence Lugas 1594* (K), *2247* (K); PINAWANTAI: 600 m, *Shea & Aban SAN 76917* (K, SING).

34.1.3. Gnetum klossii Merr. ex Markgraf, *Bull. Jard. Bot. Buitenzorg,* ser. 3, 10: 478, t. 11, f. 6–8 (1930).

Large woody climber. Hill forest. Elevation: 900 m.

Material examined. DALLAS: 900 m, *Clemens 26003* (K, L), 900 m, *26003b* (K), 900 m, *27021* (K, L), 900 m, *27022* (BM); MELANKAP TOMIS: *Lorence Lugas 2605* (K).

34.1.4. Gnetum latifolium Blume, *Tijdschr. Natuurl. Gesch. Physiol.* 1: 160 (1834).

a. var. latifolium

Large woody climber. Lower montane forest. Elevation: 1200–1500 m.

Material examined. BAMBANGAN RIVER: 1500 m, *RSNB 4629* (K, L, SING); MESILAU CAMP: 1300 m, *Meijer SAN 38567* (L); MESILAU RIVER: 1500 m, *RSNB 4229* (K, L, SING); PENIBUKAN: 1200–1500 m, *Clemens 30764* (K, L); SOSOPODON: *Sinanggul SAN 47907* (K, L).

b. var. minus (Foxw.) Markgraf, *Bull. Jard. Bot. Buitenzorg,* ser. 3, 10: 463 (1930).

Woody climber. Hill forest, lower montane forest. Elevation: 900–1800 m.

Material examined. DALLAS/BONGOL: 900 m, *Clemens 27645* (BM, L); GURULAU SPUR: 1500 m, *Clemens 50509* (BM, K, L); TENOMPOK: 1200 m, *Clemens 26826* (BM, L, SING), 1500 m, *27506* (BM, K), 1500 m, *27764* (BM, K, L); TINEKUK FALLS: 1800 m, *Clemens 40875* (BM, K, L).

34.1.5. Gnetum leptostachyum Blume, *Rumphia* 4: 5 (1848).

a. var. leptostachyum. Plate 5C.

Woody climber. Hill forest, lower montane forest. Elevation: 600–1500 m.

Material examined. DALLAS: 900 m, *Clemens s.n.* (BM), 900–1200 m, *26246* (BM, K, L), 900 m, *26264* (L), 900 m, *26284* (K), 900 m, *26429* (BM, K), 900 m, *26598* (K, L), 900 m, *26672* (BM, K, L), 900 m, *26698* (BM, K, L), 900 m, *26710* (BM, K, L), 900 m, *26848* (BM, K), 900 m, *27034* (K), 900 m, *27356* (L), 900 m, *27597* (K), 900 m, *27643* (BM, K), 900 m, *30311* (K, L); DALLAS/TENOMPOK: 1200 m, *Clemens 27597* (BM); GURULAU SPUR: 1500 m, *Clemens 51039* (BM, K, L); HIMBAAN: *Doinis Soibeh 185* (K), *231* (K); KAUNG: 800 m, *Mujin SAN 26759* (K, L); LIWAGU/MESILAU RIVERS: 1200 m, *RSNB 2595* (K, L, SING), 1200 m, *RSNB 2600* (K, L); LOHAN RIVER: 700–900 m, *Beaman 9242* (K); LOHAN/MAMUT COPPER MINE: 900 m, *Beaman 10640a* (K), 1000 m, *10645* (K); MARAI PARAI: 1500 m, *Clemens 33072* (BM, K, L); MELANGKAP TOMIS: *Lorence Lugas 703* (K); MINITINDOK GORGE: 900–1200 m, *Clemens 29670* (BM); MT. NUNGKEK: 900–1200 m, *Clemens 32728* (BM, L, SING); NALUMAD: *Daim Andau 879* (K); NAPUNG: *Sani Sambuling 526* (K); PORING HOT SPRINGS: 600 m, *Beaman 7545* (MSC); PORING HOT SPRINGS/LANGANAN WATER FALLS: 600–1000 m, *Kokawa & Hotta 4904* (L); SAYAP: *Yalin Surunda 41* (K); SINGH'S PLATEAU: 1000 m, *Meijer SAN 26418* (K, L, SING); SOSOPODON: 1200 m, *Mikil SAN 37711* (K); TEKUTAN: *Dius Tadong 619* (K); TENOMPOK: *Clemens 27532* (BM, K, L), 1500 m, *29670* (K, L).

b. var. **abbreviatum** Markgraf, *Reinwardtia* 1: 462 (1952). Type: MARAI PARAI, 1500 m, *Clemens 32990* (holotype M? n.v.; isotypes BM!, L!).

Woody climber. Lower montane forest, mostly on ultramafic substrate. Elevation: 1200–1800 m. Endemic to Mount Kinabalu.

Additional material examined. KILEMBUN RIVER HEAD: 1800 m, *Clemens 32475* (BM, K, L); LUMU-LUMU: 1600 m, *Carr SFN 2/103* (SING); MARAI PARAI: 1500 m, *Clemens 32276* (BM, L), 1400 m, *32488* (BM, L), 1800 m, *32601* (L), 1200–1500 m, *32698* (L), 1500 m, *32991* (BM, K); MARAI PARAI SPUR: 1800 m, *Clemens 32601* (BM); MARAI PARAI/NUNGKEK: 1200–1500 m, *Clemens 32698* (BM); PINOSUK PLATEAU: *Poore H 18* (K, L).

34.1.6. Gnetum neglectum Blume, *Rumphia* 4: 6, t. 175, f. 2 (1848).

Woody climber. Hill forest. Elevation: 700–1100 m.

Material examined. EASTERN SHOULDER: 1100 m, *RSNB 646* (K, SING); HEMPUEN HILL: 1000 m, *Beaman 7689* (MSC); KIPUNGIT HILL: 700 m, *Beaman 7641* (K); LOHAN RIVER: 800–1000 m, *Beaman 9051* (K); MELANGKAP KAPA: 700–1000 m, *Beaman 8788* (K).

INDEX TO NUMBERED GYMNOSPERM COLLECTIONS

Prefixes for collectors' numbers are not included.

Binideh 65143 (31.1.3); 65144 (31.1.3); 65171 (31.1.3); 65173 (33.1.2).

Boeriaatmadja 89 (33.1.2).

Burgess 25162 (33.1.1a); 25167 (32.1.1).

Carr 26437 (33.2.2); 26450 (33.5.3); 26588 (33.2.2); 27005 (31.1.3); 27010 (33.1.1a); 27103 (34.1.5b); 27241 (33.3.1); 27293 (34.1.1); 27553 (33.1.1a); 27599 (33.2.5); 27617 (32.1.1); 27632 (32.1.1); 27735 (33.1.1a); 28052 (33.1.2).

Chew & Corner 4084 (33.1.1a); 4172 (32.1.1); 4211 (33.6.1); 4229 (34.1.4a); 4249 (31.1.3); 4255 (33.5.6); 4303 (33.2.3); 4305 (33.2.3); 4330 (31.1.3); 4350 (33.5.5); 4361 (33.2.2); 4369 (33.5.3); 4457 (31.1.2); 4629 (34.1.4a); 4778 (31.1.2); 4824 (32.1.1); 4847 (33.3.1); 4878 (33.4.1); 5858 (33.6.1); 5887 (33.1.2); 7031 (33.6.1); 7102 (33.6.1); 8024 (33.2.5).

Chew, Corner & Stainton 646 (34.1.6); 710 (32.1.1); 724 (33.5.1); 756 (33.5.1); 868 (33.1.2); 1863 (33.3.1); 2595 (34.1.5a); 2600 (34.1.5a); 2657 (33.5.5).

Chow & Leopold 74513 (32.1.1); 74521 (33.1.1a).

Clemens 10004 (31.1.2); 10565 (32.1.1); 10636 (33.1.2); 10657 (33.5.1); 10685 (33.2.2); 10957 (32.1.1); 10962 (33.3.1); 26003 (34.1.3); 26003b (34.1.3); 26203 (34.1.1); 26203b (34.1.1); 26246 (34.1.5a); 26264 (34.1.5a); 26284 (34.1.5a); 26429 (34.1.5a); 26598 (34.1.5a); 26672 (34.1.5a); 26698 (34.1.5a); 26710 (34.1.5a); 26826 (34.1.4b); 26848 (34.1.5a); 27021 (34.1.3); 27022 (34.1.3); 27034 (34.1.5a); 27092 (33.1.2); 27103 (33.5.1); 27302 (31.1.1); 27356 (34.1.5a); 27506 (34.1.4b); 27532 (34.1.5a); 27597 (34.1.5a); 27643 (34.1.5a); 27645 (34.1.4b); 27764 (34.1.4b); 27825 (33.5.1); 27826 (33.5.1); 27851 (33.3.1); 27854 (33.1.2); 27930 (32.1.1); 28145 (31.1.3); 28390 (31.1.3); 28542 (33.2.5); 28631 (33.1.1a); 28729 (31.1.3); 28901 (33.5.1); 28910 (33.1.2); 28954 (33.1.1a); 29328 (32.1.1); 29670 (34.1.5a); 29743 (32.1.1); 29779 (33.1.1a); 29914 (33.1.2); 30030 (32.1.1); 30311 (34.1.5a); 30764 (34.1.4a); 30922 (33.2.2); 31838 (32.1.1); 31927 (32.1.1); 32021 (33.5.3); 32276 (34.1.5b); 32316 (33.1.2); 32316A (33.1.2); 32317 (33.1.2); 32318 (33.1.2); 32459 (32.1.1); 32475 (34.1.5b); 32488 (34.1.5b); 32502 (33.2.5); 32601 (34.1.5b); 32698 (34.1.5b); 32728 (34.1.5a); 32821 (31.1.1); 32990 (34.1.5b); 32991 (34.1.5b); 33037 (33.2.2); 33072 (34.1.5a); 33078 (33.3.1); 33618 (33.1.1a); 33671 (34.1.1); 33864 (32.1.1); 34341 (33.2.5); 34496 (31.1.2, 31.1.3); 35011 (33.1.1a); 40001 (33.5.3); 40151 (33.2.2); 40732 (31.1.3); 40875 (34.1.4b); 50051 (33.5.6); 50509 (34.1.4b); 50626 (32.1.1); 50691 (33.5.6); 50696 (33.1.1a); 50784 (32.1.1); 50790 (33.5.1); 50797 (32.1.1); 50825 (33.5.1); 51024 (33.1.1a); 51039 (34.1.5a); 51066 (33.1.2); 51201 (33.1.2); 51220 (32.1.1); 51635 (33.1.1a).

Cockburn 82780 (33.5.1); 82963 (33.2.3); 82965 (33.2.3); 82966 (33.3.1); 82988 (33.1.2).

Cockburn & Aban 82276 (33.5.1); 82771 (33.1.2); 82776 (33.5.1); 82959 (33.2.3); 82961 (33.1.1a); 82962 (33.2.3); 82967 (33.6.1); 82971 (33.2.5); 82972 (33.1.2); 82973 (32.1.1); 82978 (33.1.2); 82981 (33.1.2).

Collenette 100 (33.2.2); 542 (33.2.2); 543 (33.2.5); 579 (33.1.2); 810 (33.5.1).

Daim Andau 838 (33.5.5); 879 (34.1.5a).

de Laubenfels 619 (31.1.3); 620 (31.1.3); 621 (31.1.3); 622 (33.6.1); 623 (33.6.1); 624 (33.4.1); 625 (31.1.2); 626 (33.2.5); 627 (33.2.5); 628 (33.2.2); 629 (33.2.2); 630 (33.2.2); 631 (33.1.2); 632 (33.1.2); 633 (33.5.1); 634 (33.5.1); 635 (33.1.2); 636 (32.1.1); 637 (31.1.3); 638 (33.2.3); 639 (33.3.1); 640 (33.3.1); 642 (33.2.3); 643 (33.5.5); 644 (31.1.2); 645 (32.1.1); 646 (31.1.2); 704 (33.5.4); 705 (33.5.4); 706 (33.5.4); 707 (33.3.1); 709 (33.5.3); 710 (33.5.3); 711 (33.5.3); 712 (33.5.2); 713 (33.5.2); 714 (33.5.2); 715 (33.5.5); 716 (33.2.3); 717 (33.2.3); 718 (33.3.1); 719 (33.3.1); 720 (31.1.2).

Dius Tadong 619 (34.1.5a).

Doinis Soibeh 185 (34.1.5a); 231 (34.1.5a); 289 (34.1.2); 714 (33.5.6); 715 (31.1.1); 736 (32.1.1).

Dolois & Ansow 1941 (31.1.3).

Enriquez 18168 (33.2.2); 18169 (33.2.5).

Fosberg 44128 (33.1.1a).

Fuchs & Collenette 21430 (32.1.1).

Gibbs 4050 (33.2.2); 4067 (33.3.1); 4088 (32.1.1); 4089 (33.5.1); 4092 (33.5.3); 4152 (32.1.1); 4162 (33.2.2); 4166 (33.5.1); 4216 (33.1.2); 4238 (32.1.1); 4273 (32.1.1).

Gimpiton et al. 937 (31.1.2).

Gintus 56374 (33.5.5).

Haviland 1092 (32.1.1); 1093 (33.5.1); 1094 (33.1.2); 1095 (33.1.2); 1183 (33.2.5).

Hotta 3897 (33.2.2); 3900 (33.2.2).

Jacobs 5755 (33.1.2).

Justin 290 (33.2.4); 675 (31.1.2); 676 (33.1.2); 767 (31.1.2).

Kokawa & Hotta 3213 (33.3.1); 3217 (33.1.1a); 3379 (33.5.1); 3380 (33.1.2); 3401 (33.2.2); 3411 (33.5.1); 3451 (32.1.1); 3497 (32.1.1); 3563 (32.1.1); 3640 (33.1.2); 3642 (33.1.2); 4280 (31.1.3); 4904 (34.1.5a); 5384 (33.1.1a); 5514 (33.1.1a); 5541 (33.6.1); 5565 (33.2.5); 5676 (31.1.3); 6051 (33.2.1); 6128 (31.1.3).

Lajangah 33085 (33.1.1a); 44400 (31.1.3).

Lampangi 29290 (33.5.1).

Lee et al. 69959 (33.5.1).

Lorence Lugas 703 (34.1.5a); 1594 (34.1.2); 1962 (33.2.3); 2247 (34.1.2); 2605 (34.1.3).

Lowry 649 (31.1.3).

Madani 89369 (33.3.1); 89400 (33.5.2).

Meijer 20279 (33.3.1); 20411 (33.6.1); 20951 (33.2.4); 20952 (33.2.4); 20953 (33.3.1); 20970 (33.2.4); 21086 (33.2.3); 21097 (33.2.5); 21098 (33.2.3); 21100 (33.2.2, 33.2.3); 21968 (32.1.1); 21975 (33.5.1); 21988 (33.1.2); 22045 (33.2.2); 22065 (33.5.1); 22111 (31.1.3); 22114 (32.1.1); 23500 (33.2.5); 26418 (34.1.5a); 29153 (33.2.5); 29265 (33.1.2); 29271 (32.1.1); 38567 (34.1.4a).

Melegrito 471 (33.1.1a); 473 (31.1.3).

Mikil 33930 (33.1.1a); 37711 (34.1.5a); 56277 (32.1.1).

Mujin 26759 (34.1.5a); 33774 (33.2.3).

Nais & Dolois 3277 (32.1.1).

Nais et al. 4826 (31.1.2); 4830 (32.1.1); 4834 (33.5.1).

Native Collector 39 (32.1.1).

Nicholson 17823 (32.1.1); 17824 (33.5.1); 17825 (33.1.2); 17826 (33.2.2); 17827 (33.2.5); 39766 (33.1.2); 39768 (33.2.5).

Patrick & Gimpiton 499 (33.3.1).

Phillipps 1145 (33.3.1); 1516 (31.1.3); 1821 (33.2.2); 2165 (31.1.2); 2761 (33.5.3); 2762 (33.3.1).

Phillipps & Tan 1586 (33.5.1); 1594 (33.1.2).

Phillipps *et al.* 2388 (33.1.2).

Poore 18 (34.1.5b).

Price 183 (32.1.1); 205 (33.2.5).

Rao *et al.* 76 (32.1.1); 77 (33.5.1); 83 (33.1.2).

Rickards 153 (33.2.2); 161 (32.1.1).

Sadau 42811 (33.1.1a); 42812 (31.1.3); 42890 (33.1.1a); 49689 (33.6.1).

Sani Sambuling 526 (34.1.5a).

Sario 32246 (33.1.1a).

Sato 764 (33.1.2).

Shea & Aban 76917 (34.1.2).

Sinanggul 47907 (34.1.4a).

Sinclair *et al.* 9053 (32.1.1); 9146 (33.1.2).

Smith 453 (32.1.1); 471 (33.1.2); 474 (32.1.1); 529 (33.5.1).

Smythies 10601 (33.1.1a); 10602 (31.1.3); 10607 (33.2.5); 10614 (33.6.1); 10622 (32.1.1).

Stone 11348 (32.1.1); 11351 (33.5.1); 11368 (33.1.2); 11373 (32.1.1).

Tan & Gimpiton 507 (33.1.1a); 568 (33.3.1).

Thomas & Patrick 235 (33.1.1a).

Wong 21 (33.5.1).

Wood & Wyatt-Smith 4493 (33.1.2); 4500 (33.6.1).

Wyatt-Smith 80371 (32.1.1).

Yalin Surunda 41 (34.1.5a).

LITERATURE CITED

Beaman, J. H., & Beaman, R. S. (1993). The gymnosperms of Mount Kinabalu. *Contr. Univ. Michigan Herb.* 19: 307–340.

Cockburn, P. F. (1980). *Trees of Sabah, Vol. 2. Sabah Forest Record*, no. 10.

Corner, E. J. H. (1978). Plant life, Chapter 6, pp. 112–178 in M. Luping, Chin Wen & E. R. Dingley, eds. *Kinabalu: Summit of Borneo*. Sabah Society Monograph, Sabah Society, Kota Kinabalu, Sabah, Malaysia.

Corner, E. J. H., with minor revisions by Beaman, J. H. (1996). The plant life of Kinabalu—an introduction. Pp. 101–149 in Wong, K. M., & Phillipps, A. (eds.). *Kinabalu: Summit of Borneo, a revised and expanded edition*. The Sabah Society in association with Sabah Parks, Kota Kinabalu, Malaysia.

Kitayama, K. (1991). *Vegetation of Mount Kinabalu Park, Sabah, Malaysia. Map of physiognomically classified vegetation*. East-West Center, Honolulu, Hawaii.

Laubenfels, D. J. de. (1988). Coniferales. *Fl. Males.* I, 10(3): 337–453.

Markgraf, F. (1930). Monographie der Gattung *Gnetum. Bull. Jard. Bot. Buitenzorg*, ser. 3, 10: 407–511.

Markgraf, F. (1951). Gnetaceae. *Fl. Males.* I, 4: 336–347.

Meijer, W. (1965a). *A botanical guide to the flora of Mount Kinabalu. Symposium on ecological research on humid tropics vegetation*: 325–364. Government of Sarawak, and UNESCO.

Meijer, W. (1965b). *Plant Life on Mount Kinabalu*. Forest Department, Sabah.

Page, C. N. (1988). New and maintained genera in the conifer families Podocarpaceae and Pinaceae. *Notes Roy. Bot. Gard. Edinburgh* 45: 377–395.

Parris, B. S., Beaman, R. S. & Beaman, J. H. (1992). *The Plants of Mount Kinabalu: 1. Ferns and Fern Allies*. Royal Botanic Gardens, Kew.

Stapf, O. (1894). On the flora of Mount Kinabalu, in North Borneo. *Trans. Linn. Soc. London, Bot.* 4: 69–263, pl. 11–20.

Stapf, O. (1914). *In* L. S. Gibbs, A contribution to the flora and plant formations of Mount Kinabalu and the highlands of British North Borneo. *J. Linn. Soc., Bot.* 42: 1–240, 8 pl.

Whitmore, T. C. (1980). A monograph of *Agathis*. *Pl. Syst. Evol.* 135: 41–69.

Wood, J. J., Beaman, R. S., & Beaman, J. H. (1993). *The Plants of Mount Kinabalu: 2. Orchids.* Royal Botanic Gardens, Kew.

NON-ORCHID MONOCOTYLEDONS

INTRODUCTION

The non-orchid monocotyledons of Mount Kinabalu (hereafter generally referred to as the monocotyledons) comprise 34 families, 163 genera, and 426 fully determined species and infraspecific taxa as presently represented in the database. Additionally, the database includes various conditionally determined taxa, namely 15 taxa with 'cf.' between the generic name and the specific epithet, seven similarly designated with 'aff.,' nine referred to as sp., sp. 1, or sp. 2, and 21 genera in which a set of one or more specimens is referred to as 'indet.' (undetermined). The total number of taxon names included in the Enumeration is 478. The monocot database includes 4662 records; there are 105 undetermined specimens in the database (these include a few unnumbered duplicates).

The largest monocot families are as follows (counting fully determined and named taxa: Poaceae, 88 taxa; Arecaceae, 80 taxa; Cyperaceae, 73 taxa; Zingiberaceae, 44 taxa; and Araceae, 40 taxa. The database includes 327 fully determined taxa in five families, whereas 29 other families are represented by only 99 fully determined taxa. Although orchids are not included in this volume, it should be noted that the Kinabalu Orchidaceae enumeration (Wood *et al.* 1993) included 711 taxa; 34 additional taxa have been recorded recently by Wood and Barkman (in press).

HISTORICAL ASPECTS

The account by Stapf (1894) listed 30 non-orchid monocots in 10 families (as treated here) from Kinabalu and one from Tuaran. Ten of the 30 are still recognized under the names indicated by Stapf. The names are different for the other material treated by Stapf because of taxonomic and nomenclatural changes in some instances and redetermination of specimens in others. Ten new species and two new varieties were described by Stapf, seven of which are currently recognized, although not in all cases under the name he used. Two taxa he included were given provisional names, one of which was only to generic level. He noted that palms occur on Mount Kinabalu, but did not provide a formal treatment for any species, because, except for one incomplete specimen, not a single collection was represented in the Kew herbarium. He indicated that Burbidge had mentioned *Cocos nucifera* L. and *Oncosperma filamentosa* Blume (*O. horridum* or *O. tigillarium*) as being grown near Dusun houses. Four other palm genera also were mentioned.

Gibbs (1914) listed for Mount Kinabalu 41 non-orchid monocots in 14 families. Seventeen of the taxa she included are recognized by the same name in the present treatment. Seven new species and varieties were described along with one new genus, and one new combination was made. Additionally, she included 84 monocot taxa from other areas of Sabah, representing 17 families. Five of these were described as new and two new combinations were made. The majority of those non-Kinabalu species she listed, particularly the grasses and sedges, have now been collected on the mountain. The monocot accounts in Gibbs were authored largely by H. N. Ridley and O. Stapf, with additional contributions by O. Beccari (Arecaceae), W. B. Turrill (Centrolepidaceae and some Cyperaceae), U. Martelli (Pandanaceae), and A. B. Rendle (Hydrocharitaceae). The two largest families included in her account were the Cyperaceae with 38 taxa and Poaceae with 37.

In contrast to the other major groups (*i.e.*, pteridophytes and orchids) previously treated in this series enumerating the Kinabalu flora, there have been no earlier accounts concerning Mount Kinabalu monocotyledons except for the publications of Stapf and Gibbs. Indeed, excluding the orchids, the Kinabalu monocots have received relatively little taxonomic attention other than a few publications covering wider areas of Borneo on certain families. Noteworthy in this respect are accounts on variou palm genera (Arecaceae) by J. Dransfield and gingers (Zingiberaceae) by R. M. Smith. Eighteen families, in the sense that they are recognized here, have been treated in *Flora Malesiana*. These include the Alliaceae, Amaryllidaceae (including Hypoxidaceae), Burmanniaceae, Centrolepidaceae, Liliaceae sens. lat. (*i.e.*, Convallariaceae, Melanthiaceae, Phormiaceae), Cyperaceae, Dioscoreaceae, Flagellariaceae sens. lat. (*i.e.*, Flagellariaceae, Hanguanaceae, Joinvilleaceae), Iridaceae, Juncaceae, Limnocharitaceae (as Butomaceae p.p.), Taccaceae and Triuridaceae. Families still needing considerable taxonomic attention, and thus considered difficult for the Kinabalu area, include the Araceae, Commelinaceae, Dioscoreaceae, Dracaenaceae, Marantaceae, Musaceae, Pandanaceae (notwithstanding important papers by the late B. C. Stone), Poaceae (currently under study for *Flora Malesiana* by J. F. Veldkamp) and Smilacaceae.

In view of the fact that the Kinabalu monocotyledons include the cereal grasses, the world's most important food plants, and many other species of great economic value, including coconut, the rattans, betel nut, taro, bananas and pandans, it is remarkable that they have received virtually no attention in this area. A single monocotyledonous family, the Orchidaceae, has captured a lion's share of interest in the Kinabalu botanical literature, not because of essential useful properties, but rather because of their sometimes remarkable beauty, their extraordinary floral morphology, the rarity of some species and their attractiveness to orchid growers and fanciers. Other Kinabalu monocotyledons have beautiful flowers, but have been unable to compete for attention with the orchids, and indeed with many other components of the Kinabalu flora. It is peculiar that the otherwise marvellous chapter by Corner (1996) on the plant life of Kinabalu mentions not a single monocotyledon.

NEW COMBINATION AND NEW RECORDS

The present study includes no descriptions of new taxa. Ten groups of specimens in eight families might constitute new taxa, but they have not been sufficiently
studied to warrant publication here. One new combination, *Pycreus sanguinolentus*
(Vahl) Nees subsp. *cyrtostachys* (Miq.) D. A. Simpson, is proposed. Records for first
known occurrence in Borneo are reported for *Dioscorea sumatrana* Prain & Burkill
and *Pothos ovatifolius* Engl. The spontaneous occurrence of the tropical American
species *Paspalum virgatum* L. is reported for the first time in Malesia.

MONOCOTYLEDON COLLECTIONS

This study of the Kinabalu monocotyledons is based upon 3282 collections
represented by about 4660 specimens. These have been obtained by 121 collectors or
collecting teams and are deposited in 23 herbaria. We have attempted to examine
systematically most or all collections in BM, K and L, which provide the primary
resource for this inventory. Records from other herbaria were mostly sporadically or
serendipitously obtained.

Collectors or collecting teams from whom we have recorded 20 or more
collections are listed below with the approximate dates of their time in the field and
number of collections made and taxa obtained.

Amin Gambating *et al.*, June 1986–November 1988: 35 collections, 31 taxa
J. H. Beaman, July 1983–August 1984 (a few also in 1990, 1992 and 1994–96): 183
    collections, 90 taxa
W. L. Chew & E. J. H. Corner, January–May 1964: 107 collections, 62 taxa
W. L. Chew, E. J. H. Corner & A. Stainton, July–November 1961: 157 collections,
    103 taxa
J. & M. S. Clemens, August 1931–January 1, 1934: 1131 collections, 264 taxa
M. S. Clemens, October–December 1915: 53 collections, 46 taxa
S. Collenette (née Darnton), September 1958–March 1981: 38 collections, 31 taxa
Daim Andow, April 1995–December 1996: 54 collections, 51 taxa
S. Darnton, February–March 1954: 29 collections, 28 taxa
Dius Tadong, July 1992–November 1994: 131 collections, 70 taxa
Doinis Soibeh, September 1992–September 1994: 86 collections, 60 taxa
J. Dransfield *et al.*, April–September 1979: 65 collections, 33 taxa
L. S. Gibbs, February 1910: 56 collections, 39 taxa
G. D. Haviland, March–April 1892: 33 collections, 25 taxa
Jibrin Sibil, September 1992–November 1993: 45 collections, 34 taxa
Jusimin Duaneh, October 1992–September 1994: 38 collections, 34 taxa
Kinsun Bakia, June 1994–March 1995: 53 collections, 48 taxa
K. Kokawa & H. Hotta, January–February 1969: 21 collections, 18 taxa

Lomudin Tadong, March 1995–September 1995: 42 collections, 37 taxa
Lorence Lugas, April 1995–August 1996: 213 collections, 120 taxa
Matamin Rumutom, April 1995–September 1996: 28 collections 26 taxa
W. Meijer et al., November 1959–December 1990: 61 collections, 45 taxa
Meliden Giking, September 1992–April 1993: 25 collections, 24 taxa
Sani Sambuling, November 1993–September 1996: 150 collections, 48 taxa
D. A. Simpson, November 1989: 33 collections, 27 taxa
J. Sinclair, Kadim bin Tassim & Sisiron bin Kapis, June 1957: 24 collections, 22 taxa
J. M. B. Smith, July–August 1978: 26 collections, 20 taxa
B. C. Stone, May 1973–April 1977: 20 collections, 14 taxa

The specimens upon which the enumeration is based were collected over a period of 145 years from 1851, when Hugh Low made the first collections, through 1996. As has been the case with the other major plant groups reported upon thus far (*i.e.*, the pteridophytes and orchids), the collecting team of Joseph and Mary Strong Clemens obtained far more collections than has anyone else. The Clemenses were first on Mount Kinabalu in 1915, for a relatively short stay from October to December. At that time Joseph Clemens collected orchids, which were numbered under his name. Other taxa were numbered under Mrs. Clemens's name. The Clemenses returned to Mount Kinabalu in 1931 and collected there from August through June 1932. For most of the rest of 1932 they were in Bogor having preliminary determinations made of their specimens by the Dutch botanists van Slooten and van Steenis. They returned to Mount Kinabalu in December, 1932, and collected there until 1 January 1934. Specimens from the 1931–1934 period were numbered and labeled under both their names. A glance at the Index to Numbered Collections will give an immediate impression of the numerical dominance of Clemens collections relative to all other collectors. The enumeration includes 53 collections (71 specimens) obtained by M. S. Clemens in 1915 and 1135 collections (2039 records) obtained by both Clemenses from August 1931 through 1 January 1934. Among the 485 taxon names listed in the Enumeration (including incompletely determined taxa), 310 taxa are represented among the Clemens collections. Excluding material from the *PEK* collectors (discussed below), all collectors, including the Clemenses, have obtained 377 taxa. Therefore, it can be said that the Clemenses found 82 percent or all but 67 of the taxa of monocots recorded by botanical collectors.

Notwithstanding the enormous importance of the Clemens collections to knowledge of the flora of Mount Kinabalu, working with their material involves a number of problems. These were discussed in some detail in the treatment of the Kinabalu orchids and are not repeated here. It may be emphasized, however, that the Clemenses were not formally trained as botanists and apparently did not appreciate the importance to systematists of understanding local population variation. As a result, in their zeal to collect extensive sets of duplicates that could be sold, apparently they not infrequently obtained additional material of a particular species (or what they took to be that species) and added it to a number that had been previously collected. As a consequence a considerable number of mixed collections resulted, so one needs to be cautious in assuming that all material under a particular number is the same species and of exactly the same provenance.

IMPACT OF *PROJEK ETNOBOTANI KINABALU* (*PEK*) ON COLLECTIONS

The specimens upon which this research is based have been obtained by two distinct groups of collectors: 1) visiting botanists and others interested in the natural history of the area, referred to below as Flora of Mount Kinabalu (FMK) collectors, and 2) local people who have participated in the *Projek Etnobotani Kinabalu* (*PEK*), an ethnobotanical programme based at Kinabalu Park initiated by researchers from Sabah Parks (the State conservation agency), the Universiti Malaysia Sarawak and the People and Plants Initiative (an international ethnobotanical programme of WWF, UNESCO and the Royal Botanic Gardens, Kew).

In order to compare results from these two groups of collectors we have analysed the number and diversity of monocotyledons obtained by both groups. The results are summarised in Tables 1 and 2. We are indebted to Gary Martin for his contribution to this analysis. The FMK collectors have obtained monocotyledons represented by 374 different names, or, if collections determined only to genus are excluded, then 357 names. The *PEK* collections represent 259 different names (including specimens determined only to genus). Low-elevation, useful and mor-phologically salient monocotyledons are in general well represented in the *PEK* collections, whereas high elevation, inconspicuous plants not known to the Dusun are relatively deficient.

The *PEK* collectors, members of Dusun ethnic communities around the Park, attended an ethnobotanical training course in July 1992 and began a systematic ethnobotanical inventory immediately thereafter. Over a period of 4.5 years, 17 local collectors in nine communities have made more than 7000 collections that are being processed by the research staff of Sabah Parks. Areas in which the *PEK* collectors have worked are shown on the map in Fig. 4.

Specimens and ethnobotanical data have been collected at more than 100 locally named sites from a broad range of natural and anthropogenic vegetation types. Following *PEK* guidelines, the collectors have focused on plants named or used by Dusun people but have made general floristic collections as well. They were instructed to avoid collecting orchids, which are sometimes illegally harvested from Kinabalu Park. Sabah Parks granted them permission to collect inside Kinabalu Park only in 1995. Their access to areas at the higher elevations were thus limited during the first years of the *PEK*. In any case, except in service as guides and other tourist amenities the Dusun have rarely gone to the upper slopes, because these are traditionally considered the resting place of the dead and thus taboo to the living (Regis 1996).

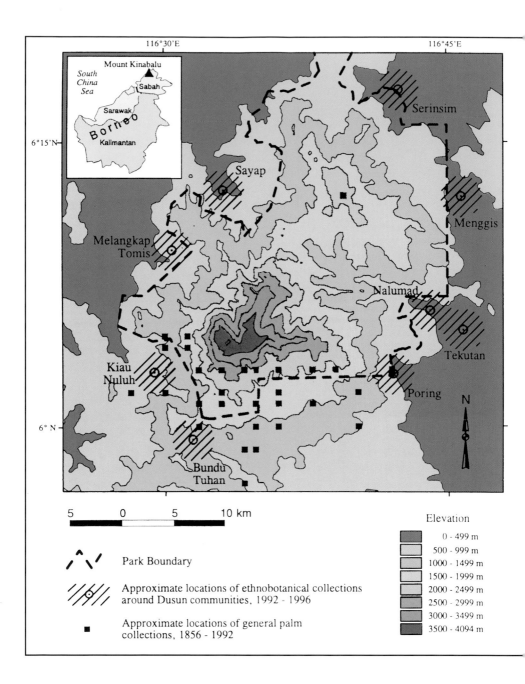

Fig. 4. Map of Mount Kinabalu, showing the approximate locations of palm collections made by botanists carrying out general surveys from 1856 to 1992, and by Dusun collectors participating in the *Projek Etnobotani Kinabalu* from 1992 to 1996.

Table 1. Comparison of the contribution of FMK and *PEK* collectors to current knowledge of genera, species and infraspecific taxa of 34 families of monocotyledons on Mount Kinabalu.

	Genera			Species and infraspecific taxa		
	FMK	*PEK*	Combined	FMK	*PEK*	Combined
Acoraceae	0	1	1	0	1	1
Agavaceae	0	1	1	0	1	1
Alliaceae	0	1	1	0	1	1
Amaryllidaceae	0	1	1	0	1	1
Araceae	14	11	14	41	15	44
Arecaceae	10	19	19	48	74	81
Burmanniaceae	3	0	3	6	0	6
Cannaceae	0	1	1	0	1	1
Centrolepidaceae	1	0	1	1	0	1
Commelinaceae	7	7	8	14	10	17
Convallariaceae	2	1	2	2	1	2
Costaceae	1	1	1	3	2	4
Cyperaceae	21	14	22	60	27	73
Dioscoreaceae	1	1	1	9	6	12
Dracaenaceae	1	1	1	3	3	3
Eriocaulaceae	1	0	1	2	0	2
Flagellariaceae	1	1	1	1	1	1
Hanguanaceae	1	1	1	1	2	2
Hypoxidaceae	1	1	1	2	2	3
Iridaceae	1	0	1	1	0	1
Joinvilleaceae	1	1	1	1	1	1
Juncaceae	2	0	1	3	0	3
Limnocharitaceae	0	1	1	0	1	1
Lowiaceae	0	1	1	0	1	1
Marantaceae	4	4	4	5	6	6
Melanthiaceae	2	0	2	2	0	2
Musaceae	1	1	1	3	1	4
Pandanaceae	2	2	2	15	8	18
Phormiaceae	1	1	1	2	1	2
Poaceae	42	32	52	67	44	88
Smilacaceae	1	1	1	8	5	8
Taccaceae	1	1	1	2	1	2

Table 1 (continued)

	Genera			Species and infraspecific taxa		
	FMK	*PEK*	Combined	FMK	*PEK*	Combined
Triuridaceae	1	0	1	4	0	4
Zingiberaceae	11	8	11	47	24	52
Totals	135	116	163	352	240	448

The Orchidaceae were excluded from the analysis in Table 1, because the present study concerns non-orchid monocotyledons and because the *PEK* collectors were not supposed to be collecting them. Despite this prohibition, *PEK* collectors made 147 orchid collections, which include 41 genera and 87 species. One genus and 13 species were previously uncollected on Mount Kinabalu. In all, *PEK* collections yielded seven families, 28 genera and 99 species of non-orchid monocotyledons previously unrecorded for the mountain. Including orchids, the monocots added by *PEK* are thus seven families, 29 genera and 112 species.

Table 2. Summary comparison of Mount Kinabalu floristic and ethnobotanical collections of monocotyledons.

	FMK collections	*PEK* collections	Combined FMK and *PEK* collections	New records in *PEK* collections
Families	27	27	34	7
Genera	135	116	163	28
Species and infraspecific taxa	353	241	449	99

Projek Etnobotani Kinabalu had an initial emphasis on collecting palms, a family of special economic and conservation importance. Because of the difficult and time-consuming process in making palm specimens, the collectors were paid a bonus for them. In order to visualize the results of palm collecting efforts by the *PEK* collectors, we have made an analysis of presently available collections in this family. These include 375 *PEK* collections now in the palm herbarium of The Royal Botanic Gardens, Kew, representing 19 genera and 74 species and infraspecific taxa. The 372 FMK palm collections included in the analysis were obtained over a 136-year period from 1856 to 1992. These have been recorded from several herbaria and belong to 10 genera and 48 species and infraspecific taxa.

Together the FMK and *PEK* collections have yielded 757 palm records that correspond to a total of 81 species and infraspecific taxa in 19 genera (Table 3).

Table 3. Comparison of palm taxa in the Mount Kinabalu floristic and ethnobotanical collections.

Genus	FMK collections (1856–1992)	PEK collections (1992–1996)	Combined FMK and PEK collections (1856–1996)
Areca	1	3	3
Arenga	2	4	4
Calamus[1]	16	20	22
Caryota	2	2	2
Ceratolobus	0	1	1
Cocos	0	1	1
Daemonorops	10	12	12
Dypsis[2]	0	1	1
Elaeis[2]	0	1	1
Eugeissona	0	1	1
Korthalsia	3	9	9
Licuala	0	3	3
Metroxylon	0	1	1
Nenga	0	1	1
Oncosperma	0	1	1
Pinanga	9	6	10
Plectocomia[3]	1	2	2
Plectocomiopsis	2	2	2
Salacca	2	3	4
Total	48	74	81

[1]One additional species of *Calamus* (*C. scipionum*) has been observed but not collected on Mount Kinabalu. Two varieties of *Calamus laevigatus* are included as distinct taxa in this summation.

[2]Introduced recently to the Mount Kinabalu area.

[3]An incompletely determined FMK collection of *Plectocomia* may be a distinct species.

Forty-one species were recorded in both sets of collections, while seven species were detected only in the floristic inventory and 33 were documented only in the ethnobotanical collections. The comparative percentages of FMK and *PEK* palm collections made during the entire history of plant collecting activities on Mount Kinabalu are shown graphically in Fig. 5.

We attribute the productivity of the *PEK* collecting of palms and other plants to the Dusun peoples' intimate knowledge of living organisms, localities and microenvironments around Mount Kinabalu, their ability to collect at a variety of locations year-round and their tendency to collect a range of cultivated, semi-cultivated and wild species. Most visiting botanists came from Europe, the United States, or Peninsular Malaysia and were often unfamiliar with the plants, ecology and geography of Mount Kinabalu. Many of them collected mainly along established trails in accessible primary and mature secondary forests of the Kinabalu massif and focused on documenting wild species. Their success in finding a diversity of plants can be attributed in part to collaboration with Dusun assistants who accompanied them to the forest and aided them in making plant collections. For example, the label data on palm specimens obtained by the Clemenses indicate that 19 collections corresponding to 10 species were made by Dusun collaborators.

In addition to new taxonomic records, the ethnobotanical inventory has enriched our knowledge of local classification, use and distribution of plants on Mount Kinabalu. The *PEK* collectors have registered local names and uses for nearly all specimens they collected. Even though many FMK plant collecting trips included Dusun helpers who gave some ethnobotanical information, visiting botanists for the most part have not systematically recorded local names and uses for the plants they collected. FMK collections come largely from the southern parts of Mount Kinabalu (as can be seen in the locations of palm collections shown in Fig. 4), generally at higher elevations within Kinabalu Park. Those made by *PEK* collectors are from sites at lower elevations that form a virtual ring around the massif, giving a more complete representation of species distributions on the mountain than was previously available. This geographical coverage is particularly useful for understanding plant distribution patterns on Mount Kinabalu, where a high percentage of plants have restricted distributions and many plant species are known from a single collection (Beaman & Beaman 1990).

Although local collectors have been more productive than visiting botanists in the overall rate of making palm collections and in the number of palm taxa recorded, our data suggest that the most complete floristic inventories will be achieved by combining community-based collecting with general floristic surveys, particularly if the collections: 1) are from vegetation types across a wide elevational gradient; 2) include the broadest possible variety of plants, useful and not, cultivated and wild, native and introduced; and 3) are obtained through plant collecting programmes carried out in collaboration with local institutions and international herbaria. In the case of *PEK*, we hypothesize that a comprehensive collection could be developed if local collectors are encouraged to obtain specimens of the flora at higher elevations,

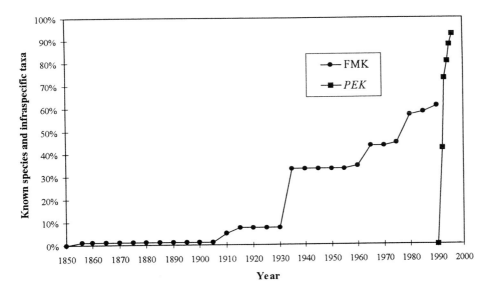

Fig. 5. Cumulative percentage of currently known Mount Kinabalu palms (species and infraspecific taxa) represented in FMK floristic collections made over 136 years (circles) and in *PEK* collections made during the first 4.5 years of *PEK* (squares). The number of new palm records in the FMK collections was counted by five-year periods from 1850 to 1990, and then yearly until 1992. The number of new palm records in the *PEK* collections was counted yearly from 1992 to 1996. (From G. J. Martin *et al.* in prep.)

and if they increased the scope of their collecting to include species regardless of whether or not they are named or used by Dusun people.

GEOGRAPHICAL DISTRIBUTIONS, ENDEMISM

In a symposium discussing results of the 1961 Royal Society expedition to Mount Kinabalu van Steenis (1964) noted that "The interest, and *uniqueness of Kinabalu lies largely with its mountain flora*." [Italics his.] He also stated that "Plant geographically the foothill zone, though extremely rich in species, yields few problems innate to Kinabalu." These observations are well borne out in the distributions of the non-orchid monocotyledons.

All of the lowland monocots have wide geographical distributions, many of them being pantropical. Few of the hill forest taxa and none of those from the lowlands have distributions localized to Mount Kinabalu or neighbouring areas. The relatively few taxa occurring in hill-forest habitats that seem to have localized distributions tend to be parts of poorly understood taxonomic complexes. Among these are *Homalomena gillii* and *Pothos kinabaluensis* (Araceae).

The monocotyledons treated in this volume constitute perhaps less than ten percent of the total flora of vascular plants of Mount Kinabalu. Therefore, we have not attempted generalizations about origins and distribution of the flora on the basis of this limited sample. Furthermore, a point about endemism made by Stapf (1894) over 100 years ago still rings true, *i.e.*, "To speak of the endemism of a district so little known and forming part of a likewise imperfectly explored flora is a very difficult task, ... " Notwithstanding this problem, we have included some brief observations on monocot distributions as presently understood.

Because we lack a ready source of data on general distributions of many of the Kinabalu monocotyledons, we have used the Cyperaceae as an exemplar family. They have been treated in Flora Malesiana (Kern 1974, Kern & Nooteboom 1979), so distribution data for the family are available. The Cyperaceae are the third largest non-orchid monocot family on Mount Kinabalu, with 73 fully determined taxa. They provide examples of lowland, mid-elevation and high-elevation occurrence, and a varied array of patterns of geographical distribution. Analyses of the Cyperaceae distributions are provided in Tables 4–7.

The predominant geographical affinity of the Cyperaceae on Mount Kinabalu is with Asia (45 taxa), as shown in Table 7. Some of these taxa are also Austral, a category that may be underrepresented in the analysis, because taxa well represented in Asia were placed in that category even when they extend to austral regions. The next most common affinity is that of taxa restricted or nearly restricted to Malesia (25). A considerable number (10) of the lowland and hill-forest taxa are pantropical, and even the lower montane forest taxa include some pantropical members (4). Austral species are fewest in the analysis (9), but 13 taxa considered Asian in affinity also extend to Austral regions. Only two species of the Cyperaceae are endemic or subendemic.

Wide disjunctions are most prominent among the upper montane/summit area Cyperaceae. A pattern repeated in several taxa is from Mount Kinabalu to southwest Sulawesi (particularly Mt. Latimodjong, ca. 3460 m) to various high mountains in New Guinea. Species with this distribution include *Carex breviculmis* var. *perciliata, Oreobolus ambiguus* and *Schoenus curvulus. Isolepis habra and I. subtilissima* likewise have this distribution, but occur also in high mountains of the Philippines and extend on to Australia and New Zealand. *Schoenus longibracteatus* skips the Philippines and Sulawesi and occurs only on Kinabalu and Mt. Doorman in Irian

Jaya, while *S. melanostachys* occurs in Australia, on Mt. Halcon in the Philippines and on Mount Kinabalu.

Table 4. Geographical affinity of lowland and hill-forest Cyperaceae on Mount Kinabalu.

Pantropical or very widespread	SE or E Asia	Australia or SW Pacific	Restricted or nearly restricted to Malesia
Cyperus compressus	*Cyperus castaneus*	*Carex graeffeana*	*Mapania caudata*
Cyperus cyperoides; introduced in Malesia?	*Cyperus compactus*	*Machaerina glomerata*	*Mapania cuspidata* var. *petiolata*
Cyperus haspan	*Cyperus diffusus*		*Mapania latifolia*
Cyperus sphacelatus; introduced in Malesia	*Cyperus laxus* var. *macrostachyus*		*Mapania palustris*
Fimbristylis complanata	*Cyperus pilosus*[1]		*Mapania squamata*
Fimbristylis dichotoma	*Fimbristylis dura*		*Pycreus sanguinolentus* subsp. *cyrtostachys*
Fumbristylis littoralis	*Fimbristylis miliacea*[1]		*Scleria motleyi*
Fuirena umbellata	*Fimbristylis obtusata*		*Scleria scrobiculata*[2]
Rhynchospora corymbosa	*Fimbristylis umbellaris*		
Rhynchospora rugosa	*Gahnia tristis*		
	Hypolytrum nemorum[1]		
	Kyllinga nemoralis		
	Rhynchospora rubra[1]		
	Schoenoplectus juncoides[1]		
	Schoenoplectus mucronatus[1]		
	Scleria ciliaris[1]		
	Scleria purpurascens var. *purpurascens*		
	Scleria terrestris[1]		

[1]Also Austral.
[2]Extending somewhat beyond Malesia to both north and southeast.

Table 5. Geographical affinity of lower-montane-forest Cyperaceae on Mount Kinabalu.

Pantropical or very widespread	SE or E Asia	Australia or SW Pacific	Restricted or nearly restricted to Malesia
Cyperus haspan	Carex cruciata var. cruciata	Machaerina glomerata	Carex perakensis var. borneensis
Fimbristylis dichotoma	Carex cruciata var. rafflesiana	Schoenus melanostachys	Costularia pilisepala
Kyllinga brevifolia	Carex cryptostachys		Machaerina aspericaulis[2]
Rhynchospora rugosa	Carex filicina		Machaerina disticha
	Carex hypolytroides		Mapania borneensis[3]
	Carex indica[1]		Mapania latifolia
	Carex perakensis var. perakensis		Mapania palustris[4]
	Carex tristachya var. pocciliformis		Schoenus delicatulus
	Cyperus pilosus[1]		Schoenus longibracteatus
	Eleocharis congesta		Scleria motleyi
	Gahnia baniensis		
	Kyllinga melanosperma		
	Lepidosperma chinense		
	Lipocarpha chinensis[1]		
	Schoenoplectus mucronatus[1]		
	Scirpus ternatanus		
	Scleria terrestris		

[1]Also Austral.
[2]Known only from Mount Kinabalu.
[3]One record from Sarawak; otherwise only known from Mount Kinabalu.
[4]Extending somewhat beyond Malesia to both north and southeast.

A few lower-montane-forest taxa, particularly in the genus *Machaerina,* also have interesting disjunctions. These include *M. disticha,* which occurs also in Kalimantan, the Philippines, Sulawesi, Maluku and Waigeo Island. *Machaerina glomerata* occurs in the Solomon and Palau Islands and in the Philippines, Sulawesi, Maluku and New Guinea. The upper montane *Machaerina falcata* has a Pacific distribution, occurring in Hawaii, Samoa, Fiji, Vanuatu and the Solomon Islands, in addition to Malesian occurrences in Sumatra, the Philippines, Sulawesi and New

Guinea. *Machaerina aspericaulis* seems to be the only strict Kinabalu endemic in the Cyperaceae. All four species of *Machaerina* on Mount Kinabalu occur in open habitats on ultramafic substrates. Another ultramafic lower montane/ upper montane species of Cyperaceae with a similar distribution is *Costularia pilisepala,* which is found on Mount Kinabalu and in Irian Jaya. The genus *Costularia* has an austral distribution centre, with nine of its approximately 20 species occurring in New Caledonia and only the one in Malesia.

Table 6. Geographical affinity of upper-montane-forest and summit-area Cyperaceae on Mount Kinabalu.

Pantropical or very widespread	SE or E Asia	Australia or SW Pacific	Restricted or nearly restricted to Malesia
	Carex breviculmis var. *breviculmis*[1]	*Isolepis habra*	*Carex breviculmis* var. *perciliata*
	Carex capillacea var. *capillacea*	*Isolepis subtilissima*	*Carex perakensis* var. *borneensis*
	Carex capillacea var. *sachalinensis*	*Machaerina falcata*	*Costularia pilisepala*
	Carex filicina	*Schoenus melanostachys*	*Oreobolus ambiguus*
	Carex perakensis var. *perakensis*	*Uncinia compacta*	*Schoenus curvulus*
	Carex verticillata		*Schoenus longibracteatus*
	Gahnia javanica		
	Lepidosperma chinense		
	Scirpus subcapitatus		

[1]Also Austral.

Table 7. Summary of geographical affinities of Cyperaceae on Mount Kinabalu.

	Pantropical or very widespread	Asian	Austral	Malesian	Endemic
Lowland and hill forest	10	19	2	8	—
Lower montane	4	17	2	11	2
Upper montane-summit area	—	9	5	6	—
Total[1]	14	45	9	25	2

[1]The total number adds up to more than the 73 taxa considered, because taxa that commonly occur in more than one habitat zone are included for each of those occurrences.

The low number of Kinabalu endemics in the Cyperaceae is characteristic of all the non-orchid monocotyledons. Table 8 lists the species thought to be endemic, subendemic or with restricted, disjunct occurrences, with the caveat that species

Table 8. Monocotyledons endemic to Mount Kinabalu or with restricted ranges and long disjunctions.

Taxon	Comments
Araceae	
Amorphophallus lambii	Also known from Tenom, the type locality
Colocasia oresbia	Recently described from Park Headquarters
Homalomena gillii	Taxonomy uncertain
Homalomena kinabaluensis	Taxonomy uncertain
Hottarum kinabaluense	Endemic
Pothos kinabaluensis	Taxonomy uncertain
Arecaceae	
Calamus mesilauensis	Possibly occurs also on Mt. Silam in eastern Sabah
Salaca dolicholepis	Endemic
Centrolepidaceae	
Centrolepis philippinensis	Also on high mountains of the Philippines, Sulawesi and New Guinea
Cyperaceae	
Costularia pilisepala	Also known from Irian Jaya
Isolepis habra	Also on high mountains in the Philippines, Sulawesi and New Guinea, and in Australia and New Zealand
Isolepis subtilissima	Also on high mountains in the Philippines, Sulawesi and New Guinea, and in Australia and New Zealand
Machaerina aspericaulis	Endemic
Mapania borneensis	Subendemic; one record from Sarawak
Oreobolus ambiguus	Also known from Sulawesi and New Guinea
Schoenus curvulus	Also known from Sulawesi and New Guinea
Schoenus delicatulus	Also known from eastern Sabah and Palawan
Schoenus longibracteatus	Also known from Irian Jaya
Eriocaulaceae	
Eriocaulon kinabaluense	Endemic
Iridaceae	
Patersonia lowii	Also in Sarawak, Sumatra (Mt. Leuser), the Philippines and New Guinea
Melanthiaceae	
Aletris foliolosa	Also known from Sumatra and the Philippines
Pandanaceae	
Freycinetia kinabaluana	Endemic
Pandanus beccatus	Endemic
Pandanus kinabaluensis	Endemic
Pandanus tunicatus	Endemic
Poaceae	
Agrostis infirma var. *borneensis*	Endemic
Agrostis infirma var. *diffusissima*	Endemic
Agrostis infirma var. *kinabaluensis*	Endemic

Table 8 (continued)

Taxon	Comments
Anthoxanthum horsfieldii var. *borneense*	Endemic
Bromus formosanus	Also known from Taiwan
Danthonia oreoboloides	Also in northern Sumatra (Mt. Leuser), Philippines, Sulawesi, New Guinea
Deschampsia flexuosa var. *ligulata*	Also in the Philippines (Mt. Pulog)
Isachne albomarginata	Also in New Guinea
Isachne clementis	Also in the Philippines
Isachne kinabaluensis	Also in Sumatra (Mt. Leuser)
Poa borneensis	Endemic
Poa epileuca	Also in Sulawesi and New Guinea
Poa papuana	Also in Sulewesi and New Guinea
Racemobambos hepburnii	Also known from the adjacent Crocker Range
Racemobambos hirsuta	Also known from eastern Sabah on Mts. Silam and Nicola
Racemobambos rigidifolia	Endemic
Sphaerobambos hirsuta	Endemic, known only from the Lohan area
Trisetum spicatum subsp. *kinabaluense*	Endemic
Yushania tessellata	Also known from Mt. Alab and in the Meligan area
Zingiberaceae	
Alpinia beamanii	Also known from the type locality in the adjacent Crocker Range
Alpinia havilandii	Also known from the Crocker Range and Mt. Trus Madi
Amomum kinabaluense	Endemic or also nearby in the Crocker Range
Amomum longipedunculatum	Endemic

presently thought to be endemic to the mountain, on further study, may turn out to be not so restricted. In the Enumeration 17 taxa are indicated as being endemic to Mount Kinabalu, a mere four percent of the non-orchid monocotyledons.

CULTIVATED AND INTRODUCED MONOCOTYLEDONS

Even though most of the Kinabalu monocotyledons have been taxonomically neglected, they constitute the majority of important cultivated plants of that area. *Projek Etnobotani Kinabalu* is now helping extend our knowledge of the cultivated plants of the mountain. For example, without the *PEK* collections there would be no herbarium records of the two most important food plants, rice (*Oryza sativa*) and maize (*Zea mays*). Many other species listed in Table 9 also have been recorded only by *PEK* collectors. Other cultivated monocots occur on Mount Kinabalu, but Table 9 includes only species documented by herbarium specimens. The extensively cultivated taro, *Colocasia esculenta*, apparently has not yet been collected. The day-

Table 9. Cultivated and introduced monocotyledons on Mount Kinabalu.

Taxon	Comments
Agavaceae	
Cordyline fruticosa	Possibly native but more likely in cultivation
Alliaceae	
Allium fistulosum	Unknown in the wild; a vegetable
Amaryllidaceae	
Pancratium zeylanicum	Possibly native but more likely in cultivation
Araceae	
Alocasia macrorrhizos	Giant taro
Arecaceae	
Areca catechu	Betel nut; extensively cultivated
Calamus caesius	Native species but cultivated around villages as a rattan source
Cocos nucifera	Coconut; extensively cultivated around villages; wild origin unknown
Dypsis lutescens	Golden bamboo palm; introduced from Madagascar; whether or not the species occurs spontaneously is not known
Elaeis guineensis	African oil palm; introduced from Africa and now extensively cultivated in plantations over much of Sabah
Eugeissona utilis	Native species but usually associated with villages and possibly a remnant of cultivation
Metroxylon sagu	Sago palm; cultivated around villages; probably of Bornean origin but wild occurrences not documented
Cannaceae	
Canna indica	Native of the New World; cultivated ornamental, possibly escaped
Cyperaceae	
Cyperus cyperoides	Widespread tropical weed; naturalized in Malesia
Cyperus sphacelatus	Widespread tropical weed; naturalized in Malesia
Pycreus pelophilus	Introduced from Africa
Limnocharitaceae	
Limnocharis flava	New World native, apparently spread since 1866 from Bogor Botanic Gardens, where cultivated at that time (van Steenis 1954)
Musaceae	
Musa acuminata	Banana; extensively grown and highly diverse on Mount Kinabalu
Pandanaceae	
Pandanus tectorius	The most commonly cultivated pandan; wild origin unknown
Poaceae	
Bambusa vulgaris	Common bamboo; introduced and probably becoming established
Coix lacryma-jobi	Job's tears; possibly native but probably mostly cultivated and escaped
Cymbopogon winterianus	Source of citronella oil; the species known only in cultivation
Gigantochloa levis	Poring bamboo; planted or naturalized
Oryza sativa	Rice; of Asiatic origin but unknown in the wild
Paspalum virgatum	Introduced from tropical America

Table 9 (continued)

Taxon	Comments
Pennisetum clandestinum	Kikuyu grass; probably recently introduced as a pasture grass on the Pinosuk Plateau
Pennisetum purpureum	Elephant grass; introduced, probably as a pasture grass
Poa annua	A recently arrived weed at high elevations around Panar Laban
Setaria italica	Foxtail millet; cultivated as both grain and forage plant; whether or not it occurs spontaneously on Kinabalu is unknown
Sorghum bicolor	Sorghum; cultivated as both grain and forage plant; possibly naturalized
Zea mays	Maize; introduced from the New World; probably mostly used as a vegetable; possibly occurring spontaneously as an escape
Zingiberaceae	
Alpinia galanga	Unknown in the wild
Curcuma zedoaria	Natural origin not precisely established; easily becomes naturalized
Etlingera elatior	Torch ginger; widely cultivated

lily, *Hemerocallis*, perhaps *H. fulva* or one of its hybrids, is abundantly planted around Park Headquarters; it appears in numerous photographs of the area, and many tourists must think it is native. Apparently it has not escaped the Headquarters area.

Ecological Associations

Our research on the flora of Mount Kinabalu has a primarily taxonomic orientation. Research on vegetation structure and composition is presently being carried out by Kanahiro Kitayama. Three of his publications are cited in the section on Vegetation. The herbarium specimens upon which our study is based have been rather randomly and serendipitously obtained, but they can be used to give some insight into common components of the vegetation and the habitats in which species occur. We have therefore attempted to obtain a somewhat objective measure of frequency of the monocotyledons in the major vegetation zones by counting the number of collections by which each taxon is represented. First we determined which taxa have been collected eight or more times; these 135 taxa are listed in Table 10. For each of the zones, *i.e.*, lowlands, hill forest, lower montane forest, upper montane forest and summit area, it is indicated whether or not a species occurs in that habitat. The method provides an imperfect means for determining species frequency but is the only approach available from our data. This procedure likely will bias the results toward particularly conspicuous, attractive, easy to collect, or taxonomically interesting taxa, or perhaps toward species that have extended flowering and fruiting times.

Although we have arbitrarily used eight collections as the criterion for recognizing common Kinabalu monocotyledons, it is noteworthy that some taxa are

Table 10. Monocotyledons collected eight or more times on Mount Kinabalu and the habitats in which they occur.

	low-lands	hill forest	lower montane forest	upper montane forest	summit area	ultra mafic	dis-turbed areas
Araceae							
Aglaonema simplex		X	X				
Amydrium medium	X	X	X				
Arisaema filiforme		X	X				
Arisaema umbrinum	X	X					
Homalomena gillii		X	X				
Homalomena propinqua		X				X	
Piptospatha elongata		X	X				
Rhaphidophora korthalsii		X	X				
Rhaphidophora sylvestris		X	X				
Schismatoglottis calyptrata		X	X				
Scindapsus borneensis	X		X				
Scindapsus curranii	X						
Scindapsus pictus		X	X				
Scindapsus rupestris		X	X				
Arecaceae							
Areca kinabaluensis			X				
Arenga undulatifolia	X	X					X
Calamus acuminatus	X	X					
Calamus amplijugus	X	X					
Calamus blumei		X					
Calamus gibbsianus			X	X			
Calamus javensis		X	X	X		X	
Calamus marginatus		X	X			X	
Calamus mesilauensis			X				
Calamus pogonacanthus	X	X	X				
Calamus tenompokensis		X	X				
Caryota mitis	X	X					X
Daemonorops korthalsii	X	X					
Daemonorops longipes	X	X	X				
Daemonorops longistipes		X	X				
Daemonorops microstachys	X	X					
Daemonorops sabut	X	X				X	

Table 10 (continued)

	low-lands	hill forest	lower montane forest	upper montane forest	summit area	ultra-mafic	dis-turbed areas
Daemonorops							
sparsiflora	X	X					
Korthalsia echinometra	X	X					
Korthalsia jala	X	X					
Korthalsia rigida	X	X					
Licuala valida	X	X					
Nenga pumila var.							
pachystachya	X	X					
Pinanga capitata		X	X	X			
Pinanga pilosa		X	X				
Salacca dolicholepis		X	X				
Burmanniaceae							
Burmannia longifolia			X			X	
Centrolepidaceae							
Centrolepis							
philippinensis					X		
Commelinaceae							
Amischotolype glabrata		X	X				
Amischotolype							
marginata		X					
Pollia secundiflora		X	X				
Convallariaceae							
Ophiopogon caulescens		X	X				
Costaceae							
Costus speciosus	X	X					
Cyperaceae							
Carex cruciata							
var. *cruciata*		X	X			X	
Carex filicina			X	X	X	X	
Carex perakensis var.							
borneensis		X	X	X		X	
Carex verticillata				X	X		
Costularia pilisepala			X	X		X	
Gahnia javanica			X	X			
Machaerina falcata				X		X	
Mapania latifolia		X	X				
Mapania palustris		X	X			X	
Schoenus curvulus				X	X	X	
Schoenus							
melanostachys			X	X		X	
Scirpus subcapitatus			X	X	X	X	
Scleria purpurascens var.							
purpurascens	X	X					

Table 10 (continued)

	low-lands	hill forest	lower montane forest	upper montane forest	summit area	ultra-mafic	dis-turbed areas
Scleria terrestris		X	X				
Dioscoreaceae							
Dioscorea pyrifolia		X	X				
Dracaenaceae							
Dracaena angustifolia	X	X					
Dracaena elliptica		X	X				
Eriocaulaceae							
Eriocaulon hookerianum var. *hookerianum*				X		X	
Eriocaulon kinabaluense				X	X		
Hanguanaceae							
Hanguana major		X	X				
Hypoxidaceae							
Curculigo latifolia	X	X	X				
Iridaceae							
Patersonia lowii			X	X		X	
Joinvilleaceae							
Joinvillea ascendens subsp.*borneensis*		X	X				
Marantaceae							
Donax canniformis	X	X					
Phacelophrynium maximum		X				X	
Phrynium placentarium	X	X					
Melanthiaceae							
Aletris foliolosa			X	X		X	
Petrosavia stellaris			X	X			
Pandanaceae							
Freycinetia kinabaluana		X	X				
Freycinetia rigidifolia		X	X				
Pandanus kinabaluensis			X				
Pandanus pectinatus		X	X				
Pandanus tunicatus		X	X			X	
Phormiaceae							
Dianella ensifolia		X	X				
Dianella javanica		X	X	X			
Poaceae							
Agrostis infirma var. *borneensis*					X		
Agrostis infirma var. *diffusissima*				X	X		
Agrostis infirma var. *kinabaluensis*					X		

Table 10 (continued)

	low-lands	hill forest	lower montane forest	upper montane forest	summit area	ultra-mafic	dis-turbed areas
Anthoxanthum horsfieldii var. *borneense*				X	X		X
Centotheca lappacea	X	X					X
Coix lacryma-jobi	X	X					
Cyrtococcum accrescens		X	X				
Danthonia oreoboloides				X	X		
Deschampsia flexuosa var. *ligulata*				X	X		
Dinochloa sublaevigata	X	X					X
Eleusine indica	X	X					
Gigantochloa levis	X	X					
Ichnanthus pallens		X	X				
Imperata conferta	X						X
Isachne albens		X	X	X			
Isachne albomarginata		X	X			X	
Isachne clementis				X		X	
Isachne kinabaluensis		X	X				
Lophatherum gracile		X					
Microstegium geniculatum		X	X				
Miscanthus floridulus var. *malayanus*		X	X	X			
Oplismenus compositus	X	X	X				
Oplismenus hirtellus		X	X			X	
Panicum notatum		X					X
Panicum sarmentosum	X	X					
Paspalum conjugatum		X					
Poa epileuca				X	X		
Poa papuana				X	X		
Racemobambos gibbsiae			X	X			
Racemobambos hepburnii			X				
Setaria palmifolia	X	X	X				X
Thysanolaena latifolia		X					
Trisetum spicatum subsp. *kinabaluense*				X	X		X
Smilacaceae							
Smilax borneensis	X	X					
Smilax corbularia		X	X				
Smilax lanceifolia		X	X	X			
Smilax leucophylla		X	X	X		X	
Smilax odoratissima	X	X					

Table 10 (continued)

	low-lands	hill forest	lower montane forest	upper montane forest	summit area	ultra-mafic	dis-turbed areas
Smilax sp. 1		X	X				
Taccaceae							
Tacca palmata		X				X	
Zingiberaceae							
Alpinia aquatica		X					
Alpinia havilandii			X				
Alpinia ligulata	X						X
Alpinia nieuwenhuizii	X	X	X				
Amomum kinabaluense			X				
Amomum sceletescens		X	X				
Burbidgea schizocheila		X	X				
Globba franciscii		X	X				
Globba pendula		X					
Globba propinqua	X	X	X				
Globba tricolor var. gibbsiae		X					
Hedychium cylindricum		X	X			X	
Hedychium muluense		X	X				

represented by far greater numbers. The species that apparently has been collected more times than any other is *Smilax lanceifolia,* with 61 collections; *S. leucophylla* is close with 56. Why the greenbriars have been so popular with collectors is an open question, but they may have been tangled up with them. The other most frequently collected taxa include five palms, namely *Daemonorops longipes* (54), *Calamus gibbsianus* (49), *C. javensis* (47), *Pinanga pilosa* (45) and *P. capitata* (41). All of the frequently collected taxa occur in at least two vegetation zones, and most of them occur in three. For example, the two species of *Smilax* noted above occur in hill forest, lower montane forest and upper montane forest; *Daemonorops longipes* occurs in the lowlands, hill forest and lower montane forest and *Calamus gibbsianus* occurs in lower and upper montane forest. The latter species on Mount Kinabalu occurs at a higher elevation than any other palm in Southeast Asia (Uhl & Dransfield 1987). Thirty-three taxa are represented in the database by 20 or more collections, and the palms are the lead family in this respect with ten abundantly collected taxa.

LOWLANDS

Among all Kinabalu monocotyledons the palms are the most predominantly lowland family. Over 60 percent (*i.e.*, about 50 species) of the species in this family are recorded from the lowlands, and most of the rest occur in hill forest at only slightly higher elevations. The rattans are particularly prominent among the lowland palms, with more than half of the species of *Calamus*, three-fourths of *Daemonorops* and all of *Korthalsia*, *Plectocomia* and *Plectocomiopsis* lowland species, some of which may also occur at higher elevations.

In addition to the palms two other families with large numbers of lowland species are the Poaceae and Cyperaceae. Many of these are widespread weedy taxa. Lowland grasses include the bamboos *Bambusa vulgaris*, *Dinochloa sublaevigata*, *Gigantochloa levis* and *Schizostachyum latifolium* and other species, such as *Dactyloctenium aegyptium*, *Digitaria junghuhniana*, *D. setigera*, *D. violascens*, *Echinochloa colona*, *Eleusine indica*, *Eragrostis unioloides*, *Imperata conferta*, *I. cylindrica*, *Isachne miliacea*, *Ischaemum barbatum*, *I. polystachyum*, *Panicum sarmentosum*, *Pogonatherum crinitum*, *Schizachryium brevifolium*, *Scrotochloa urceolata*, *Setaria palmifolia*, *S. sphacelata* and *Sorghum propinquum*. Lowland sedges include *Cyperus castaneus*, *C. compactus*, *C. cyperoides*, *C. diffusus*, *C. laxus*, *C. pilosus*, *C sphacelatus*, *Fimbristylis complanata*, *F. dichotoma*, *F. dura*, *F. littoralis*, *F. miliacea*, *F. obtusata*, *F. umbellaris*, *Fuirena umbellata*, *Kyllinga nemoralis*, *Pycreus sanguinolentus* subsp. *cyrtostachys*, *Rhynchospora rubra*, *Scleria ciliaris* and *S. scrobiculata*.

A number of gingers (Zingiberaceae), including *Alpinia ligulata*, *Amomum uliginosum*, *Boesenbergia pulchella*, *Etlingera elatior*, *F. fimbriobracteata*, *E. littoralis*, *Globba propinqua* and *Hornstedtia havilandii*, are lowland species, as are several species of the Commelinaceae: *Amischotolype mollissima*, *Belosynapsis ciliata*, *Commelina diffusa*, *Floscopa scandens*, *Murdannia japonica* and *Rhopalephora vitiensis*.

Miscellaneous other noteworthy lowland monocotyledons include the large herb *Amorphophallus lambii* (Araceae), *Costus speciosus* (Costaceae), *Dioscorea hispida* (Dioscoreaceae), *Flagellaria indica* (Flagellariaceae), *Curculigo latifolia* (Hypoxidaceae), *Donax canniformis*, *Phrynium placentarium* and *Stachyphrynium cylindricum* (Marantaceae), *Pandanus gibbsianus* and *P. leuconotis* (Pandanaceae) and *Smilax borneensis* and *S. odoratissima* (Smilacaceae).

Data from *PEK* have provided a considerably better basis for knowledge of the lowland flora of Mount Kinabalu than was previously available. Nevertheless, there still is not much information about the original lowland primary forest around the mountain, but collections from remnant stands and scattered individual plants, particularly on the north side of the mountain, help fill some gaps in understanding of the lowland forests.

HILL FOREST

Over half of the Kinabalu monocotyledons (285 taxa), including three-fourths (60 taxa) of palms, are recorded from hill forest. This is a significantly higher number of taxa than recorded from any of the other major vegetation types including the lowlands, for which 159 records are listed in the enumeration, lower montane forest with 165 records, upper montane forest with 48 records, and the summit area with 19 taxa recorded. Thus, we see a higher diversity of non-orchid monocotyledons at lower elevations than was the case for the pteridophytes (Parris *et al.* 1992), orchids (Wood *et al.* 1993) and gymnosperms (p. 11), in which the highest diversity was in the lower montane forest at about 1500 m. It appears that the highest diversity for non-orchid monocotyledons may be at about 1000 m.

A few more taxa (95) are common to hill forest and the lowlands than occur in both hill forest and lower montane forest (92 taxa). Relatively few taxa (9) occur in both hill forest and upper montane forest, and none are found in both hill forest and the summit area.

As noted above, the Poaceae are the largest non-orchid monocot family on Mount Kinabalu, but the Arecaceae have more taxa (60) recorded in hill forest than any other family. The Poaceae are a close second with 54 hill-forest taxa recorded. Three other families also have relatively large numbers of taxa occurring in hill forest, *i.e.*, the Zingiberaceae with 36, Araceae with 31 and Cyperaceae with 30. Other families with significant numbers of hill-forest taxa include the Pandanaceae with 12, the Smilacaceae with seven and the Marantaceae with five.

LOWER MONTANE FOREST

At the lower elevations on Mount Kinabalu in lowland and hill forest habitats, particularly at 1000 m and below, much of the primary forest, at least in areas where substantial numbers of collections have been made, already was destroyed before botanical collectors visited the mountain. Consequently, in hill forest and below the present flora is represented in large part by collections of secondary species. In contrast, in the lower montane forest and above, herbarium specimens represent mostly the original floristic components.

Lower montane forest habitats have more monotyledonous taxa of Cyperaceae (33) than of any other family. The Poaceae rank second, with 28 taxa. The Araceae are third, with 25 taxa and the Zingiberaceae are fourth with 21. The palms, a primarily lowland family, which has the most taxa in both the lowlands and hill forest, are fifth with 19 taxa. The Pandanaceae are the sixth largest lower-montane-forest family, with 10 taxa. Other families have four or fewer taxa in lower montane forest.

UPPER MONTANE FOREST

The predominant monocotyledonous families in upper montane forest are the Cyperaceae and Poaceae, each with ten taxa. Six other families are represented by one to three species, as follows: Arecaceae, three species; Eriocaulaceae, two species;

Iridaceae, one species; Melanthiaceae, two species; Phormiaceae, one species; and Smilacaceae, three species. The upper montane monocotyledons, in spite of having rather low taxonomic diversity, are fairly conspicuous components of this vegetation. Although hardly achieving a dominance anywhere, the grasses and sedges are commonly seen in this zone.

SUMMIT AREA

In the summit area only 10 monocot taxa have been frequently collected. These include five species of grasses, three of sedges and one species from each of two other families, Centrolepidaceae and Eriocaulaceae. The majority of the summit area monocotyledons are low mat-formers, a life-form appropriate for the harsh windy and cold conditions of the summit.

ULTRAMAFIC SUBSTRATES

A distinctly lower percentage (15%) of non-orchid monocots on Mount Kinabalu occurs on ultramafic substrates than Wood *et al.* (1993) reported for the orchids (40%). In the habitat data presented in the enumeration a total of 70 taxa of the non-orchids is characterized as occurring on ultramafic substrates, with various qualifiers for some of the species (such as mostly, usually, often, sometimes and probably). The Cyperaceae have 18 taxa on ultramafics, more than any other family. This includes five in *Carex* and four each in *Machaerina* and *Schoenus*. The Arecaceae have 12 ultramafic-occurring taxa, including five species of *Calamus*. Ten taxa in the Poaceae occur on this substrate. *Sciaphila* (Triuridaceae), a genus of tiny saprophytes, has all four of its determined species occurring exclusively on ultramafics, and the likewise predominantly saprophytic Burmanniaceae have three ultramafic taxa, *Burmannia longifolia* (non-saprophytic), *Gymnosiphon aphyllus* and *Thismia ophiurus*.

Most of the ultramafic outcrops around Mount Kinabalu are above the lowlands. Consequently, relatively few (10) lowland taxa are recorded from this substrate. Noteworthy among these are six palms including the rheophytic *Areca rheophytica, Calamus zonatus, Daemonorops sabut, Korthalsia concolor, Pinanga lepidota,* and *Plectocomiopsis mira*. The largest number of monocots on ultramafic substrates (45 taxa in 14 families) occurs in hill forest, perhaps in part because extensive areas of hill forest are ultramafic. The principal families in hill forest with taxa recorded as occurring on ultramafic substrates are the Arecaceae with nine taxa, Cyperaceae with six and Poaceae with seven. The lower montane forest has 36 taxa on ultramafic substrates, including the Cyperaceae with 14 and the Poaceae with six. Seventeen taxa are recorded as occurring on ultramafic substrates in upper montane forest, and the predominant family is the Cyperaceae with 10. Species that occur on ultramafic substrates in the upper montane forest sometimes are also characteristic of the granitic rocks of both upper montane forest and the summit area, but there are no ultramafic outcrops on the summit.

WET HABITATS

No submerged aquatic monocotyledons are known from Mount Kinabalu. The subclass Alismatidae, known as the aquatic monocotyledons, is notably absent from the mountain, being represented only by the introduced species *Limnocharis flava*. The Kinabalu aquatics are thus merely plants rooted in wet situations such as swamps, bogs, ditches, stream banks, and, at high elevations, rivulets and seepage areas. These subaquatic and wetland plants are listed in Table 11.

Table 11. Monocotyledons of wet habitats on Mount Kinabalu.

Taxon	Habitat
Acoraceae	
Acorus calamus	Hill forest in disturbed, open, wet situations
Arecaceae	
Licuala paludosa	Lowland swamp forest
Nenga pumila var. *pachystachya*	Lowlands, infrequently also in hill forest, often in swampy habitats
Commelinaceae	
Amischotolype glabrata	Hill forest, lower montane forest, in wet areas
Amischotolype marginata	Hill forest in wet areas
Belosynapsis ciliata	Lowlands, hill forest; rocky areas near water
Commelina paludosa	Hill forest, near rivers
Cyperaceae	
Carex hypolytroides	Lower montane forest in wet places
Carex indica	Hill forest, wet places
Carex perakensis var. *borneensis*	Hill forest, lower montane forest, upper montane forest, sometimes on ultramafic substrate, in wet places
Costularia pilisepala	Lower montane forest, upper montane forest, sometimes on ultramafic substrate, in marshy or boggy areas
Cyperus haspan	Hill forest, lower montane forest, in drainage ditches, margins of shallow standing water, and paddy fields
Cyperus laxus var. *macrostachyus*	Lowland primary forest near streams, sometimes on ultramafic substrate
Cyperus pilosus	Lowlands, hill forest, lower montane forest, in drainage ditches in open areas
Eleocharis congesta	Lower montane forest, in the open in drainage ditches
Fimbristylis littoralis	Lowlands, in drainage ditches
Fimbristylis umbellaris	Lowlands and hill forest in open, wet secondary situations
Fuirena umbellata	Lowlands and hill forest in open, wet areas
Hypolytrum nemorum	Hill forest, in swampy places

Table 11 (continued)

Taxon	Habitat
Isolepis habra	Upper montane forest in boggy areas
Isolepis subtilissima	Upper montane forest in boggy areas with sphagnum
Lipocarpha chinensis	Lower montane forest, drainage ditches, damp roadside verges in open areas
Mapania palustris	Hill forest (rarely), lower montane forest, sometimes on ultramafic substrate
Oreobolus ambiguus	Summit area, around shrubs and on wet rocks sheltered by boulders, and in damp hollows on granitic rocks
Pycreus sanguinolentus subsp. *cyrtostachys*	Lowlands in drainage ditches in open places
Rhynchospora rugosa	Lower montane forest, in open places on banks of streams and drainage ditches
Schoenoplectus mucronatus	Lower montane forest, in open wet or swampy areas
Schoenus curvulus	Upper montane forest, summit area, in wet places in the open or under shrubs, sometimes on ultramafics
Scirpus subcapitatus	Lower montane forest (rarely), upper montane forest, summit area, sometimes on ultramafic substrate, in open boggy places
Scirpus ternatanus	Lower montane forest, in open, wet situations and drainage ditches
Eriocaulaceae	
Eriocaulon kinabaluense	Open boggy, rocky areas in upper montane forest and the summit area
Juncaceae	
Juncus bufonius	In disturbed situations at upper edge of upper montane forest
Juncus effusus	Upper montane forest?
Limnocharitaceae	
Limnocharis flava	Open wet situations at low elevations, weed in old rice paddys
Poaceae	
Agrostis infirma var. *diffusissima*	Upper montane forest, summit area, on stream banks and in granitic depressions and crevices
Aniselytron treutleri	Lower or upper montane forest among river boulders
Coix lacryma-jobi	Lowlands, hill forest in wet situations by rice paddys or along streams
Danthonia oreoboloides	Upper montane forest, summit area, in open wet places, often in granitic crevices
Isachne albomarginata	Lower montane forest, on wet rocks by rivers and waterfalls
Oplismenus hirtellus	Hill forest, lower montane forest, at edge of streams, sometimes on ultramafic substrate
Oryza sativa	Cultivated, apparently occasionally occurring spontaneously

Table 11 (continued)

Taxon	Habitat
Panicum humidorum	Hill forest in old rice paddys
Poa borneensis	Upper montane forest, summit area, in open scrub and on wet rocks
Poa papuana	Upper montane forest, summit area, in marshy areas and in wet granitic crevices and depressions
Schizachyrium brevifolium	Lowlands, hill forest, in paddy fields
Trisetum spicatum subsp. *kinabaluense*	Upper montane forest to summit area, in seepage areas and disturbed places, sometimes weedy
Zingiberaceae	
Etlingera elatior	Lowlands and hill forest, in wet places in primary and secondary situations
Plagiostachys crocydocalyx	Hill forest along streams

Probably the reason that there are no submerged or floating aquatics on Mount Kinabalu is that there are no appropriate habitats for them. The topography is so precipitous that small or large bodies of water do not accumulate. The rivers do not provide a satisfactory habitat for aquatics, because of their periodic torrential flows associated with heavy downpours in which ordinary aquatics would be washed away. Hence, only the rheophytes are adapted to stream proximities.

DISTURBED HABITATS

In the lowlands two families, the Cyperaceae and Poaceae, are predominant in open, disturbed habitats. In areas that formerly were hill forest many of the same genera and species also are prominent. The tall clumped grass *Miscanthus floridulus* var. *malayanus,* with terminal panicles of silvery white spikelets, is commonly seen along open roadsides in hill forest and lower montane forest habitats. The grass *Imperata cylindrica* is an extremely serious weed in areas formerly covered by hill forest. Considering its present dominance over vast areas, the species is greatly underrepresented in herbarium collections. *Globba franciscii* (Zingiberaceae) is one of the few monocots of disturbed habitats that is not a grass or sedge.

Most of the disturbed areas in the lower montane forest and below are the result of anthropogenic activities. In the upper montane forest and summit area, in contrast, human activities have had relatively little impact on the vegetation, with most of the disturbance occurring along the main summit trail and around Panar Laban. The principal natural disturbances in upper montane forest are landslides, of which there are many. Some particularly interesting species occur on old landslides, and, as with disturbed areas at lower elevations, the grasses and sedges predominate. Common plants of open areas in upper montane forest and the summit area include: *Carex verticillata* and *Schoenus curvulus* (Cyperaceae) and *Anthoxanthum horsfieldii* var. *borneensis*, *Danthonia oreoboloides* and *Poa borneensis* (Poaceae).

HABITAT DATA IN THE DUSUN LANGUAGE

Projek Etnobotani Kinabalu specimen-label data generally designate the habitat for the collection in the Dusun language. It may be useful, therefore, to have the meaning of these terms in English. The Dusun words *puru, talun* and *timbaan* mean primary forest. *Timbaan* implies a primary forest with large trees, whereas *puru* and *talun* do not have that implication. The word *geuten* is used for a dense, relatively impenetrable vegetation type that probably best translates as thicket. *Temulek* means secondary vegetation, usually a secondary forest, although it can imply disturbed areas in successional stages prior to what could be thought of as forest, and originates from abandoned cultivated land. Disturbed areas along roadsides are often referred to as *temulek*. *Butur* is a grassy area, either a pasture or within a house compound. *Tume* is a cultivated area. *Liwan* (in Bundu Tuhan and Serinsim) and *natad* are equivalent words for house garden. *Bawang* means river. *Labak* or *labak-labak* is a boggy or permanently muddy area.

LIFE-FORMS OF THE MONOCOTYLEDONS

Tomlinson (1995) noted that the most distinctive feature of monocotyledons is the organization of the vascular system, which is completely different from that of dicotyledons and is a distinctive morphogenetic type within all vascular plants. The monocotyledonous vascular system lacks a secondary cambium, with the result that the tree habit is uncommon, and when it does occur, as in palms and pandans, there are specializations in the stem and root systems that compensate for the lack of water-conducting vessels produced by a cambium as found in the dicotyledons. The primary root of monocotyledons is too small to provide the water requirements of a large crown, so crown enlargement, when it occurs, is associated with various modifications, such as the growth of adventitious and stilt roots or the occurrence of large obconical bases produced by a special establishment growth.

Among the 34 monocotyledonous families included in this treatment, 30 are herbaceous or essentially so and four, the Agavaceae, Arecaceae, Dracaenaceae and Pandanaceae, are woody. The Dioscoreaceae, Flagellariaceae and Smilacaceae include vines with tough, wiry stems that approach being woody. The bamboos (Poaceae), with hard culms and hollow internodes, are sometimes tree-like, but have a unique life-form.

TREES

Palms are one of the exceptional monocotyledonous families in which trees are frequent, with about 17 Kinabalu species having this habit. Some of these, however, such as the species of *Pinanga* and *Nenga* are better referred to as treelets rather than trees, and others, such as *Arenga* and *Salacca*, are very short-stemmed. Excluding the palms, the only Kinabalu monocots with the tree habit are two species of *Dracaena* and the cultivated *Pandanus tectorius*.

SHRUBS

Above-ground branching systems in monocotyledons are infrequent. This is also related to limitations in the nature of the vascular system (Tomlinson 1995). Consequently, few Kinabalu species are recorded as shrubs, and these are principally in the genus *Pandanus,* for which six species are characterized as shrubs. *Cordyline fruticosa* is also considered a shrub; an unnamed species of *Dracaena* likewise has the shrub habit as do the aroid species *Aglaonema nitidum* and *A. simplex*.

HERBS

The organization of the vascular system with lack of secondary growth through a cambium is a major factor in the predominance of the herbaceous life-form in monocotyledons (Tomlinson 1995, p. 597). Perennial herbs are most common, and there are numerous examples of the rhizomatous habit; tubers or corms are also frequent, and bulbs occur in some families, particularly the Alliaceae and Amaryllidaceae. Examples of tuberous perennials include members of the Araceae such as *Alocasia, Amorphophallus, Arisaema* and *Colocasia* and *Musa* (Musaceae) and *Tacca* (Taccaceae). The rhizomatous habit is found in *Acorus* (Acoraceae), many Araceae, Cannaceae, Commelinaceae, Convallariaceae, Costaceae, Cyperaceae, Hypoxidaceae, Iridaceae, Juncaceae, Lowiaceae, Marantaceae, Melanthiaceae, Phormiaceae, Taccaceae and Zingiberaceae.

All of the Zingiberaceae, which have a distinctive "ginger habit," are rhizo-matous perennials. Some members of the family are very large herbs reaching four or more metres in height. Two principal modes of inflorescence presentation occur in the family. The genera *Alpinia, Boesenbergia, Burbidgea, Globba, Hedychium* and *Plagiostachys* have terminal inflorescences on leafy stems. In contrast, the inflorescences emanate from the base in *Amomum, Etlingera, Hornstedtia* and *Zingiber*. Some species of *Amomum* and *Hornstedtia* bear rhizomes on stilt roots.

The grass-sedge habit occurs in the Poaceae and Cyperaceae as well as in the Joinvilleaceae and Juncaceae. Grasses and sedges are often tufted or caespitose, and may also be rhizomatous or stoloniferous. One of the best known and least collected rhizomatous grasses is *Imperata cylindrica* (lalang), which has tough, fire-resistant rhizomes that enable it to survive and thrive with frequent fires. It now forms a near single-species dominance in large areas of Borneo, including some of the drier areas to the north of Mount Kinabalu. This grass is unpalatable to cattle and otherwise difficult to eradicate, with the result that areas where it dominates are virtually useless for agriculture.

The stoloniferous habit, also associated with weediness, is likewise common among the grasses, as in *Dactyloctenium aegyptium, Digitaria setigera, D. violascens, Echinochloa colona, Eleusine indica, Oplismenus compositus, O. hirtellus* and *Schizachrium brevifolium.*

In the Cyperaceae the most common life-form is that of tufted perennial with short rhizome. Among the numerous species with this habit are *Carex cruciata* (both varieties), *C. filicina, C. indica, C. perakensis* (both varieties), *C. tristachya, Costularia pilisepala, Cyperus compactus, C. cyperoides, C. diffusus, C. laxus, Fimbristylis complanata, F. dichotoma, F. dura, Fuirena umbellata, Gahnia tristis, Hypolytrum nemorum, Lepidosperma chinense, Machaerina aspericaulis, M. disticha, M. falcata, M. glomerata, Pycreus sanguinolentus* subsp. *cyrtostachys, Rhynchospora corymbosa, R. rubra, R. rugosa, Schoenoplectus mucronatus, Schoenus curvulus, S. delicatulus, S. longibracteatus, S. melanostachys, Scirpus subcapitatus* and *Uncinia compacta.* A few Cyperaceae are stoloniferous, notably *Carex hypolytroides, Cyperus pilosus* and *Mapania latifolia.*

A number of dense, low, tufted or tussock grasses occur in the upper montane forest and summit area, including *Agrostis infirma* (all three varieties), *Aniselytron treutleri, Anthoxanthum horsfieldii* var. *borneense, Bromus formosanus, Deschampsia flexuosa* var. *ligulata, Isachne kinabaluensis, Poa borneensis* and *Trisetum spicatum.* The densely tufted sedge species *Carex breviculmis* (both varieties) and *C. capillacea* (both varieties) are also found at high elevations.

Several cushion-plant or mat-forming species occur in the summit area, including *Centrolepis philippinensis* (Centrolepidaceae), *Isolepis subtilissima* and *Oreobolus ambiguus* (Cyperaceae), *Eriocaulon kinabaluense* (Eriocaulaceae), and *Danthonia oreoboloides, Poa epileuca* and *P. papuana* (Poaceae).

The annual habit in monocotyledons is rare (Tomlinson 1995, p. 597), and most of the herbs on Mount Kinabalu are perennials. Nevertheless, some annuals do occur, particularly in the families Cyperaceae and Poaceae. Among the annual sedges (Cyperaceae) are *Cyperus castaneus, C. compressus, C. sphacelatus, Fimbristylis dichotoma, F. littoralis, F. miliacea, F. obtusata, Lipocarpha chinensis, Pycreus sanguinolentus* subsp. *cyrtostachys* and *Schoenoplectus juncoides.* Annual grasses (Poaceae) include *Dactyloctenium aegypticum, Dimeria ornithopoda, Paspalum scrobiculatum* var. *bispicatum, Sacciolelpis indica, Setaria italica,* and the cultivated species *Coix lacryma-jobi, Oryza sativa, Sorghum bicolor* and *Zea mays.* Some of these species may be either annual or perennial, depending on conditions under which they are growing. Two additional annuals are the autotrophic *Burmannia coelestis* (Burmanniaceae) and *Juncus bufonius* (Juncaceae).

CLIMBERS, SCRAMBLERS, CREEPERS

Climbers or vines are recorded particularly among the rattan palms, of which the 48 Kinabalu taxa have this habit. These generally climb by means of reflexed spines on a cirrus (whiplike extension of the leaf rachis) or flagellum (whiplike climbing organ derived from a sterile inflorescence). Other families in which climbers, scramblers or vines occur are the Araceae, Cyperaceae, Dioscoreaceae, Flagellariaceae, Poaceae and Smilacaceae. The Araceae have about 14 climbing taxa. The Araceae which climb do so by adventitious roots. Most of them are herbaceous, although the genus *Pothos* can have tough, wiry stems. Other vining Araceae occur in the genera *Amydrium, Rhaphidophora* and *Scindapsus*. Only one species of Cyperaceae, *Scleria terrestris*, is a scrambler. The 11 species of *Dioscorea* (Dioscoreaceae) are all vines, which consistently twine either left or right. *Flagellaria indica* (Flagellariaceae) climbs by tendrils at the leaf apices. The nine species of *Freycinetia* (Pandanacae) are woody climbers (lianas). The Poaceae include various grasses such as *Isachne albens, Ottochloa nodosa, Panicum brevifolium, P. notatum,* and *P. sarmentosum* that are characterized as scramblers. Among the bamboos four genera, *Dinochloa, Kinabaluchloa, Racemobambos* and *Sphaerobambos,* are scramblers. The four Kinabalu species of *Racemobambos* occur at relatively high elevations, and are noted for forming near-impenetrable thickets. The genus *Smilax* (Smilacaceae) includes only tough, wiry stemmed vines among the eight Kinabalu species. Some of these are armed with prickles while others are unarmed. A few collections made by the Clemenses have label data indicating that the plants do not climb. Some Kinabalu *Smilax* have tendrils while other species lack them.

In addition to the vines, climbers and scramblers, a group of creepers which stay close to the ground includes *Scindapsus borneensis* and *S. rupestris* (Araceae), *Belosynapsis ciliata, Commelina paludosa, Floscopa scandens, Murdannia loriformis, Rhopalephora scaberrima* and *R. vitiensis* (Commelinaceae) and *Paspalum scrobiculatum* and *Pennisetum clandestinum* (Poaceae).

BAMBOOS

A particularly distinctive life-form among monocotyledons is that of the bamboos. In addition to the scrambling bamboos noted above, erect bamboos occur in the genera *Bambusa, Gigantochloa, Kinabaluchloa, Schizostachyum, Sphaerobambos* and *Yushania. Kinabaluchloa nebulosa* and *Sphaerobambos hirsuta* may be either erect or scrambling. The common name poring, from which Poring Hot Springs gets its name, is often associated with the large bamboo, *Gigantochloa levis.*

EPIPHYTES

As noted by Wood *et al.* (1993) the epiphytic habit is predominant among Kinabalu orchids. In contrast, it occurs infrequently in other Kinabalu monocotyledons, most notably in *Burbidgea schizocheila* (Zingiberaceae). Two aroids, *Amydrium medium* and *Rhaphidophora sylvestris,* are infrequently secondary hemiepiphytes (cf. Putz & Holbrook 1986), *i.e.,* they begin their development rooted in the ground, grow as vines on supporting vegetation, and ultimately lose their connection with the ground. Various other species may be facultatively epiphytic, but true epiphytes are nevertheless rare.

RHEOPHYTES

Rheophytes are plants that grow in or along rivers and are adapted to withstand periodic inundation, sometimes of extremely swiftly flowing water. They generally have tough, narrow leaves and roots that are tightly attached to boulder surfaces. Most of the Kinabalu rheophytes are in the Araceae, in the genera *Hottarum, Piptospatha* and *Schismatoglottis,* all of which are in the tribe Schismatoglottidae. Additionally, *Scindapsus rupestris* (tribe Monstereae) and some species of *Homalomena* (tribe Homalomeneae) are rheophytes, but from the habitat data available it is not certain that any of the Kinabalu species in the latter genus are rheophytic.

The well documented rheophytic aroids on Kinabalu include *Hottarum kinabaluense, Piptospatha elongata, P. grabowskii* and *P.* cf. *marginata.* Some Kinabalu species of *Schismatoglottis* must also be rheophytes, but many of the specimens in this poorly understood taxonomic complex are not identified.

In addition to the araceous rheophytes, two palm species, *Areca rheophytica* and *Pinanga tenella,* also have this habit. The type locality of *Pinanga tenella* is Mount Kinabalu, but it has not been recollected on the mountain since Lobb made the type collection at Bungol in 1856. That specimen has the distinction of being the first palm collected on Kinabalu. The species was subsequently collected at the neighbouring Kelawat by Mrs. Clemens in 1915, but that locality is not considered to be part of Mount Kinabalu.

SAPROPHYTES

Three families include some or all members that are saprophytic. Thus the Burmanniaceae include both saprophytic and autotrophic members in the genus *Burmannia,* while the species of *Gymnosiphon* and *Thismia* are saprophytic. The genus *Petrosavia* (here placed in the Melanthiaceae) on Mount Kinabalu includes a single saprophytic species, *P. stellaris.* The family Triuridaceae is represented by four minute, rarely collected saprophytic species. All of the Kinabalu saprophytes are small herbs.

LIFE-FORMS IN THE DUSUN LANGUAGE

The Dusun language has a series of expressions for plant life-form that have been used on the specimen labels of the *PEK* collections. Thus the word *saket* means herb, although it can also mean weed, and is applied to grasses and sedges as well as the types of herbs that would be designated forb in English life-form expressions. The word *kayu* means tree, whether large or small, and sometimes what the botanist would consider a large herb is referred to as a *kayu*. In some localities (kampungs or communities) different words are used to express certain plant habits than are used in other localities. For example, in Poring, Nalumad, and Tekutan a climber or liana is called *tangau* and a rattan is *wakau*. In Bundu Tuhan, Kiau, Melangkap Tomis, and Serinsim a climber is called *wakau* and rattan is *tuai*. A general word for bamboo is tuluh (Bundu Tuhan, Kiau, Melangkap Tomis and Serinsim) or *wuluh* (Poring, Nalumad, Tekutan). *Poring* is used for a few large bamboos, particularly *Gigantochloa levis,* and is the basis for the place name Poring. The term *liar* is a borrowed Malay word for wild, which is used for any wild relatives of cultivated species (for example *Pandanus*). The Malay word *rumpun* means a cluster, and is used sometimes for clumped grasses such as *Sorghum* and some bamboos as well as certain pandans such as *Pandanus rusticus. Parai* is a commonly used term, referring exclusively to rice. *Parai-parai* or *marai-parai*, however, mean life-forms resembling rice. In expressing habits in English the Dusun collectors have often used the word shrub for what the botanist would call an herb or grass. The word *syrub* appears to have been borrowed from English. It is sometimes used for what the botanist would call a shrub, but rarely appears in the Dusun part of the *PEK* specimen labels. Other life-form terms that have been borrowed include *bambu* and *palma* (English) and *sayur* (vegetable; Malay).

CLASSIFICATION AND EVOLUTION

Classification of the monocotyledons has undergone revolutionary changes in the past quarter century. Nevertheless, earlier classifications, particularly those from the latter part of the 19th and early 20th centuries, have had a lasting impact in the way monocot families have been arranged. Among these are the works of Bentham and Hooker (1883) and Hutchinson (1934). Older classifications used a few conspicuous morphological characters, particularly in the structure of the flowers, to distinguish major groups. In a number of the more recent classifications, particularly those of Takhtajan (1980), Cronquist (1981) and Thorne (1976), four or five major groups of monocotyledons have been recognized. These treatments used more characters but still have relied primarily on morphological data and subjective opinions of phylogenetic relationships. Major changes in classification of the monocotyledons began with the work of Dahlgren and Clifford (1982), Dahlgren and Rasmussen (1983) and Dahlgren *et al.* (1985). They were among the first botanists to use cladistic methodology as an aid in classification of the higher categories of seed plants. Cladistics is having continuing major impact on the understanding of monocotyledon relationships, and, together with newly obtained molecular data, is most responsible for the recent revolution in classifications of the monocotyledons.

The 35 families of monocotyledons that occur on Mount Kinabalu include most of the evolutionary lines that are currently recognized for the class. Several earlier classification put great emphasis on an aquatic ancestry for the mono-cotyledons, with the subclass Alismatidae considered to be the basal group and to have relationships with the dicotyledons. Recent studies, however, point to an advanced position for the Alismatidae and lack of support for relationship through them with the dicotyledons. Of the major groups of monocotyledons on Mount Kinabalu, the Alismatidae are conspicuous by their absence, with only the genus *Limnocharis* (Limnocharitaceae) represented, and it not native.

The morphological and molecular data sets presented in a recent symposium on monocotyledons (Rudall *et al.* 1995) indicate considerable agreement as to relationships among various families and orders. It now appears that the genus *Acorus* (formerly included in the Araceae) is particularly noteworthy among mono-cotyledons in being basal (Duvall *et al.* 1993, Les & Haynes 1995, p. 357, Mayo *et al.* 1995 p. 278) and is even suggested as a link to such dicotyledonous families as the Piperaceae and Aristolochiaceae.

In the Dahlgren *et al.* (1985) classification the Dioscoreaceae are in a basal position, *i.e.*, closest to the basal dicotyledonous orders Annonales and Piperales. Chase *et al.* (1995, p. 132) indicate that the Dioscoreales *sensu* Dahlgren *et al.* are polyphyletic. Bouman (1995) likewise indicates that it is difficult to relate the Discorealeales *sensu* Dahlgren *et al.* (1985) with other monocotyledonous orders. Chase *et al.* (1995, p. 697, fig. 4b) present one randomly selected tree, among 96 equally parsimonious possibilities, of the monophyletic Lilianae that includes a clade which has the Dioscoreaceae, Taccaceae, Burmanniaceae, Thismiaceae (here included in the Burmanniaceae) and Pandanaceae in the order Dioscoreales.

The Arales are another near-basal group of monocotyledons, as placed in the Dahlgren *et al.* (1985) classification, and may have relationships with the Alisma-tidae, as suggested by Mayo *et al.* (1995), and supported by the combined morpho-logical and molecular data sets of Chase *et al.* (1995). French *et al.* (1995) note that the Araceae are among the most diverse of all monocotyledons in stem, leaf and flower structure and in pollination biology.

The Dahlgren *et al.* (1985) classification had the order Triuridales (including only the family Triuridaceae) in an isolated position in the general neighborhood of the Alismatidae and Arales. Chase *et al.* (1995) likewise have it in a clade with aralian and alismatid families. The genus *Petrosavia* in most recent classifications has been included in the Melanthiaceae, but Stevenson and Loconte (1995, Table 4, p. 560), among others, recognize it in its own family, Petrosaviaceae, and include it in the Triuridales. Chase (pers. comm.) has told us that recent DNA data for *Petrosavia* indicate that it is among the basal monocots. The Triuridaceae have not yet yielded DNA data, because of their lack of chlorophyll.

The order Asparagales in the Dahlgren *et al.* (1985) classification is in a relatively central position with respect to other higher monocots and appears reasonably monophyletic. It is also the largest order of the subclass Liliifloreae, and among the Kinabalu monocot families includes the Convallariaceae, Dracaenaceae, Hypoxidaceae, Phormiaceae, Alliaceae and Amaryllidaceae, at least according to the combined morphological and molecular data sets of Chase *et al.* (1995). This grouping is supported by the data of Rudall *et al.* (1997) who place the Agavaceae, Alliaceae, Amaryllidaceae, Convallariaceae and Lomandraceae (including *Cordyline*) in the "higher asparagoids." Rudall *et al.* have the Hypoxidaceae, Iridaceae and Phormiaceae (for which the older name is Hemerocallidaceae) in the "lower asparagoids."

In the Chase *et al.* (1995) analysis the Dracaenaceae, Asparagaceae, and *Cordyline* (here included in the Agavaceae) are in a clade of the order Asparagales. The Agavaceae and Alliaceae are part of another clade in the Asparagales near the clade that includes the Dracaenaceae. The Iridaceae are in a clade close to the one with the Agavaceae, etc. The Hypoxidaceae and Orchidaceae are in still another neighbouring asparagalean clade.

Dahlgren *et al.* included the Hanguanaceae in the Asparagales, but this family has been excluded by Rudall and Cutler (1995) and others, *e.g.*, Chase *et al.* (1995), who found a tentative relationship of the family with the palms. Stevenson and Loconte (1995) placed it in its own order, Hanguanales, and suggested it to be the sister group of the Zingiberales.

The order Liliales was formerly one of the largest orders of monocotyledons. Efforts of many authors, however, to make the Liliales monophyletic have resulted in drastic reduction in the number of families represented, which, among the Kinabalu monocots, would include only the Smilacaceae and Melanthiaceae, based on the data of Chase *et al.* (1995). It should be noted, however, that Rudall *et al.* (1997) place the Smilacaceae among the lower asparagoids.

A group of monocotyledons whose hypothesized relationships have met with considerable agreement are centred around the Commelinales. Commelinid monocots are characterized by starchy endosperm and cell walls with p-coumarins. The Commelinales include the Kinabalu families Commelinaceae and Eriocaulaceae.

Commelinid monocotyledons also include the Zingiberales, an order with six of its eight families represented on Mount Kinabalu. These are the Musaceae,

Lowiaceae, Zingiberaceae, Costaceae, Cannaceae and Marantaceae. In contrast to the rather ambiguous relationships of various other monocot families, the Zingiberales are an uncontested monophyletic order (Kress 1990, 1995), even if there is not general agreement as to relationships among the included families.

Kellogg and Linder (1995) found that the Poales, including the Kinabalu families Poaceae, Joinvilleaceae, Centrolepidaceae, and Flagellariaceae, are mono-phyletic and have the Typhales as a sister group. Furthermore, they consider this clade to be the sister group to the Cyperales. The Cyperales have many characters in common (Simpson 1995), and two of its three families, Cyperaceae and Juncaceae, are represented on Mount Kinabalu.

The Arecaceae (palms) were considered by Kellogg and Linder (1995) to be a sister group to the clade including the Zingiberales, Poales, and Cyperales. Uhl *et al.* (1995) indicated that the palms are probably the most morphologically diverse of all monocot families. It was noted above that the Araceae are another highly diverse monocot group.

Enumeration of the Monocotyledons

35. ACORACEAE

35.1. ACORUS L.

Engler, A. (1905). Araceae-Pothoideae. *Pflanzenr*. IV. 23B (Heft 21): 1–330 [*Acorus* on pp. 308–313].

35.1.1. Acorus calamus L., *Sp. Pl.*: 324 (1753).

Grasslike perennial herb with long, thick rhizome. Hill forest in disturbed, open, wet situations.

Material examined. KIAU: *Jusimin Duaneh 169* (K).

36. AGAVACEAE

36.1. CORDYLINE Adans.

36.1.1. Cordyline fruticosa (L.) A. Chev., *Cat. Pl. Jard. Bot. Saigon:* 66 (1919).

Erect, few-branched shrub. Hill forest (cultivated?). Introduced?

Material examined. BUNDU TUHAN: *Doinis Soibeh 93* (K); KIAU: *Jusimin Duaneh 112* (K).

37. ALLIACEAE

Buijsen, J. R. M. (1993). Alliaceae. *Fl. Males*. I, 11: 375–384.

37.1. ALLIUM L.

37.1.1. Allium fistulosum L., *Sp. Pl.*: 301 (1753).

Tufted perennial herb with small bulb and relatively wide leaves. Hill forest, cultivated and possibly escaped into secondary situations. Native to Siberia and China and not known in the wild.

Material examined. KIAU: *Jusimin Duaneh 181* (K), *342* (K).

38. AMARYLLIDACEAE

Geerinck, D. J. L. (1993). Amaryllidacae (including Hypoxidaceae). *Fl. Males*. I, 11: 353–373.

38.1. PANCRATIUM L.

38.1.1. Pancratium zeylanicum L., *Sp. Pl.*: 290 (1753).

Bulbous, glabrous perennial herb. Lowlands in open, grassy areas.

Material examined. MENGGIS: *Matamin Rumutom 77* (K).

39. ARACEAE

In collaboration with P. C. Boyce (K)

Engler, A. (1920). Araceae: Pars generalis et Index familiae generalis. *Pflanzenr.* IV. 23A (Heft 74): 1–71. Hay, A., Bogner, J., Boyce, P. C., Hetterscheid, W. L. A., Jacobsen, N. and Murata, J. (1995). Checklist & Botanical Bibliography of the Aroids of Malesia, Australia, and the tropical western Pacific. *Blumea* Suppl. 8, 210 pp. Mayo, S. J., Bogner, J., and Boyce, P. C. (1997). *The Genera of Araceae.* Royal Botanic Gardens, Kew. xii + 370 pp.

39.1. AGLAONEMA Schott

Engler, A. (1915). Araceae-Philodendroideae-Anubiadeae, Aglaonemateae, Dieffenbachieae, Zante-deschieae, Typhonodoreae, Peltandreae. *Pflanzenr.* IV. 23Dc (Heft 64): 1–78 [*Aglaonema* on pp. 10–34]. Nicolson, D. H. (1964). *A taxonomic revision of the genus* Aglaonema *(Araceae).* Ph. D. thesis, Cornell Univ. 329 pp. Univ. Microfilms. Nicolson, D. H. (1969). A revision of the genus *Aglaonema. Smithsonian Contr. Bot.* 1: i–iv, 11–69.

39.1.1. Aglaonema nitidum (Jack) Kunth, *Enum. Pl.* 3: 56 (1841).

Shrub, sub-creeping to more or less erect, up to 45 cm tall. Hill forest. Elevation: 900 m. Widespread in Borneo.

Material examined. DALLAS: 900 m, *Clemens 26142* (BM, K), 900 m, *26703* (BM, SING n.v.); MELANGKAP TOMIS: *Lorence Lugas 1903* (K); TEKUTAN: *Dius Tadong 797* (K).

39.1.2. Aglaonema simplex Blume, *Rumphia* 1: 152, t. 65 (1837). Plate 5D.

Shrub to 1 m or more tall. Hill forest, lower montane forest. Elevation: 900–2100 m. Distinguished from *A. nitidum* by the thinner leaves with conspicuous primary veins.

Material examined. DALLAS: 900 m, *Clemens 26820* (BM, K, L); KIAU: *Clemens, M. S. 10003* (BO n.v., PNH n.v.); KUNDASANG: 1200 m, *RSNB 1434* (K); LUBANG GORGE: *Clemens, M. S. 10442* (PNH n.v.); MENGGIS: *Matamin Rumutom 281* (K); MT. NUNGKEK: 1200 m, *Darnton 481* (BM); PENIBUKAN: 1200–1500 m, *Clemens 31149* (BM, SING n.v.), 2100 m, *40985* (SING n.v.), 1000 m, *Nooteboom & Aban 1544* (L); TENOMPOK: 1500 m, *Clemens 29133* (BM, K, L); TINEKEK FALLS: 2100 m, *Clemens s.n.* (BM).

39.2. ALOCASIA (Schott) G. Don

Engler, A., and Krause, K. (1920). Araceae-Colocasioideae. *Pflanzenr.* IV. 23E (Heft 71): 3–139 [*Alocasia* on pp. 71–115].

39.2.1. Alocasia beccarii Engl., *Bull. Soc. Tosc. Ortic.* 4: 300 (1879).

Terrestrial perennial herb. Hill forest, lower montane forest, on ultramafic substrate. Elevation: 1200–1500 m.

Material examined. PENIBUKAN: 1200 m, *Clemens s.n.* (BM, BM), 1500 m, *31548* (BM).

39.2.2. Alocasia cuprea (C. Koch & Bouché) C. Koch, *Wochenschr. Vereines Beförd. Gartenbaues Königl. Preuss. Staaten* 4: 141 (1861). Plate 5E, F.

Terrestrial perennial herb. Hill forest.

Material examined. HIMBAAN: *Doinis Soibeh 689* (K).

39.2.3. Alocasia cf. cuprea (C. Koch & Bouché) C. Koch, *Wochenschr. Vereines Beförd. Gartenbaues Königl. Preuss. Staaten* 4: 141 (1861).

Terrestrial perennial herb. Hill forest. Elevation: 1100 m.

Material examined. TAHUBANG RIVER: 1100 m, *Clemens 40588* (BM).

39.2.4. Alocasia longiloba Miq., *Bot. Zeit.* 14: 564 (1856). Plate 6A.

Robust perennial herb. Hill forest, lower montane forest. Elevation: 900–1700 m.

Material examined. DALLAS: 900 m, *Clemens 26213* (K), 900 m, *26268* (K); LIWAGU/MESILAU RIVERS: 1500 m, *RSNB 1954* (K); PENIBUKAN: 1200 m, *Clemens 31161* (K); TENOMPOK: 1500 m, *Clemens 29148* (K); WEST MESILAU RIVER: 1600–1700 m, *Beaman 7451* (K, MSC, US), 1600–1700 m, *7453* (MSC, US).

39.2.5. Alocasia macrorrhizos (L.) Schott in Schott & Endl., *Melet. Bot.*: 18 (1832). Plate 6B.

Robust perennial herb. Hill forest. Elevation: 800 m. Most likely introduced.

Material examined. EASTERN SHOULDER: 800 m, *RSNB 574* (K).

39.3. AMORPHOPHALLUS Blume ex Decne.

Engler, A. (1911). Araceae-Lasioideae. *Pflanzenr.* IV. 23C (Heft 48): 1–130 [*Amorphophallus* on pp. 61–109].

39.3.1. Amorphophallus hottae Bogner & Hett., *Blumea* 36: 470 (1992).

Large perennial herb from a tuberous base. Hill forest, lower montane forest.

Material examined. KINATEKI RIVER (NEAR MINITINDUK): *Holttum SFN 25590* (SING); MESILAU RIVER: *Furtado s.n.* (SING).

39.3.2. Amorphophallus lambii Mayo & Widjaja, *Bot. Mag.* 184(2): 62, t. 852 (1982). Plate 6C.

Large tuberous perennial herb. Lowlands. Elevation: 400 m. Reported in the original description as occurring at Ranau, but no specimen from there seen.

Material examined. PORING HOT SPRINGS: 400 m, *Idzushi & Togashi s.n.* (TI, fide Hetterscheid).

39.3.3. Amorphophallus indet.

Material examined. DALLAS: 900 m, *Clemens 30454* (BM); LOHAN/MAMUT COPPER MINE: 900 m, *Beaman 10605* (K, MSC, US); MELANGKAP TOMIS: *Lorence Lugas 1611* (K), *2608* (K); PENIBUKAN: 1200–1500 m, *Clemens 35001* (BM).

39.4. AMYDRIUM Schott

Engler, A., and Krause, K. (1908). Araceae-Monsteroideae. *Pflanzenr.* IV. 23B (Heft 37): 4–138 [*Amydrium* on p. 118]. Engler, A., and Krause, K. (1908). Araceae-Monsteroideae. *Pflanzenr.* IV. 23B (Heft 37): 4–138 [*Amydrium* on p. 118]. Nicolson, D. H. (1968). A revision of *Amydrium* (Araceae). *Blumea* 16: 123–127.

39.4.1. Amydrium medium (Zoll. & Moritzi) Nicolson, *Blumea* 16: 124 (1968). Plate 7A.

Slender to more or less robust climber, occasionally a facultative epiphyte. Lowlands, hill forest, lower montane forest. Elevation: 500–1500 m. Widespread in Borneo.

Material examined. BAMBANGAN RIVER: *Amin Gambating SAN 121562* (K); BUNDU TUHAN: *Aban Gibot SAN 74128* (K, SAN); DALLAS: 900 m, *Clemens 26733* (K, L, UC); GURULAU SPUR: *Clemens, M. S. 10773* (UC); KG. PAKA: *Julius K. SAN 124867* (K); KIPUNGIT HILL: 700–1000 m, *Beaman 8221* (K, US, US); KUNDASANG: 1200 m, *RSNB 1430* (K); LOHAN RIVER: *Amin & Jarius 121417* (K, SAN); MEKEDEU RIVER: 500 m, *Shea & Aban SAN 77263* (K, L); MELANGKAP TOMIS: *Lorence Lugas 802* (K), *2599* (K); NALUMAD: *Daim Andau 791* (K); PAKA: *Julius K. SAN 124867* (E); PENIBUKAN: 1200–1500 m, *Clemens 51420* (K, L), 1100 m, *Nooteboom & Aban 1583* (L); PINOSUK PLATEAU: 1400–1500 m, *Beaman 9200* (MSC); TENOMPOK: 1500 m, *Clemens 29144* (K, L).

39.5. ANADENDRUM Schott

Engler, A. (1905). Araceae-Pothoideae. *Pflanzenr.* IV. 23B (Heft 21): 1–330 [*Anadendrum* on pp. 46–50].

39.5.1. Anadendrum indet.

The genus is taxonomically chaotic; thus the specimens have not been determined to species.

Material examined. BUNDU PAKA: *Lorence Lugas 2637* (K); DALLAS: 900–1200 m, *Clemens s.n.* (BM), 900–1200 m, *26741* (BM), 900 m, *27376* (K); HEMPUEN HILL: 800–1200 m, *Beaman 7693* (MSC); KULUNG HILL: 800 m, *Beaman 7786* (US); MELANGKAP TOMIS: 900–1000 m, *Beaman 8976* (K, US), *Lorence Lugas 1453* (K); MINITINDUK GORGE: 900–1200 m, *Clemens s.n.* (BM); NALUMAD: *Daim Andau 856* (K); PENIBUKAN: 1200 m, *Clemens s.n.* (BM), 1500 m, *31017* (BM, K), 1200 m, *32179* (BM), 1200 m, *40104* (BM); PINAWANTAI: 700 m, *Shea & Aban SAN 76956* (K); PORING: *Meijer & Abu Bakar SAN 131913* (SAN); TENOMPOK: 1500 m, *Clemens 26217* (BM, K).

39.6. ARISAEMA Mart.

Engler, A. (1920). Araceae-Aroideae und Araceae-Pistioideae. *Pflanzenr.* IV. 23F (Heft 73): 1–274 [*Arisaema* on pp. 149–220].

39.6.1. Arisaema filiforme Blume, *Rumphia* 1: 102, t. 28 (1836). Plate 7B.

Tuberous perennial herb. Hill forest, lower montane forest. Elevation: 1200–1900 m.

Material examined. EAST MESILAU RIVER: 1600 m, *Bailes & Cribb 503* (K); GOLF COURSE SITE: 1700 m, *Beaman 8552* (MSC); MARAI PARAI: 1500 m, *Clemens 32477* (BM); MENERINTOG: 1800 m, *Clemens 29156* (BM, K); MENGGIS: *Matamin Rumutom 183* (K); MESILAU CAMP: 1500 m, *RSNB 6010* (K); MESILAU CAVE: 1900 m, *Wood 856* (K); MESILAU RIVER: 1500 m, *RSNB 4081* (K), 1500 m, *RSNB 4921* (K), 1400 m, *RSNB 7052* (K); MT. NUNGKEK: 1200 m, *Darnton 461* (BM); PENIBUKAN: 1800 m, *Clemens s.n.* (BM), 1200 m, *30508* (BM, K), 1200–1500 m, *35000* (BM), 1200–1500 m, *35153* (BM); PINOSUK PLATEAU: 1700 m, *RSNB 1857* (K), 1500 m, *Collenette 639* (K); TENOMPOK: 1500 m, *Clemens s.n.* (BM), 1500 m, *26840* (K); WEST MESILAU RIVER: 1600–1700 m, *Beaman 7450* (MSC, US), 1600–1700 m, *8638* (US).

39.6.2. Arisaema laminatum Blume, *Rumphia* 1: 99, t. 27 (1836). Plate 7C.

Tuberous perennial herb. Hill forest. Elevation: 1300 m.

Material examined. HEMPUEN HILL: 1300 m, *Wood 601* (K); LUGAS HILL: 1300 m, *Beaman 10588* (K, US).

39.6.3. Arisaema umbrinum Ridl., *J. Straits Branch Roy. Asiat. Soc.* 44: 171 (1905). Plate 7D.

Arisaema simplicifolium Ridl., *J. Linn. Soc., Bot.* 42: 171 (1914). Type: KADAMAIAN RIVER, 1100 m, *Gibbs 4099* (holotype BM!; isotype K!).

Tuberous perennial herb. Lowlands, hill forest. Elevation: 400–1200 m.

Additional material examined. DALLAS: 900–1200 m, *Clemens 26628* (BM, K); KADAMAIAN RIVER: 500 m, *Collenette A 132* (BM); LOHAN RIVER: 700–900 m, *Beaman 9245* (US); MELANGKAP KAPA: 600–700 m, *Beaman 8587* (MSC, US); MELANGKAP TOMIS: 900–1000 m, *Beaman 8401a* (MSC); TAHUBANG RIVER: 500 m, *Collenette A 8* (BM); TINEKEK/TAHUBANG RIVERS: 400 m, *Collenette A 7* (BM).

39.7. COLOCASIA Schott

Engler, A., and Krause, K. (1920). Araceae-Colocasioideae. *Pflanzenr.* IV. 23E (Heft 71): 3–139 [*Colocasia* on pp. 62–71].

39.7.1. Colocasia oresbia A. Hay, *Sandakania* 7: 39 (1996). Type: PARK HEADQUARTERS, *Hay 10046* (holotype SNP n.v.; isotype K!).

Robust to moderately robust perennial herb. Lower montane forest. Elevation: 1400–1500 m.

Additional material examined. KIBAMBANG LUBANG: 1400 m, *Clemens 33884* (BM); TENOMPOK: 1500 m, *Clemens 29141* (K).

39.8. HOMALOMENA Schott

Engler, A., and Krause, K. (1912). Araceae-Philodendroideae-Philodendreae. *Pflanzenr.* IV. 23Da (Heft 55): 1–134 [*Homalomena* on pp. 25–81]. Furtado, C. X. (1939). Araceae Malesicae II. Notes on some Indo-Malaysian *Homalomena* species. *Gard. Bull. Straits Settlem.* 10: 183–238.

39.8.1. Homalomena gillii Furtado, *Gard. Bull. Straits Settlem.* 10: 221 (1939). Type: DALLAS, 1100 m, *Clemens 29138* (holotype SING n.v.; isotype BM!).

Perennial herb. Hill forest, lower montane forest. Elevation: 900–1500 m. Possibly not distinct from *H. kinabaluensis* Furtado.

Additional material examined. DALLAS: 900–1200 m, *Clemens 26226* (BM, K), *26730* (SING n.v.), 900 m, *26730A* (BM), *29140* (SING n.v.); MT. NUNGKEK: 1100 m, *Clemens 35006* (BM); PENIBUKAN: 1200 m, *Clemens 40104 bis* (BM); TAHUBANG RIVER: 1500 m, *Clemens 31023* (BM, K); TENOMPOK: 1400 m, *Clemens 29142* (BM, SING n.v.), 1500 m, *29145* (BM, SING n.v.).

39.8.2. Homalomena humilis (Jack) Hook. f.

a. var. pumila (Hook. f.) Furtado, *Gard. Bull. Straits Settlem.* 10: 203 (1939).

Perennial herb. Hill forest. Elevation: 900–1200 m.

Material examined. DALLAS: 900 m, *Clemens 26199* (BM, K), 900 m, *27230* (K), 1100 m, *29139* (BM, K, L); LANGANAN FALLS: 1200 m, *Beaman 10964* (K); MARAI PARAI: 1200 m, *Clemens 32508* (BM, K, L); MOUNT KINABALU: 900 m, *Burbidge s.n.* (K).

39.8.3. Homalomena kinabaluensis Furtado, *Gard. Bull. Straits Settlem.* 10: 222 (1939). Type: TENOMPOK, 1400 m, *Clemens 29137* (holotype SING n.v.; isotypes BM!, K!).

Perennial herb. Lower montane forest. Elevation: 1400–1500 m.

Additional material examined. GURULAU SPUR: 1500 m, *Clemens 51019* (BM, K, L).

39.8.4. Homalomena ovata Engl., *Bull. Soc. Tosc. Ortic.* 4: 296 (1879).

Perennial herb. Hill forest. Elevation: 900 m.

Material examined. DALLAS: 900 m, *Clemens 29134* (BM, K, L).

39.8.5. Homalomena propinqua Schott in Miq., *Ann. Mus. Bot. Lugduno-Batavum* 1: 280 (1863).

Perennial herb. Hill forest, often on ultramafic substrate. Elevation: 500–1700 m.

Material examined. DALLAS: 900 m, *Clemens 26086* (BM, K, L), 900 m, *26732* (BM, K), 900 m, *26986* (BM, L), 900 m, *29152* (BM, K, L); EASTERN SHOULDER: 1000 m, *RSNB 966* (K, L); MT. NUNGKEK: 900 m, *Darnton 424* (BM); NALUMAD: *Shea & Aban SAN 77311* (K, L); PENATARAN RIVER: 500 m, *Beaman 9293* (MSC, US, US); PENIBUKAN: 1200 m, *Clemens 32084* (BM, L), 1700 m, *50279* (BM, SING n.v.).

39.8.6. Homalomena rubra Hassk. in Hoeven & de Vriese, *Tijdschr. Natuurl. Gesch. Physiol.* 9: 162 (1842).

Perennial herb. Lower montane forest. Elevation: 1400 m.

Material examined. TENOMPOK: 1400 m, *Clemens 29151* (BM, SING n.v.).

39.8.7. Homalomena sp. sect. Cyrtocladon

Perennial herb. Hill forest.

Material examined. MELANGKAP TOMIS: *Lorence Lugas 668* (K), *2622* (K).

39.8.8. Homalomena indet.

Material examined. LOHAN/MAMUT COPPER MINE: 900 m, *Beaman 10638* (K, MSC, US); SAYAP: 800–1000 m, *Beaman 9768* (K, MSC, US).

39.9. HOTTARUM Bogner & Nicolson

39.9.1. Hottarum kinabaluense Bogner, *Pl. Syst. Evol.* 145: 161 (1984). Type: MESILAU CAVE, 2000 m, *Collenette 21634* (holotype L!; isotype K!).

Perennial herb. Lower montane forest; rheophytic. Elevation: 1400–2100 m.

Additional material examined. GOLF COURSE SITE: *Madani SAN 111616* (K); MENERINTOG: 2100 m, *Clemens 29135* (BM, K); PARK HEADQUARTERS: 1400 m, *Edwards 2162* (K), 1500 m, *Ogata 11083* (L).

39.10. PIPTOSPATHA N. E. Br.

Engler, A., and Krause, K. (1912). Araceae-Philodendroideae-Philodendreae. *Pflanzenr.* IV. 23Da (Heft 55): 1–134 [*Piptospatha* on pp. 124–128].

39.10.1. Piptospatha elongata (Engl.) N. E. Br., *Bot. Mag.* sub t. 7410 (1895). Plate 8A.

Perennial herb. Hill forest, lower montane forest, mostly a rheophyte on rocks and muddy banks. Elevation: 500–1500 m.

Material examined. KIAU: *Jusimin Duaneh 461* (K); KUNDASANG: 1200 m, *Clemens 29136* (A, BM, K, L); LIWAGU/MESILAU RIVERS: 1200 m, *RSNB 2501* (K, L); MARAI PARAI: 1500 m, *Clemens 32290* (BM, GH); MELANGKAP KAPA: 600–700 m, *Beaman 8594* (K, MSC, US), 700–1000 m, *8811* (MSC); PARK HEADQUARTERS: 1400 m, *Price 165* (K); RANAU: 500 m, *Darnton 187* (BM); TAHUBANG RIVER: 900 m, *Clemens 31876* (BM, GH), 900 m, *Nooteboom & Aban 1508* (K, L); UPPER LIWAGU RIVER: *Amin et al. 123353* (K).

39.10.2. Piptospatha grabowskii Engl., *Pflanzenr.* IV. 23Da (Heft 55): 125 (1912).

Perennial herb. Lower montane forest; rheophyte on rocks and muddy banks. Elevation: 2000 m.

Material examined. EASTERN SHOULDER: 2000 m, *RSNB 708* (K, L).

39.10.3. Piptospatha cf. marginata (Engl.) N. E. Br., *Bot. Mag.* sub t. 7410 (1895).

Perennial herb. Lower montane forest; rheophyte on rocks. Elevation: 1800 m.

Material examined. MESILAU CAVE TRAIL: 1800 m, *Beaman 7473* (K, MSC, US).

39.11. POTHOS L.

Engler, A. (1905). Araceae-Pothoideae. *Pflanzenr.* IV. 23B (Heft 21): 1–330 [*Pothos* on pp. 21–44].

39.11.1. Pothos beccarianus Engl. in DC., *Monogr. Phan.* 2: 92 (1879).

Slender root climber. Lower montane forest. Elevation: 1500 m.

Material examined. MESILAU RIVER: 1500 m, *RSNB 4977* (K, L, LE, SING, US).

39.11.2. Pothos borneensis Furtado, *Gard. Bull. Straits Settlem.* 8: 148 (1935).

Herbaceous climber. Hill forest. Elevation: 900–1200 m. The species may not be distinct from *P. insignis* Engl.

Material examined. DALLAS: 900–1200 m, *Clemens 26542* (BM, SING).

39.11.3. Pothos cf. **brevistylus** Engl., *Bull. Soc. Tosc. Ortic.* 4: 267 (1879).

Herbaceous climber. Hill forest. Elevation: 700–1300 m.

Material examined. KULUNG HILL: 700–800 m, *Beaman 8361* (MSC, US); LAKANG: *Daim Andau 579* (K); LUGAS HILL: 1300 m, *Beaman 8463* (K, MSC, US).

39.11.4. Pothos kinabaluensis Furtado, *Gard. Bull. Straits Settlem.* 8: 149 (1935). Type: TENOMPOK, 1500 m, *Clemens 29155* (holotype SING!; isotypes A!, BM!, K!).

Herbaceous climber. Hill forest on ultramafic substrate. Elevation: 1200–1500 m.

Additional material examined. PENIBUKAN: 1200 m, *Clemens 31126* (A, BM, K, SING), 1200 m, *35003* (BM), 1500 m, *50710* (BM, SING); TAHUBANG RIVER: 1200 m, *Clemens s.n.* (BM); TENOMPOK: 1500 m, *Clemens 28514* (BM, K, SING).

39.11.5. Pothos ovatifolius Engl., *Pflanzenr.* IV. 23B (Heft 21): 40 (1905).

Clinging climber appressed to tree trunks. Lowlands. The species was previously known only from the Philippines.

Material examined. MELANGKAP TOMIS: *Lorence Lugas 320* (K).

39.11.6. Pothos scandens L., *Sp. Pl.*: 968 (1753).

Slender climber. Hill forest. Elevation: 500–1100 m.

Material examined. DALLAS/TENOMPOK: 1100 m, *Clemens 26808* (A, BM, K, L, SING, UC); PINAWANTAI: 500 m, *Shea & Aban SAN 76765* (K, KEP, L); PORING: *Sani Sambuling 29* (K), *421* (K).

39.12. RHAPHIDOPHORA Hassk.

Engler, A., and Krause, K. (1908). Araceae-Monsteroideae. *Pflanzenr.* IV. 23B (Heft 37): 4–138 [*Rhaphidophora* on pp. 17–53].

39.12.1. Rhaphidophora korthalsii Schott in Miq., *Ann. Mus. Bot. Lugduno-Batavum* 1: 129 (1863). Plate 8B, C.

Robust climber, occasionally lithophytic. Hill forest, lower montane forest. Elevation: 900–1800 m.

Material examined. DALLAS: 900 m, *Clemens 26495* (BM, K, L, UC); KULUNG HILL: *Meijer SAN 122414* (SAN), *SAN 122429* (SAN); MENERINTOG: *Clemens 29146* (BM, BO, K, L, SING, UC); MESILAU: 1200 m, *Mikil SAN 38691* (K); MESILAU BASIN: 1800 m, *Clemens 29146a* (K, L, SING); MESILAU CAMP: *Poore H 131* (K); NALUMAD: *Daim Andau 632* (K); PENATARAN RIVER: 900–1200 m, *Clemens 34298* (BM, BO, K, L, UC); PENIBUKAN: 1200–1500 m, *Clemens 31150* (BM, K, SING); TENOMPOK: 1500 m, *Clemens 28813* (BM, K).

39.12.2. Rhaphidophora puberula Engl., *Bot. Jahrb.* 1: 180 (1881).

Climbing aroid with long pendent shoots. Hill forest, lower montane forest. Elevation: 900–1700 m. Some specimens have been identified previously as *R. peepla* (Roxb.) Schott.

Material examined. DALLAS: 900 m, *Clemens 26453* (BM, K, L), 900 m, *26731* (BM), 900 m, *26876* (A, BM, BO, K, L, SING, UC); GURULAU SPUR: *Clemens, M. S. 10772* (UC); KIAU: *Clemens, M. S. 10142* (UC); PARK HEADQUARTERS/TENOMPOK: 1500–1700 m, *Kokawa & Hotta 3089* (KYO); TENOMPOK: 1400 m, *Clemens 26900* (BM, BO, K, L, UC).

39.12.3. Rhaphidophora sylvestris (Blume) Engl. in DC., *Monogr. Phan.* 2: 239 (1879).

Slender climber, occasionally a semi-epiphyte. Hill forest, lower montane forest. Elevation: 900–2000 m.

Material examined. DALLAS: 900 m, *Clemens s.n.* (BM), 900 m, *26720* (K, SING), 900 m, *26921* (BO, K, L, SING), 900 m, *27030* (BM, BO, K, L, UC), 900–1200 m, *27312* (BM, BO, K); DALLAS/ TENOMPOK: 1100 m, *Clemens s.n.* (BM); HEMPUEN HILL: *Amin & Jarius SAN 121155* (K, SAN); MENERINTOG: 1800 m, *Clemens 28487* (BM, K, L, SING); MESILAU CAMP/MESILAU CAVE: 1600–2000 m, *Kokawa & Hotta 3995* (KYO); MESILAU VALLEY: 1500 m, *Cockburn SAN 70108* (SAN); PAHU: *Doinis Soibeh 722* (K); SAYAP: *Yalin Surunda 36* (K); TENOMPOK: 1500 m, *Clemens 29227* (BM, BO, K, L, SING, UC).

39.13. SCHISMATOGLOTTIS Zoll. & Moritzi

Engler, A., and Krause, K. (1912). Araceae-Philodendroideae-Philodendreae. *Pflanzenr.* IV. 23Da (Heft 55): 1–134 [*Schismatoglottis* on pp. 82–122].

39.13.1. Schismatoglottis calyptrata (Roxb.) Zoll. & Moritzi ex Moritzi, *Syst. Verz.*: 83 (1846). Plate 8D.

Perennial herb. Hill forest, lower montane forest. Elevation: 800–1500 m.

Material examined. DALLAS: 900 m, *Clemens 26024* (BM, K, L), 900 m, *26025* (BM), 900 m, *29183* (BM, K), 900 m, *29184* (BM, K, L), 900 m, *29188* (BM, K, L); LOHAN/MAMUT COPPER MINE: 900 m, *Beaman 10632* (K, MSC, US); MESILAU RIVER: 1500 m, *RSNB 4899* (K); SAYAP: 800–1000 m, *Beaman 9769* (MSC, US); SINGH'S PLATEAU: 900 m, *RSNB 1026* (K); TENOMPOK: *Amin et al. SAN 123499* (K), 1400 m, *Clemens 29185* (BM, K).

39.13.2. Schismatoglottis cf. **calyptrata** (Roxb.) Zoll. & Moritzi ex Moritzi, *Syst. Verz.*: 83 (1846).

Perennial herb. Lowlands.

Material examined. SERINSIM: *Kinsun Bakia 81* (K), *232* (K); TEKUTAN: *Lomudin Tadong 228* (K), *299* (K).

39.13.3. Schismatoglottis caulescens Ridl., *J. Straits Branch Roy. Asiat. Soc.* 44: 182 (1905).

Perennial herb. Hill forest. Elevation: 1200 m.

Material examined. KADAMAIAN RIVER: 1200 m, *Gibbs 4100* (BM).

39.13.4. Schismatoglottis cf. **hastifolia** Hallier f. ex Engl., *Pflanzenr.* IV. 23Da (Heft 55): 116 (1912).

Perennial herb. Lower montane forest. Elevation: 1500 m.

Material examined. TENOMPOK: 1500 m, *Clemens 29493* (BM, K, L).

39.13.5. Schismatoglottis lancifolia Hallier f. & Engl., *Pflanzenr.* IV. 23Da (Heft 55): 88 (1912).

Perennial herb. Lower montane forest. Elevation: 1500 m.

Material examined. TENOMPOK: 1500 m, *Clemens 29272* (BM, K, L).

39.13.6. Schismatoglottis retinervia Furtado, *Gard. Bull. Straits Settlem.* 8: 157 (1935). Type: TENOMPOK, 1500 m, *Clemens 29153* (holotype SING n.v.; isotypes BM!, K!, L!).

Perennial herb. Lower montane forest. Elevation: 1500–1700 m.

Additional material examined. LUMU-LUMU: 1700 m, *Clemens 29186* (BM, K), 1700 m, *29187* (BM, K, L); TENOMPOK: 1500 m, *Clemens 29154* (BM, K).

39.13.7. Schismatoglottis indet.

The genus is not well understood taxonomically; thus a large number of collections are undetermined.

Material examined. BAMBANGAN RIVER: *Amin & Jarius SAN 116539* (K); DALLAS: 900 m, *Clemens 29157* (BM, K, L); EAST MESILAU RIVER: 1400 m, *Meijer SAN 38078* (K, L); EASTERN SHOULDER: 1100 m, *RSNB 642* (K, L); KILEMBUN BASIN: 1200 m, *Clemens s.n.* (BM); NALUMAD: *Daim Andau 495* (K),

715 (K); PENIBUKAN: 1200 m, *Clemens 32205* (BM), 1200 m, *35007* (BM), 1200–1500 m, *51694* (BM, K, L); PINOSUK PLATEAU: 1400–1600 m, *Kokawa & Hotta 4481* (K); SOSOPODON: 1400–1500 m, *Kokawa & Hotta 4513* (K); TAHUBANG RIVER: 1200–1500 m, *Clemens 31345* (BM), 1200 m, *50079* (BM); TEKUTAN: *Lomudin Tadong 32* (K); TENOMPOK: 1500 m, *Clemens 28552* (BM), 1500 m, *28647* (BM), 1400 m, *29143* (BM); TINEKEK FALLS: 1800 m, *Clemens 40992* (BM).

39.14. SCINDAPSUS Schott

Engler, A., and Krause, K. (1908). Araceae-Monsteroideae. *Pflanzenr.* IV. 23B (Heft 37): 4–138 [*Scindapsus* on pp. 67–80].

39.14.1. Scindapsus borneensis Engl., Pflanzenr. IV. 23B (Heft 37): 74 (1908).

Rhaphidophora kinabaluensis Furtado, *Gard. Bull. Straits Settlem.* 8: 152 (1935). Type: TENOMPOK, 1400 m, *Clemens 26875* (holotype SING!; isotypes B!, BM!, K!, L!, UC!).

Climber. Rarely in the lowlands, mostly in lower montane forest. Elevation: 400–2400 m.

Additional material examined. BAMBANGAN RIVER: *Amin et al. SAN 121008* (K, KEP); DALLAS: 900 m, *Clemens 26121* (BM, K, L, UC), 900–1200 m, *26243* (BM, K, L), 900 m, *28142* (BM, K, L, UC); EAST MESILAU/MENTEKI RIVERS: 1700 m, *Beaman 8747* (K, US), 1700–2000 m, *9582* (MSC, US); GOLF COURSE SITE: 1700–1800 m, *Beaman 7201* (MSC), *Jamili Nais SNP 4287* (SAN); GURULAU SPUR: 1800 m, *Clemens 51128* (BM); KAUNG: 400 m, *Clemens 26121* (BM); KEMBURONGOH: 1800 m, *Clemens 26713* (BM, K, L, UC); KILEMBUN RIVER: 1400 m, *Clemens 35128* (BM, L); LOHAN RIVER: 800–1000 m, *Beaman 9984* (K, MSC, US); MAMUT COPPER MINE: 1600–1700 m, *Beaman 9925* (K, MSC, US); MESILAU CAMP/MESILAU CAVE: 1600–2000 m, *Kokawa & Hotta 4152* (KYO); MESILAU CAVE TRAIL: 1800 m, *Beaman 7492* (MSC), 1700–1900 m, *9107* (MSC, US); MESILAU RIVER: 1500 m, *RSNB 4864* (K, L, LE); PARK HEADQUARTERS: 1600 m, *Stone 11414* (US); PENIBUKAN: 1200–1500 m, *Clemens 35004* (BM, BO); TENOMPOK: 1300 m, *Lau Choon Teng 8/LCT* (UKMB).

39.14.2. Scindapsus curranii Engl. & K. Krause, *Pflanzenr.* IV. 23B (Heft 37): 138 (1908).

Vine. Lowlands. Elevation: 500–1500 m. The species has long been overlooked in Borneo.

Material examined. DALLAS: 900 m, *Clemens s.n.* (BM); KIAU/LUBANG: *Topping 1596* (K); KIPUNGIT RIVER: 500 m, *Meijer et al. SAN 122513* (A); LIWAGU/MESILAU RIVERS: 1400 m, *RSNB 2770* (K), 1200 m, *RSNB 2940* (K, L); MINITINDUK GORGE: 900–1200 m, *Clemens 29617* (BM, BO, K, L, UC); PENIBUKAN: 1200–1500 m, *Clemens 31250* (BM); PORING: *Sani Sambuling 343* (K); SAYAP: 800–1000 m, *Beaman 9802* (K, US); SERINSIM: *Kinsun Bakia 446* (K); SOSOPODON: 1400–1500 m, *Kokawa & Hotta 5177* (KYO); TENOMPOK: 1400 m, *Clemens 29182* (BM), 1500 m, *29893* (BM, BO, K, SING, UC).

39.14.3. Scindapsus longistipitatus Merr., *Philipp. J. Sci.* 29: 353 (1926).

Climber. Lowlands, hill forest. Elevation: 700–800 m.

Material examined. PORING: 700–800 m, *Beaman 10933* (K); TEKUTAN: *Dius Tadong 162* (K).

39.14.4. Scindapsus perakensis Hook. f., *Fl. Brit. India* 6: 542 (1893).

Climber. Hill forest. Elevation: 800–1800 m.

Material examined. DALLAS: 900 m, *Clemens s.n.* (BM), 1100 m, *26171* (BM, BO, K, L, UC); KULUNG HILL: 800 m, *Beaman 7807* (MSC); LIWAGU/MESILAU RIVERS: 1200 m, *RSNB 2941* (K, L); MENERINTOG: 1800 m, *Clemens 29181* (BM, K, L); PORING: *Meijer SAN 131924* (A); SAYAP: 800–1000 m, *Beaman 9799* (K, US).

39.14.5. Scindapsus pictus Hassk. in Hoeven & de Vriese, *Tijdschr. Natuurl. Gesch. Physiol.* 9: 164 (1842). Plate 9A.

Climber with juvenile leaves tightly appressed to tree trunks; mature plants with much-branched free stems. Hill forest, lower montane forest. Elevation: 1100–1700 m.

Material examined. DALLAS: 1100 m, *Clemens 29335* (BM, K); GURULAU SPUR: 1700 m, *Clemens s.n.* (BM); KILEMBUN BASIN: 1700 m, *Clemens 34099* (BM, BO); MELANGKAP TOMIS: *Lorence Lugas 174* (K); PAHU (NEAR BUNDU TUHAN): *Doinis Soibeh 724* (K); PENIBUKAN: 1400 m, *Clemens s.n.* (BM), 1200 m, *31315* (BM), 1200–1500 m, *31637* (BM, K), 1200–1500 m, *40601* (K, L); TENOMPOK: 1300 m, *Lau Choon Teng 15/LCT* (UKMB).

39.14.6. Scindapsus rupestris Ridl., *J. Straits Branch Roy. Asiat. Soc.* 44: 184 (1905).

Creeper, often terrestrial, sometimes occurring as a rheophyte. Hill forest, lower montane forest. Elevation: 1200–2100 m.

Material examined. BAMBANGAN RIVER: 1500 m, *RSNB 4562* (K); GURULAU SPUR: 1800 m, *Clemens 51128* (K), 1500 m, *Gibbs 4010* (BM); KUNDASANG: 1400 m, *RSNB 1383* (K); MARAI PARAI SPUR: 1500 m, *Clemens 35005* (BM); MESILAU CAVE: 1800 m, *RSNB 4801* (K), MESILAU RIVER: 1500 m, *RSNB 4083* (K), 1500 m, *RSNB 4353* (K); PARK HEADQUARTERS: 1400 m, *Price 124* (K), 1400 m, *168* (K); PENIBUKAN: 1200–1500 m, *Clemens 31094* (BM, K); SOSOPODON: 1500 m, *Meijer SAN 29020* (K, L); TINEKEK FALLS: 2100 m, *Clemens s.n.* (BM).

39.14.7. Scindapsus indet.

Material examined. BAMBANGAN RIVER: 1600 m, *RSNB 4939* (K); TAHUBANG RIVER: 1100 m, *Nooteboom & Aban 1582* (L).

40. ARECACEAE

In collaboration with J. Dransfield (K), R. J. Carrington (K)
and G. J. Martin (Marrakesh)

Uhl, N. W., and Dransfield, J. (1987). *Genera Palmarum, a Classification of Palms Based on the Work of Harold E. Moore, Jr.* L. H. Bailey Hortorium and the International Palm Society, Lawrence, Kansas.

40.1. ARECA L.

Dransfield, J. (1984). The genus *Areca* (Palmae: Arecoideae) in Borneo. *Kew Bull.* 39: 1–22. Furtado, C. X. (1933). The limits of *Areca* and its sections. *Repert. Spec. Nov. Regni Veg.* 33: 217–239.

40.1.1. Areca catechu L., *Sp. Pl.*: 1189 (1753). Plate 9B.

Moderately robust single-stemmed palm to 10 m tall. Lowlands, hill forest. The betel palm; widely cultivated.

Material examined. HIMBAAN: *Doinis Soibeh 778* (K); KIAU: *Jusimin Duaneh 255* (K); MELANGKAP TOMIS: *Lorence Lugas 442* (K); PORING: *Sani Sambuling 8* (K), *76* (K); SERINSIM: *Jibrin Sibil 152* (K), *Kinsun Bakia 416* (K).

40.1.2. Areca kinabaluensis Furtado, *Repert. Spec. Nov. Regni Veg.* 33: 233 (1934). Plate 9C. Type: LUMU-LUMU, 1700 m, *Clemens 28761* (holotype SING!; isotypes K!, L!).

Slender, solitary undergrowth palm to 5 m tall. Lower montane forest. Elevation: 1200–1800 m.

Additional material examined. HIMBAAN: *Doinis Soibeh 257* (K), *328* (K), *329* (K), *330* (K); LIWAGU RIVER TRAIL: *Dransfield, J. et al. JD 5708* (K); LUGAS HILL: *PEK 12* (K), *16* (K); LUMU-LUMU: 1700 m, *Clemens 28761a* (SING); MESILAU RIVER: 1500 m, *RSNB 7018* (K), 1400 m, *RSNB 7051* (K); PAHU: *Doinis Soibeh 729* (K), *731* (K); PARK HEADQUARTERS: 1600 m, *Stevens et al. 619* (L); PARK HEADQUARTERS/POWER STATION: 1500–1800 m, *Jacobs 5705* (L), 1500–1800 m, *5705A* (L); SOSOPODON: 1400 m, *Badak SAN 32388* (K), 1500 m, *Meijer SAN 29015* (K); SOSOPODON/KUNDASANG: 1200 m, *Carson SAN 28035* (K); TENOMPOK: 1500 m, *Clemens 29202* (BM, K, L, SING); TINEKEK FALLS: 1800 m, *Clemens 40932* (BM, SING).

40.1.3. Areca rheophytica J. Dransf., *Kew Bull.* 39: 18 (1984).

Slender single-stemmed palm with finely divided leaves. Lowlands, restricted to river banks on ultramafic substrates.

Material examined. NALUMAD: *Daim Andau 148* (K); TEKUTAN: *Dius Tadong 10* (K).

40.2. ARENGA Labill. in DC.

Dransfield, J., and Mogea, J. P. (1984). The flowering behaviour of *Arenga* (Palmae: Caryotoideae). *Bot. J. Linn. Soc.* 88: 1–10.

40.2.1. Arenga brevipes Becc., *Malesia* 3: 95 (1889).

Robust short-stemmed palm with massive leaves to 8 m long, leaflets rigid. Hill forest, lower montane forest. Elevation: 900–1500 m.

Material examined. BUNDU TUHAN: *Doinis Soibeh 355* (K); DALLAS: 900 m, *Clemens 26879* (BM, K); KIAU: *Jusimin Duaneh 182* (K); PENIBUKAN: 1200–1500 m, *Clemens 31237* (BM, L); SAYAP: *Jatin Tungking 4* (K); TENOMPOK: 1500 m, *Clemens 28333* (BM, K, L).

40.2.2. Arenga "distincta" Mogea ined.

Slender, clustering undergrowth palm to 1.5 m tall. Lowlands, generally occurring on slopes and in valley bottoms.

Material examined. TEKUTAN: *Dius Tadong 417* (K), *433* (K).

40.2.3. Arenga retroflorescens H. E. Moore & Meijer, *Principes* 9: 100 (1965).

Densely clustering short-stemmed palm forming thickets of leaves to 3.5 m long. Lowlands.

Material examined. SERINSIM: *Akungsai Bakia 2* (K), *Kinsun Bakia 2* (K); TEKUTAN: *Jibrin Sibil 102* (K).

40.2.4. Arenga undulatifolia Becc., *Malesia* 3: 95 (1889). Plates 9D, 10A.

Robust short-stemmed palm to 6 m tall with leaves to 8 m long, leaflets undulate. Disturbed areas in the lowlands and hill forest. Elevation: 500 m.

Material examined. MELANGKAP TOMIS: *Lorence Lugas 571* (K), *1601* (K); NALUMAD: *Daim Andau 85* (K); PORING: *Meliden Giking 124* (K), *Smith & Everard 145* (K); SAYAP: *Jatin Tungking 9* (K); TEKUTAN: *Dius Tadong 198* (K), *586* (K), *Jibrin Sibil 100* (K), 500 m, *Shea & Aban SAN 77153* (K, L).

40.3. CALAMUS L.

Dransfield, J. (1982). Notes on rattans (Palmae: Lepidocaryoideae) occurring in Sabah, Borneo. *Kew Bull.* 36: 783–815. Dransfield, J. (1984). *The rattans of Sabah*. Sabah Forest Records, No. 13. Forest Dept., Sabah.

40.3.1. Calamus acuminatus Becc., *Ann. Roy. Bot. Gard. (Calcutta)* 11, Suppl.: 16 (1913).

Slender, clustering, flagellate rattan. Lowlands, hill forest. Elevation: 700–1100 m.

Material examined. BAMBANGAN: *Amin & Jarius SAN 121138* (K); BUNDU TUHAN: *Doinis Soibeh 611* (K); DALLAS: 900 m, *Clemens 27320* (BM, K, L); EASTERN SHOULDER, CAMP 1: 1100 m, *Meijer SAN 26413* (K); KAULUAN: *Sani Sambuling 168* (K); KIAU: *Jusimin Duaneh 119* (K); MURUK: *Sani Sambuling 171* (K); NAAPONG: *Sani Sambuling 59* (K); NALUMAD: *Daim Andau 505* (K); NAMPASAN BARU: *Sani Sambuling 179* (K); NAPONG: *Sani Sambuling 56* (K), *241* (K); PORING: 700 m, *Dransfield, J. et al. JD 5561* (K), *Sani Sambuling 72* (K), *119* (K), *137* (K), *183* (K); PORING/LIPOSU: *Sani Sambuling 27* (K); SAYAP: *Tungking Simbayan 24* (K).

40.3.2. Calamus amplijugus J. Dransf., *Kew Bull.* 36: 787 (1982).

Slender, clustering, flagellate rattan. Lowlands, hill forest. Elevation: 700–900 m.

Material examined. Bongkud: *Sani Sambuling 178* (K); Hempuen Hill: 700 m, *Dransfield, J. et al. JD 5577* (K), 800–900 m, *Meijer SAN 20268* (K); Kauluan: *Sani Sambuling 163* (K); Langanan Falls: *Sani Sambuling 258* (K); Liposu: *Sani Sambuling 253* (K); Lohan: *Sani Sambuling 173* (K); Melangkap Tomis: *Lorence Lugas 1600* (K); Muruk: *Sani Sambuling 172* (K); Naapong: *Sani Sambuling 58* (K); Nampasan: *Sani Sambuling 180* (K), *249* (K); Napong: *Sani Sambuling 239* (K); Narawang: *Sani Sambuling 229* (K); Perancangan: *Sani Sambuling 146* (K); Poring: *Sani Sambuling 13* (K), *68* (K); Serinsim: *Kinsun Bakia 178* (K); Tekutan: *Dius Tadong 11* (K), *320* (K), *522* (K), *716* (K), *Lomudin Tadong 197* (K); Tensungoi: *Sani Sambuling 201* (K).

40.3.3. Calamus blumei Becc., *Rec. Bot. Surv. India* 2: 209 (1902).

Calamus penibukanensis Furtado, *Gard. Bull. Singapore* 15: 79 (1956). Type: Penibukan, 1200 m, *Clemens 40520* (holotype SING!; isotypes K!, L!).

Moderate, clustering, high-climbing, flagellate rattan. Hill forest. Elevation: 700–1500 m.

Additional material examined. Kauluan: *Sani Sambuling 159* (K); Langanan Falls: *Sani Sambuling 256* (K), *315* (K); Liposu: *Sani Sambuling 252* (K); Muruk: *Sani Sambuling 219* (K); Nalumad: *Daim Andau 229* (K); Nampasan: *Sani Sambuling 251* (K); Napong 1: *Sani Sambuling 150* (K); Narawang: *Sani Sambuling 230* (K); Penibukan: 1200–1500 m, *Clemens 31587* (BM); Poring: *Aban Gibot SAN 85732* (K), 700 m, *Dransfield, J. et al. JD 5562* (K, L); Sayap: *Tungking Simbayan 23* (K); Tenompok: *Jibrin Sibil 1* (K); Tensungoi: *Sani Sambuling 204* (K).

40.3.4. Calamus caesius Blume, *Rumphia* 3: 57 (1849).

Densely clustering slender cirrate rattan climbing to 50 m or more. Lowlands; widely cultivated on village margins and sometimes apparently wild.

Material examined. Melangkap Tomis: *Lorence Lugas 582* (K), *1571* (K); Menggis: *Matamin Rumutom 290* (K); Nalumad: *Daim Andau 108* (K); Tekutan: *Dius Tadong 236* (K), *291 p.p.* (K), *476* (K).

40.3.5. Calamus comptus J. Dransf., *Kew Bull.* 45: 98 (1990).

Slender solitary or clustering flagellate rattan with neat fine leaflets. Lowlands on slopes and in valley bottoms.

Material examined. Tekutan: *Jibrin Sibil 96* (K).

40.3.6. Calamus convallium J. Dransf., *Kew Bull.* 36: 800 (1982).

Short clustering rattan with subcirrate leaves with distant irregular leaflets. Hill forest, restricted to valley bottoms at the lower elevations, rarely on slopes in the uplands.

Material examined. BUNDU TUHAN: *Doinis Soibeh 617* (K); KIAU: *Jusimin Duaneh 120* (K).

40.3.7. Calamus elopurensis J. Dransf., *Kew Bull.* 36: 787 (1982).

Slender clustering rattan, rarely exceeding 8 m tall. Hill forest on ultramafic substrate, but usually in alluvial forest in the lowlands of eastern Sabah.

Material examined. MELANGKAP TOMIS: *Lorence Lugas 1626* (K).

40.3.8. Calamus gibbsianus Becc., *Ann. Roy. Bot. Gard. (Calcutta)* 11, Suppl.: 58 (1913). Plate 10B. Type: PAKA-PAKA CAVE, 2400–2700 m, *Gibbs 4348* (holotype BM!).

Calamus dachangensis Furtado, *Gard. Bull. Straits Settlem.* 8: 247 (1935). Type: DACHANG, 3400 m, *Clemens 29198* (holotype SING!; isotypes K!, L!).

Slender, clustering, flagellate rattan. Lower montane forest, upper montane forest; occurring at higher elevations than any other SE Asian palm. Elevation: 1200–3400 m. A very polymorphic species; forms from exposed ridges on ultramafic rock have short, stiff, crowded leaflets.

Additional material examined. BAMBANGAN RIVER: 1500 m, *RSNB 4461* (K, L, SING); BUNDU TUHAN: *Doinis Soibeh 614* (K); DACHANG: 3400 m, *Clemens 29198a* (BM, K, L, SING); EASTERN SHOULDER: 2400 m, *RSNB 199* (K, SING), 2900 m, *RSNB 788* (K, L, SING); GOLF COURSE: *Amin Gambating SAN 114236* (K); GURULAU SPUR: 2100 m, *Clemens 50756* (BM), 2700–3000 m, *50805* (BM, SING), 2300 m, *51416* (BM, SING); KEMBURONGOH: 2400 m, *Clemens 29196* (BM, K, L, SING), 2700 m, *29196a* (SING), 2400 m, *29197* (BM, K, L), 2400 m, *29197b* (SING), 2400 m, *30524* (K, L), 2200 m, *Dransfield, J. et al. JD 5677* (K, L), 2500 m, *JD 5682* (K, L), 2200 m, *Meijer SAN 20360* (K), 2100 m, *Price 201* (K); KIAU VIEW TRAIL: 1600 m, *Cockburn SAN 76812* (K); KINATEKI RIVER: 1200–1500 m, *Clemens 31438* (BM, K); KINATEKI RIVER HEAD: 2700 m, *Clemens 31722* (BM, K, L), 2400 m, *31785* (BM, K, L), 2700 m, *31910* (BM, K, L); KUNDASANG: 1200–1500 m, *Meijer SAN 21039* (K); LAYANG-LAYANG: 3000 m, *Dransfield, J. et al. JD 5683* (K); LIWAGU RIVER TRAIL: 2700 m, *Allen AK 66 - 4* (K); LUMU-LUMU: 1800 m, *Clemens 29201* (BM), 1800 m, *29201a* (BM); MARAI PARAI: 1500 m, *Clemens 32600* (BM, L); MENERINTOG: 2100 m, *Clemens 29200* (BM, K, L, SING); MESILAU CAVE: 2000 m, *Dransfield, J. et al. JD 5670* (K, L); MESILAU RIVER: 1500 m, *RSNB 4159* (K, L, SING); MT. NUNGKEK: 1700 m, *Clemens 32672* (K), 1700 m, *35035* (BM); MT. TEMBUYUKEN: 2000 m, *Argent 1153* (K), 2600 m, *Justin Jukian SNP 196* (K), 2300 m, *Meijer SAN 34615* (K, L); MURUTURA RIDGE: 1500–1800 m, *Clemens 34351* (BM); PANAR LABAN: 2400 m, *Sidek bin Kiah S. 52* (SING); PARK HEADQUARTERS: 1800 m, *Dransfield, J. et al. JD 5668* (K, L); PARK HEADQUARTERS/POWER STATION: 1500–1800 m, *Jacobs 5704* (L); PENIBUKAN: 1200–1500 m, *Clemens 31438* (SING); PIG HILL: 2300 m, *RSNB 4520* (K, L, SING); PINOSUK PLATEAU: 2000 m, *Dransfield, J. et al. JD 5669* (K); POWER STATION: 2000 m, *Dransfield, J. et al. JD 5674* (K, L); SUMMIT TRAIL: 2400 m, *Jacobs 5711* (K, L); TENOMPOK: 1500 m, *Clemens 28800* (BM, SING), 1500 m, *29199* (BM, K, L, SING).

40.3.9. Calamus gonospermus Becc., *Rec. Bot. Surv. India* 2: 202 (1902).

Clustering, very slender, flagellate rattan. Lowlands, hill forest. Elevation: 700 m.

Material examined. HEMPUEN HILL: 700 m, *Dransfield, J. et al. JD 5580* (K); HIMBAAN: *Doinis Soibeh 424* (K); PAHU: *Doinis Soibeh 728* (K).

40.3.10. Calamus javensis Blume, *Rumphia* 3: 63 (1847).

Clustering, very slender, flagellate rattan. Hill forest, lower montane forest, rarely upper montane forest, often on ultramafic substrate. Elevation: 500–2600 m. A highly polymorphic species.

Material examined. BAMBANGAN RIVER: 1600 m, *RSNB 4938* (K); BONGKUD: *Sani Sambuling 176* (K); DALLAS: 900 m, *Clemens 27320* (BM); EASTERN SHOULDER: 2000 m, *RSNB 168* (K, SING); HEMPUEN HILL: 800 m, *Meijer SAN 21064* (K, L, SING), 800 m, *SAN 21350* (K), 700 m, *Singh SAN 28335* (K); HIMBAAN: *Doinis Soibeh 137* (K); KIAU: *Clemens, M. S. 10025* (K); KIBAMBANG LOBANG: 2300 m, *Clemens 32511* (BM); KINATEKI RIVER: 1200–1500 m, *Clemens 31433* (K, SING); LANGANAN FALLS: *Sani Sambuling 257* (K); LIWAGU RIVER TRAIL: 1700 m, *Dransfield, J. et al. JD 5699* (K, L), 1700 m, *JD 5701* (K); LOHAN: *Sani Sambuling 174* (K); LUGAS HILL: *PEK 2* (K), *6* (K), *11* (K); LUMU-LUMU: 2000 m, *Clemens 27835* (BM, SING); MARAI PARAI: 1500 m, *Clemens 32269* (BM), 1200 m, *32396A* (L), 1500–1800 m, *32520* (L); MARAI PARAI SPUR: 1500 m, *Clemens 32466* (BM), 1500–1800 m, *32520* (BM); MELANGKAP TOMIS: *Lorence Lugas 1627* (K); MENERINTOG: 1500 m, *Clemens 28564* (BM, SING), 1500 m, *28564a* (BM, SING); MESILAU CAVE: 1800 m, *RSNB 4723* (K), 1800 m, *RSNB 4835* (K, L); MESILAU CAVE TRAIL: *Meijer SAN 48116* (K); MESILAU RIVER: 1500 m, *RSNB 4345* (K), 1500 m, *RSNB 4920* (K), 1500 m, *RSNB 4995* (K), 1500 m, *RSNB 7025* (K); MT. NUNGKEK: 1200 m, *Clemens 32396A* (BM); MURUK: *Sani Sambuling 220* (K); NALUMAD: *Sani Sambuling 43* (K); PAHU: *Doinis Soibeh 785* (K); PENATARAN RIVER: 2300 m, *Clemens 32511* (L), 2600 m, *33663* (BM); PENIBUKAN: 1200 m, *Clemens 30830 bis* (SING), 1200–1500 m, *31433* (BM); PORING: *Sani Sambuling 17* (K), *110* (K), *140* (K), *182* (K), *184* (K); POWER STATION: 2000 m, *Dransfield, J. et al. JD 5672* (K, L), 2000 m, *JD 5673* (K, L); RANAU: 500 m, *Darnton 177* (BM); SOSOPODON: 1700 m, *Mikil SAN 38896* (K).

40.3.11. Calamus kiahii Furtado, *Gard. Bull. Straits Settlem.* 8: 251 (1935). Type: LUMU-LUMU, 1800 m, *Clemens 29195* (holotype SING!; isotypes K!, L!).

Moderate, clustering, cirrate rattan to 10 m. Hill forest, lower montane forest. Elevation: 1000–1800 m.

Additional material examined. BAMBANGAN RIVER: 1500 m, *RSNB 4440* (K); EASTERN SHOULDER: 1000 m, *RSNB s.n.* (K); LIWAGU RIVER TRAIL: *Benedict Busin 2b* (K), 1800 m, *Dransfield, J. et al. JD 5691* (K); MAMUT RIVER: 1200 m, *RSNB 1735* (K, L, SING); MESILAU RIVER: 1500 m, *RSNB 4280* (K).

40.3.12. Calamus laevigatus Mart., *Hist. Nat. Palm.* 3: 339 (1853).

a. var. laevigatus

Single-stemmed slender cirrate rattan; basal leaflets on each leaf swept back across the stem forming a chamber sometimes occupied by ants. Lowlands to lower montane forest, widespread in a variety of habitats.

Material examined. TEKUTAN: *Dius Tadong 7* (K), *63* (K), *321* (K); TENOMPOK: *Doinis Soibeh 1* (K).

b. var. **mucronatus** (Becc.) J. Dransf., *Bot. J. Linn. Soc.* 81: 8 (1980).

Slender, solitary, cirrate rattan, climbing to 40 m. Lowlands, hill forest. Elevation: 700 m.

Material examined. HEMPUEN HILL: 700 m, *Dransfield, J. et al. JD 5579* (K); SERINSIM: *Jibrin Sibil 161* (K).

40.3.13. Calamus marginatus (Blume) Mart., *Hist. Nat. Palm.* 3: 342 (1853).

Calamus rostratus Furtado, *Gard. Bull. Straits Settlem.* 8: 257 (1935). Type: TENOMPOK, 1500 m, *Clemens 28650* (holotype SING n.v.; isotypes BM!, K!, L!).
Calamus regularis Burret, *Notizbl. Bot. Gart. Mus. Berlin-Dahlem* 15: 816 (1943). Type: PENIBUKAN, 1200–1500 m, *Clemens 31497* (holotype B†; isotypes BM!, K!, SING!).

Robust, very spiny, solitary, flagellate rattan, climbing to 20 m. Hill forest, lower montane forest, often on ultramafic substrate. Elevation: 700–2400 m.

Additional material examined. DALLAS: 900 m, *Clemens 26360* (BM, K); EASTERN SHOULDER: 1000 m, *RSNB 21* (K, L, SING); GURULAU SPUR: 1500–2100 m, *Clemens 50757* (BM, K, L, SING), 1500 m, *50794* (BM, K, SING), 2400 m, *50985* (BM, K, L, SING); HEMPUEN HILL: 700 m, *Dransfield, J. et al. JD 5582* (K); MARAI PARAI, 1500 m, *Clemens 32327* (BM, L); MESILAU RIVER: 1500 m, *RSNB 4356* (K); MT. NUNGKEK: 1100–1400 m, *Clemens 32008* (BM); PENIBUKAN: 1200–1500 m, *Clemens 31498* (BM, K, SING); TENOMPOK: 1500 m, *Clemens 28375* (BM, K, L, SING), 1500 m, *28565* (BM, SING), 1500 m, *28566* (BM, K), 1500 m, *28566 bis* (SING), 1400 m, *28844* (BM, K, L, SING), *Danson Kandaong 1* (K).

40.3.14. Calamus mesilauensis J. Dransf., *Kew Bull.* 36: 797 (1982). Type: MESILAU RIVER, 1500 m, *Dransfield JD 5556* (holotype K!; isotype L!).

Slender, clustering, cirrate rattan with unusually large reddish brown fruit. Lower montane forest. Elevation: 1200–1700 m.

Additional material examined. LIWAGU RIVER TRAIL: 1700 m, *Dransfield, J. et al. JD 5696* (K), 1700 m, *JD 5698* (K); LIWAGU/MESILAU RIVERS: 1500 m, *RSNB 1957* (K); MESILAU RIVER: *RSNB 4125 p.p.* (K), 1400 m, *RSNB 7069* (K), 1500 m, *Dransfield, J. et al. JD 5557* (K); PENATARAN RIVER: 1200 m, *Clemens 34164* (BM); PINOSUK PLATEAU: 1600 m, *RSNB 1817* (K, SING).

40.3.15. Calamus muricatus Becc., *Nelle Foreste di Borneo*: 609 (1902).

Solitary slender flagellate rattan with horizontally ridged leaf sheaths. Lowlands.

Material examined. BOTONG: *Sani Sambuling 75* (K); TEKUTAN: *Dius Tadong 477* (K).

40.3.16. Calamus optimus Becc., *Nelle Foreste di Borneo*: 610 (1902).

Calamus stramineus Furtado, *Gard. Bull. Straits Settlem.* 8: 258 (1935). Type: DALLAS, 900 m, *Clemens 27010* (holotype SING!; isotypes BM!, K!).
Calamus stramineus Furtado var. *megalocarpus* Furtado, *Gard. Bull. Straits Settlem.* 8: 259 (1935). Type: DALLAS, 900 m, *Clemens 27009* (holotype SING!; isotype K!).

Moderate, clustering, cirrate rattan, climbing to 40 m. Lowlands and hill forest. Elevation: 900–1200 m. One of the best canes in Borneo.

Additional material examined. DALLAS: 900 m, *Clemens 26496* (BM, K), 900–1200 m, *26755* (BM).

40.3.17. Calamus ornatus Blume ex Schult. & Schult. f., *Syst. Veg.* 7: 1326 (1830). Plate 10C.

Robust, clustering, flagellate rattan, climbing to 40 m. Lowlands and hill forest. Elevation: 600–1100 m.

Material examined. LOHAN: *Dius Tadong 1* (K); PAHU: *Doinis Soibeh 732* (K); PENIBUKAN: 1100 m, *Clemens 40589* (BM, SING); PORING: 600 m, *Dransfield, J. et al. JD 5570* (K), *Meliden Giking 121* (K); TEKUTAN: *Dius Tadong 717* (K).

40.3.18. Calamus pogonacanthus Becc. in H. J. P. Winkl., *Bot. Jahrb. Syst.* 48: 91 (1912).

Moderate, clustering, cirrate rattan, climbing to 20 m. Lowlands, hill forest, lower montane forest. Elevation: 600–1500 m.

Material examined. BAMBANGAN CAMP: 1500 m, *RSNB 4581* (K); BUNDU TUHAN: *Doinis Soibeh 612* (K); HEMPUEN HILL: *Dransfield, J. et al. JD 5578* (K); HIMBAAN: *Doinis Soibeh 136* (K), *341* (K), *343* (K); KAGAPON HILL: 600 m, *Brand & Anak SAN 25302* (K); KAULUAN: *Sani Sambuling 162* (K); KIAU: *Jusimin Duaneh 114* (K); LUGAS HILL: *PEK 15* (K); NARAWANG: *Sani Sambuling 44* (K); PAHU: *Doinis Soibeh 786* (K); PORING: 700 m, *Dransfield, J. et al. JD 5560* (K), *Meliden Giking 40* (K), *Sani Sambuling 24* (K), *70* (K), *135* (K), *581* (K); SAYAP: *Jatin Tungking 1* (K); TEKUTAN: *Dius Tadong 225* (K), *506* (K), *584* (K); TENSUNGOI: *Sani Sambuling 205* (K), *217* (K).

40.3.19. Calamus praetermissus J. Dransf., *Kew Bull.* 36: 802 (1982).

Moderately robust flagellate clustering rattan with broad plicate leaflets. Lowlands, hill forest, generally on lower slopes. Elevation: 900 m.

Material examined. BAMBANGAN (SERINSIM AREA): *Akungsai Bakia 15* (K); DALLAS: 900 m, *Clemens 26683* (BM); TEKUTAN: *Dius Tadong 166* (K).

40.3.20. Calamus tenompokensis Furtado, *Gard. Bull. Straits Settlem.* 8: 260 (1935). Type: Tenompok, 1500 m, *Clemens 28408* (holotype SING!; isotypes BM!, K!).

Calamus nanus Burret, *Notizbl. Bot. Gart. Mus. Berlin-Dahlem* 15: 818 (1943). Type: Gurulau Spur, 1800 m, *Clemens 50397* (holotype B†; isotypes BM!, SING!).

Short-stemmed, erect, clustering palm with short flagella. Hill forest, lower montane forest. Elevation: 1200–1800 m.

Additional material examined. Dallas/Tenompok: 1200 m, *Clemens 27339* (BM, SING); Himbaan: *Doinis Soibeh 353* (K); Liwagu River Trail: 1700 m, *Dransfield, J. et al. JD 5706* (K), 1700 m, *JD 5707* (K, L); Lugas Hill: *PEK 8* (K); Mamut River: 1200 m, *RSNB 1653* (K, L, SING); Mesilau River: 1500 m, *RSNB 4056* (K), 1500 m, *Dransfield, J. et al. JD 5554* (K), 1500 m, *JD 5555* (K); Penataran Basin: 1500 m, *Clemens 34163* (BM, L), 1400 m, *40206* (BM); Penibukan: 1200 m, *Clemens 31028* (SING); Pinosuk Plateau: 1700 m, *RSNB 1892* (K, L, SING); Sayap: *Jatin Tungking 11* (K); Tenompok: 1500 m, *Clemens 27899* (BM, SING), 1500 m, *29203* (BM, K, SING).

40.3.21. Calamus zonatus Becc., *Nelle Foreste di Borneo*: 609 (1902).

Very slender, clustering, flagellate rattan with ridged sheaths. Lowlands, hill forest, sometimes on ultramafic substrate. Elevation: 700–1000 m.

Material examined. Hempuen Hill: 700 m, *Dransfield, J. et al. JD 5576* (K); Nalumad: *Daim Andau 82* (K); Poring/Nalumad: 1000 m, *Meijer SAN 18823* (K, L); Tekutan: *Dius Tadong 9* (K), *190* (K).

40.3.22. Calamus indet.

Material examined. Mamut Ridge: 1700–1800 m, *Hotta 20194* (L); Marai Pakai: 1200 m, *Clemens 32396* (L); Tekutan: *Dius Tadong 520* (K).

40.4. CARYOTA L.

Beccari, O. (1877). Palmae della Nuova Guinea. *Malesia* 1: 8–96. Dransfield, J. (1974). Notes on *Caryota no* Becc. and other Malesian *Caryota* species. *Principes* 18: 87–93.

40.4.1. Caryota mitis Lour., *Fl. Cochinch.* 2: 569 (1790).

Clustering, moderate, fish-tail palm with stems to 10 m tall. Lowlands, hill forest, in disturbed sites. Elevation: 900 m.

Material examined. Dallas: 900 m, *Clemens 26224* (BM, K, L), *27219* (BM, K, L), 900 m, *27226* (BM, K, L); Himbaan: *Doinis Soibeh 235* (K); Kauluan: *Sani Sambuling 160* (K); Melangkap Tomis: *Lorence Lugas 176* (K); Nalumad: *Daim Andau 194* (K); Pahu: *Doinis Soibeh 787* (K); Poring: *Meliden Giking 17* (K); Serinsim: *Jibrin Sibil 193* (K), *Kinsun Bakia 338* (K); Tekutan: *Dius Tadong 59* (K), *355* (K).

40.4.2. Caryota no Becc., *Nuovo Giorn. Bot. Ital.* 3: 12 (1871). Plate 10D.

Immense, solitary, fish-tail palm with stems to 20 m tall. Lowlands, hill forest. Elevation: 600–800 m. Virtually uncollected on Kinabalu, but observed at Kaung and Langanan Falls, in addition to the localities cited below.

Material examined. BUNDU TUHAN: *Benedict Busin 4* (K); MENGGIS: 700–800 m, *Beaman 10948* (MSC); PORING HOT SPRINGS: *Beaman 10950* (K, MSC).

40.5. CERATOLOBUS Blume in Schult. & Schult. f.

40.5.1. Ceratolobus concolor Blume, *Rumphia* 2: 165, t. 130 (1843).

Slender clustering rattan with spiculate leaf sheaths and diamond-shaped praemorse leaflets, reminiscent of *Korthalsia*. Lowlands, widespread but infrequently collected on Mount Kinabalu. The taxon is immediately distinguished from *Korthalsia* by its sheath knee.

Material examined. PORING: *Sani Sambuling 112* (K); TEKUTAN: *Dius Tadong 502* (K).

40.6. COCOS L.

40.6.1. Cocos nucifera L., *Sp. Pl.*: 1188 (1753).

Single-stemmed tree palm with pinnate leaves. Lowlands, hill forest; cultivated. The coconut.

Material examined. MELANGKAP TOMIS: *Lorence Lugas 516* (K); PORING: *Meliden Giking 123* (K), *Sani Sambuling 74* (K).

40.7. DAEMONOROPS Blume in Schult. & Schult. f.

Dransfield, J. (1982). Notes on rattans (Palmae: Lepidocaryoideae) occurring in Sabah, Borneo. *Kew Bull.* 36: 783–815. Dransfield, J. (1984). *The rattans of Sabah.* Sabah Forest Records, No. 13. Forest Dept., Sabah.

40.7.1. Daemonorops didymophylla Becc. in Hook. f., *Fl. Brit. India* 6: 468 (1893).

Clustering, moderate, cirrate rattan, climbing to 10 m. Lowlands, hill forest. Elevation: 700 m.

Material examined. HIMBAAN: *Doinis Soibeh 688* (K); KIAU: *Jusimin Duaneh 129* (K); MELANGKAP TOMIS: *Lorence Lugas 1658* (K); PORING: 700 m, *Dransfield, J. et al. JD 5564* (K); SAYAP: *Jatin Tungking 3* (K); TEKUTAN: *Dius Tadong 61* (K).

40.7.2. Daemonorops elongata Blume, *Rumphia* 3: 16 (1847).

Short clustering rattan with distant, somewhat irregular leaflets. Hill forest on ultramafic substrate. The present material, not identical to the typical form of the species, seems closely related.

Material examined. MELANGKAP TOMIS: *Lorence Lugas 572* (K), *1628* (K).

40.7.3. Daemonorops fissa Blume, *Rumphia* 3: 17 (1847). Plate 11A.

Moderate, clustering, cirrate rattan, climbing to 10 m. Lowlands, hill forest, in disturbed sites. Elevation: 600 m.

Material examined. HIMBAAN: *Doinis Soibeh 827* (K); PORING: 600 m, *Dransfield, J. et al. JD 5569* (K), 600 m, *JD 5571* (K, L); TEKUTAN: *Dius Tadong 12* (K), *583* (K).

40.7.4. Daemonorops ingens J. Dransf., *Bot. J. Linn. Soc.* 81: 20 (1980).

Robust, clustering, "stemless" palm. Hill forest in valley bottoms. Elevation: 900 m.

Material examined. DALLAS: 900 m, *Clemens 26972* (BM), 900 m, *29189* (BM, K, L); MELANGKAP TOMIS: *Lorence Lugas 1558* (K); SAYAP: *Tungking Simbayan 5* (K).

40.7.5. Daemonorops korthalsii Blume, *Rumphia* 3: 23 (1847).

Moderate, clustering, cirrate rattan, climbing to 10 m. Lowlands, hill forest. Elevation: 700–800 m.

Material examined. EASTERN SHOULDER: 800 m, *RSNB 218* (K); LUGAS HILL: *PEK 7* (K); NAAPONG: *Sani Sambuling 60* (K); PORING: *Beaman 11019* (K), 700 m, *Dransfield, J. et al. JD 5565* (K), *Sani Sambuling 73* (K), *138* (K); SERINSIM: *Jibrin Sibil 104* (K), *Kinsun Bakia 173* (K); TEKUTAN: *Dius Tadong 29* (K), *206* (K), *521* (K), *587* (K).

40.7.6. Daemonorops longipes (Griff.) Mart., *Hist. Nat. Palm.* 3 (ed. 2): 205 (1845).

Daemonorops calothyrsa Furtado, *Gard. Bull. Straits Settlem.* 8: 345 (1935). Type: TENOMPOK, 1200–1500 m, *Clemens 29194* (holotype SING n.v.; isotypes BM!, K!, L!).
Daemonorops longipedunculata Furtado, *Gard. Bull. Straits Settlem.* 8: 353 (1935). Type: PENIBUKAN, 1200–1500 m, *Clemens 31280* (holotype SING!; isotypes BM!, K!, L!).

Very variable, moderate to robust, clustering, cirrate rattan. Lowlands, hill forest, lower montane forest. Elevation: 900–1700 m.

Additional material examined. BAMBANGAN RIVER: 1500 m, *RSNB 4435* (K, L); BUNDU TUHAN: *Doinis Soibeh 615* (K), *616* (K); DALLAS: 900–1200 m, *Clemens s.n.* (BM, SING), 900 m, *26793* (BM, SING), 900 m, *26807* (BM, K, L, SING), 900 m, *27156* (BM, K, SING), 900 m, *27156A* (SING), 900–1200 m, *27188* (BM, SING), 900 m, *27268* (SING), 900 m, *27269* (BM, K); EASTERN SHOULDER: 1200 m, *RSNB 256* (K, SING); GURULAU SPUR: 1500 m, *Clemens 50547* (BM, SING); HIMBAAN: *Doinis Soibeh 256* (K), *345* (K); KIAU: *Jusimin Duaneh 113* (K); LAKANG: *Daim Andau 30* (K); LIWAGU RIVER TRAIL: 1700 m, *Dransfield, J. et al. JD 5697* (K, L); LIWAGU/MESILAU RIVERS: 1200 m, *RSNB 2517* (K, SING); LUGAS HILL: *PEK 3* (K), *5* (K); MAMUT RIVER: 1400 m, *RSNB 1669* (K, L, SING); MELANGKAP TOMIS: *Lorence Lugas 80* (K), *144* (K), *1625* (K), *1645* (K); MESILAU RIVER: 1500 m, *RSNB 4325* (K, L); PENIBUKAN: 1200–1500 m, *Clemens 30887* (BM, K, L, SING), 1200–1500 m, *31280a* (BM, K, L, SING), 1200–1500 m, *31581* (BM, K, L, SING), 1200 m, *31758* (BM), 1200 m, *32166* (BM, L), 1200–1500 m, *35034* (BM), 1200 m, *40421* (BM, SING), 1200 m, *40754* (BM, SING), 1500 m, *50343* (BM, SING); PINOSUK PLATEAU: 1500 m, *Dransfield, J. et al. JD 5551* (K), 1500 m, *JD 5552* (K); PORING: *Dransfield, J. et al. JD 5559* (K); PORING/LOHAN: *Sani Sambuling 25* (K); SAYAP: *Jatin Tungking 6* (K), *7* (K), *Tungking Simbayan 8* (K); SERINSIM: *Akungsai Bakia 1* (K), *Jibrin Sibil 103* (K), *Kinsun Bakia 1* (K); SOSOPODON/KUNDASANG: 1500 m, *Carson SAN 26786* (K); TAHUBANG RIVER: 1200 m, *Clemens 50213* (SING); TEKUTAN: *Dius Tadong 4* (K), *189* (K), *Lomudin Tadong 158* (K); TENOMPOK: 1200–1500 m, *Clemens 29193* (BM, K, L, SING).

40.7.7. Daemonorops longistipes Burret, *Notizbl. Bot. Gart. Mus. Berlin-Dahlem* 15: 798 (1943). Plate 11B. Type: PENIBUKAN, 1200 m, *Clemens 30830* (holotype B†; isotype K!).

Daemonorops elongata Blume var. *montana* Becc. in Gibbs, *J. Linn. Soc., Bot.* 42: 169 (1914). Type: GURULAU SPUR, *Gibbs 3983* (lectotype (Dransfield, 1982) BM!).

Daemonorops pleioclada Burret, *Notizbl. Bot. Gart. Mus. Berlin-Dahlem* 15: 797 (1943). Type: MOUNT KINABALU, *Clemens s.n.* (holotype B†).

Slender to moderate, clustering, cirrate rattan, climbing to 8 m. Hill forest, lower montane forest. Elevation: 900–1800 m.

Additional material examined. BAMBANGAN RIVER: 1500 m, *RSNB 4385* (K), 1500 m, *RSNB 4391* (K); DALLAS: 1200 m, *Clemens s.n.* (BM, BM); EASTERN SHOULDER: 1100 m, *RSNB 235* (K, SING), 1200 m, *RSNB 1543* (K); GURULAU SPUR: 1500 m, *Clemens 50462* (BM, K, L), 1500 m, *50743* (BM); KILEMBUN RIVER HEAD: 1800 m, *Clemens 32500 bis* (BM, L); LIWAGU RIVER TRAIL: 1800 m, *Dransfield, J. et al. JD 5690* (K, L), 1700 m, *JD 5693* (K, L), 1700 m, *JD 5694* (K, L), 1700 m, *JD 5695* (K), 1700 m, *JD 5700* (K, L); LUGAS HILL: *PEK 1* (K); MENERINTOG: 1800 m, *Clemens 29192* (BM, K, SING); MESILAU CAVE: 1800 m, *RSNB 4722* (K); MESILAU RIVER: 1500 m, *RSNB 4125 p.p.* (K); MT. NUNGKEK: 1200 m, *Clemens 32865* (BM); NALUMAD: *Daim Andau 81* (K); PENIBUKAN: 1200–1500 m, *Clemens 30672* (BM), 1200–1500 m, *31284* (BM, K), 1200–1500 m, *31604* (BM, L), 1200 m, *40380* (BM), 1100 m, *50140* (BM, K), 1700 m, *50244* (BM); TAHUBANG RIVER: 1100 m, *Clemens 40939* (BM, K); TENOMPOK: 1500 m, *Clemens 28351* (BM, K, SING), 1500 m, *28560* (BM, K, SING), 1500 m, *28566* (SING), 1200–1500 m, *29191* (BM, K, L, SING), 1500 m, *29191a* (BM, K, L, SING), 1500 m, *29192* (L, SING); TINEKEK RIVER, *Jusimin Duaneh 136* (K).

40.7.8. Daemonorops microstachys Becc., *Rec. Bot. Surv. India* 2: 225 (1902).

"Stemless," clustering, undergrowth rattan. Lowlands, hill forest. Elevation: 1300 m.

Material examined. HIMBAAN: *Doinis Soibeh 325* (K), *327* (K), *342* (K), *347* (K), *348* (K); LUGAS HILL: 1300 m, *Beaman 10555* (K, L, MSC), *PEK 20* (K), *46* (K); PORING: *Sani Sambuling 2* (K); SERINSIM: *Jibrin Sibil 164* (K).

40.7.9. Daemonorops periacantha Miq., *Fl. Ned. Ind., Eerate Bijv.*: 256, 592 (1861).

Coarse clustering cirrate rattan with densely golden-spiny sheaths and grouped leaflets. Lowlands, widespread in a variety of habitats.

Material examined. PORING: *Meliden Giking 122* (K).

40.7.10. Daemonorops ruptilis Becc., *Rec. Bot. Surv. India* 2: 230 (1902).

a. var. ruptilis

Robust, clustering, very spiny, cirrate rattan, climbing to 15 m. Lowlands, hill forest. Elevation: 600 m.

Material examined. PORING: 600 m, *Dransfield, J. et al. JD 5567* (K), 600 m, *JD 5568* (K); TEKUTAN: *Dius Tadong 8* (K).

40.7.11. Daemonorops sabut Becc. in Hook. f., *Fl. Brit. India* 6: 469 (1893).

Moderate, clustering, cirrate ant-rattan, climbing to 20 m. Lowlands, hill forest, sometimes on ultramafic substrate. Elevation: 700 m.

Material examined. HEMPUEN HILL: 700 m, *Dransfield, J. et al. JD 5575* (K); HIMBAAN: *Doinis Soibeh 686* (K); NALUMAD: *Daim Andau 84* (K); NAPONG: *Sani Sambuling 240* (K); PORING: *Sani Sambuling 107* (K), *232* (K), *248* (K); SERINSIM: *Jibrin Sibil 163* (K), *Kinsun Bakia 240* (K); TEKUTAN: *Dius Tadong 92* (K), *292* (K), *478* (K), *Lomudin Tadong 157* (K).

40.7.12. Daemonorops sparsiflora Becc., *Rec. Bot. Surv. India* 2: 224 (1902).

Moderate, clustering, cirrate rattan, climbing to 15 m. Lowlands, hill forest. Elevation: 400–1000 m.

Material examined. DALLAS: 900 m, *Clemens 26888* (BM, K), 900 m, *27012* (BM, K), 900 m, *27157A* (BM, SING), 900 m, *27216* (BM, K); EASTERN SHOULDER: 1000 m, *RSNB 24* (K); HIMBAAN: *Doinis Soibeh 206* (K), *346* (K); KAUNG: 400 m, *Darnton 269* (BM); KIAU: *Jusimin Duaneh 121* (K); LUGAS HILL: *PEK 4* (K), *10* (K); MELANGKAP TOMIS: *Lorence Lugas 1678* (K), *1679* (K); MURUK: *Sani Sambuling 170* (K), *221* (K); NALUMAD: *Daim Andau 83* (K); NAMPASAN: *Sani Sambuling 51* (K); NARAWANG: *Sani Sambuling 57* (K), *67* (K), *158* (K), *231* (K); PORING: *Meliden Giking 17* (K), *Sani Sambuling 71* (K), *134* (K), *238* (K), *247* (K), *580* (K), *687* (K); PORING/LIPOSU: *Sani Sambuling 41* (K); RANAU/PORING ROAD: 700 m, *Dransfield, J. et al. JD 5507* (K); TARAWAS: *Sani Sambuling 78* (K); TEKUTAN: *Dius Tadong 93* (K), *504* (K); TENSUNGOI: *Sani Sambuling 203* (K).

40.8. DYPSIS Noronha ex Mart.

40.8.1. Dypsis lutescens (H. Wendl.) Beentje & J. Dransf., *Palms of Madagascar*: 212 (1995).

A clustering species with finely pinnate leaves, reminiscent of *Pinanga*. Introduced from Madagascar; probably cultivated. Golden bamboo palm.

Material examined. HIMBAAN: *Doinis Soibeh 777* (K).

40.9. ELAEIS Jacq.

40.9.1. Elaeis guineensis Jacq., *Select. Stirp. Amer. Hist.*: 280 (1763).

Thick-trunked tree palm with pinnate leaves and persistent leaf bases. Lowlands and hill forest; cultivated. The African oil palm or *kelapa sawit*.

Material examined. MELANGKAP TOMIS: *Lorence Lugas 89* (K), *1595* (K).

40.10. EUGEISSONA Griff.

40.10.1. Eugeissona utilis Becc., *Nuovo Giorn. Bot. Ital.* 3: 26 (1871). Plate 11C.

Densely clustering tree palm with fiercely black-spiny sheaths, and spire-like compound terminal inflorescences. On Kinabalu restricted to lowlands and hill forest, usually associated with villages. Possibly a remnant of cultivation.

Material examined. RAGANG TANAH: *Benedict Busin 2a* (K).

40.11. KORTHALSIA Blume

Dransfield, J. (1981). A synopsis of *Korthalsia* (Palmae: Lepidocaryoideae). *Kew Bull.* 36: 163–194. Dransfield, J. (1984). *The rattans of Sabah.* Sabah Forest Records, No. 13. Forest Dept., Sabah.

40.11.1. Korthalsia cheb Becc., *Malesia* 2: 67 (1884).

Robust clustering fish-tailed rattan with swollen ocreas occupied by ants. Usually in the lowlands but collected in hill forest on Mount Kinabalu.

Material examined. HIMBAAN: *Doinis Soibeh 683* (K).

40.11.2. Korthalsia concolor Burret, *Notizbl. Bot. Gart. Mus. Berlin-Dahlem* 15: 736 (1942).

Slender clustering high-climbing rattan branching in the forest canopy. Lowlands, usually restricted to ultramafic substrates up to 400 m elevation.

Material examined. SERINSIM: *Kinsun Bakia 241* (K).

40.11.3. Korthalsia debilis Blume, *Rumphia* 2: 169 (1843).

Slender clustering high-climbing rattan. Lowlands up to 500 m.

Material examined. TEKUTAN: *Dius Tadong 323* (K).

40.11.4. Korthalsia echinometra Becc., *Malesia* 2: 66 (1884). Plate 11D.

Moderately robust clustering high-climbing rattan, with lanceolate leaflets; ocreas swollen, densely spiny and usually filled with ants. Widespread in the lowlands and hill forest to 900 m.

Material examined. KAULUAN: *Sani Sambuling 169* (K); NARAWANG: *Sani Sambuling 45* (K), *55* (K); PORING: *Meliden Giking 22* (K), *Sani Sambuling 144* (K); TEKUTAN: *Dius Tadong 19* (K), *318* (K), *Lomudin Tadong 179* (K); TENSUNGOI: *Sani Sambuling 202* (K).

40.11.5. Korthalsia furtadoana J. Dransf., *Kew Bull.* 36: 185 (1981).

Slender clustering high-climbing ant rattan. Widespread in lowlands and uplands but rather infrequently seen on Mount Kinabalu.

Material examined. NAMPASAN: *Sani Sambuling 250* (K); NARAWANG: *Sani Sambuling 228* (K); PORING: *Sani Sambuling 109* (K), *114* (K), *117* (K); TENSUNGOI: *Sani Sambuling 198* (K).

40.11.6. Korthalsia hispida Becc., *Malesia* 2: 72 (1884).

Moderate clustering rattan, high climbing, with inrolled ocreas filled with bellicose noisy ants. Lowlands, hill forest, usually in damp sites.

Material examined. KIAU: *Jusimin Duaneh 118* (K); SERINSIM: *Jibrin Sibil 162* (K).

40.11.7. Korthalsia jala J. Dransf., *Kew Bull.* 36: 183 (1981). Type: PORING, 600 m, *Dransfield JD 5574* (holotype K!).

Robust, clustering, cirrate rattan with curious net-like ocreas. Lowlands, hill forest. Elevation: 600–700 m.

Additional material examined. NALUMAD: *Sani Sambuling 42* (K); PORING: *Meliden Giking 21* (K), *Sani Sambuling 77* (K); RANAU/PORING ROAD: 700 m, *Dransfield, J. et al. JD 5505* (K, L); TEKUTAN: *Dius Tadong 111* (K), *207* (K), *511* (K), *585* (K), *600* (K).

40.11.8. Korthalsia rigida Blume, *Rumphia* 2: 167 (1843).

Slender to moderate, cirrate rattan, branching high in the canopy. Lowlands, hill forest. Elevation: 700 m.

Material examined. HIMBAAN: *Doinis Soibeh 344* (K), *685* (K); KAULUAN: *Sani Sambuling 165* (K); KIAU: *Clemens, M. S. 10052* (K), *Jusimin Duaneh 123* (K); PORING: 700 m, *Dransfield, J. et al. JD 5566* (K, L); SAYAP: *Tungking Simbayan 25* (K); TEKUTAN: *Dius Tadong 323* (K).

40.11.9. Korthalsia robusta Blume, *Rumphia* 2: 170 (1843). Plate 12A.

Moderate, clustering, cirrate rattan, ocreas filled with bellicose ants. Lowlands, hill forest, in disturbed sites. Elevation: 700 m.

Material examined. HIMBAAN: *Doinis Soibeh 828* (K); PORING: 700 m, *Dransfield, J. et al. JD 5563* (K), *Sani Sambuling 3* (K), *133* (K), *164* (K); SERINSIM: *Kinsun Bakia 339* (K).

40.12. LICUALA Thunb.

Saw, L. G. (1997). A revision of *Licuala* (Palmae) in the Malay Peninsula. *Sandakania* 10: 1–95.

40.12.1. Licuala bidentata Becc., *Malesia* 3: 80 (1886).

Clustering undergrowth fan palm, usually more or less "stemless." Lowlands.

Material examined. SERINSIM: *Jibrin Sibil 86* (K), *Kinsun Bakia 138* (K).

40.12.2. Licuala paludosa Griff., *Hist. Nat. Palm.* 3: 237 (1838).

Clustering undergrowth fan palm with stems to 3 m tall. Lowland swamp forest.

Material examined. SERINSIM: *Jibrin Sibil 118* (K), *Kinsun Bakia 253* (K).

40.12.3. Licuala valida Becc. in H. J. P. Winkl., *Bot. Jahrb. Syst.* 48: 90 (1912). Plate 12B.

Robust "stemless" fan palm with large leaf blades borne on spiny petioles. Lowlands, hill forest.

Material examined. HIMBAAN: *Doinis Soibeh 208* (K); LOHAN: *Sani Sambuling 175* (K); NAMPASAN BARU: *Sani Sambuling 181* (K); NARAWANG: *Sani Sambuling 53* (K); PORING: *Meliden Giking 23* (K), *Sani Sambuling 33* (K), *627* (K); TEKUTAN: *Dius Tadong 5* (K), *480* (K).

40.12.4. Licuala indet.

Material examined. BONGKUD: *Sani Sambuling 177* (K); PAHU: *Doinis Soibeh 730* (K); TEKUTAN: *Dius Tadong 322* (K).

40.13. METROXYLON Rottb.

Rauwerdink, J. B. (1986). An essay on *Metroxylon*, the sago palm. *Principes* 30: 165–180.

40.13.1. Metroxylon sagu Rottb., *Nye Saml. Kongel. Danske Vidensk. Selsk. Skr.* 2: 527 (1783).

Massive clustering tree palm. Lowlands, cultivated on village margins. The true sago palm.

Material examined. MELANGKAP TOMIS: *Lorence Lugas 2652* (K); SAYAP: *Tungking Simbayan 10* (K); SERINSIM: *Jibrin Sibil 150* (K).

40.14. NENGA H. Wendl. & Drude

Fernando, E. S. (1983). A revision of the genus *Nenga*. *Principes* 27: 55–70.

40.14.1. Nenga pumila (Mart.) H. Wendl.

a. var. **pachystachya** (Blume) Fernando, *Principes* 27: 61 (1983).

Clustering undergrowth palm with stems to 4 m tall; fruit brick-red. Lowlands, infrequently also in hill forest, often in swampy habitats.

Material examined. LANGANAN FALLS: *Sani Sambuling 255* (K); MELANGKAP TOMIS: *Lorence Lugas 1644* (K); MURUK: *Sani Sambuling 218* (K); NAMPASAN: *Sani Sambuling 246* (K); NAKAWANG: *Sani Sambuling 227* (K); PORING: *Meliden Giking 25* (K), *Sani Sambuling 237* (K); TENSUNGOI: *Sani Sambuling 197* (K).

40.15. ONCOSPERMA Blume

40.15.1. Oncosperma horridum Scheff., *Natuurk. Tijdschr. Ned.-Indië* 32: 191 (1873).

Massive spiny pinnate tree palm with clustering spiny stems to 20 m tall. Lowlands, hill forest to 900 m.

Material examined. PORING: *Sani Sambuling 1* (K).

40.16. PINANGA Blume

Dransfield, J. (1980). Systematic notes on *Pinanga* (Palmae) in Borneo. *Kew Bull.* 34: 769–788.

40.16.1. Pinanga angustisecta Becc., *Malesia* 3: 119 (1886).

Squat, undergrowth, "stemless" palm with narrow leaflets. Lowlands, hill forest. Elevation: 1100 m.

Material examined. EASTERN SHOULDER: 1100 m, *RSNB 591* (K).

40.16.2. Pinanga aristata (Burret) J. Dransf., *Kew Bull.* 34: 775 (1980). Plate 12C.

Pseudopinanga aristata Burret, *Notizbl. Bot. Gart. Mus. Berlin-Dahlem* 13: 193 (1936). Type: DALLAS, 900 m, *Clemens 27259* (holotype B†; isotypes BM!, L!, SING!).

Slender, clustering, undergrowth palm with mottled leaves. Hill forest, lower montane forest. Elevation: 900–1500 m.

Additional material examined. KINATEKI RIVER: 1400–1500 m, *Bailes & Cribb 846* (K).

40.16.3. Pinanga brevipes Becc., *Malesia* 3: 121 (1886).

Squat, undergrowth stemless palm with broad leaflets. Hill forest. Elevation: 1200 m.

Material examined. EASTERN SHOULDER: 1200 m, *RSNB 287* (K); MELANGKAP TOMIS: *Lorence Lugas 1542* (K).

40.16.4. Pinanga capitata Becc. in Gibbs, *J. Linn. Soc., Bot.* 42: 168 (1914). Plate 12D. Type: KEMBURONGOH/PAKA-PAKA CAVE, 2400–3000 m, *Gibbs 4219* (holotype BM!).

Pinanga gibbsiana Becc. in Gibbs, *J. Linn. Soc., Bot.* 42: 168 (1914). Type: KIAU, 800 m, *Gibbs 3968* (holotype BM!).
Pinanga clemensii Furtado, *Repert. Spec. Nov. Regni Veg.* 35: 279 (1934). Type: MENERINTOG, 2100 m, *Clemens 27901* (holotype SING!; isotypes BM!, K!, L!).
Pinanga dallasensis Furtado, *Repert. Spec. Nov. Regni Veg.* 35: 278 (1934). Type: DALLAS, 900 m, *Clemens 26686* (holotype SING!; isotype BM!).
Pinanga lumuensis Furtado, *Repert. Spec. Nov. Regni Veg.* 35: 280 (1934). Type: LUMU-LUMU, 1800 m, *Clemens 27896* (holotype SING!; isotype BM!).

Polymorphic, slender, undergrowth palm, solitary or clustered, up to 5 m tall. Hill forest, lower montane forest, upper montane forest. Elevation: 800–3000 m.

Additional material examined. BAMBANGAN RIVER: 1500 m, *RSNB 4487* (K), 1500 m, *RSNB 4488* (K), 1500 m, *RSNB 4589* (K); DALLAS: 900 m, *Clemens 26682* (BM, SING), 900 m, *26684* (BM, K, L, SING), 900 m, *27276* (BM, K, L, SING), 900–1200 m, *28407* (K); EASTERN SHOULDER: 2400 m, *RSNB 202* (K, SING); GURULAU SPUR: 2400 m, *Clemens 50836* (BM, K, L, SING), 2400 m, *51118* (BM, K, SING); HIMBAAN: *Doinis Soibeh 207* (K), *681* (K); KEMBURONGOH: 2500 m, *Dransfield, J. et al. JD 5681* (K); KIAU: 1100 m, *Dransfield, J. et al. JD 5549* (K), *Jusimin Duaneh 126* (K), *397* (K); KILEMBUN BASIN: 3000 m, *Clemens 33795* (BM, L); KILEMBUN RIVER: 2900 m, *Clemens 33837* (BM, L); LIWAGU RIVER TRAIL: 1700 m, *Dransfield, J. et al. JD 5702* (K, L), 1700 m, *JD 5705* (K, L); LIWAGU/MESILAU RIVERS: 1200 m, *RSNB 2599* (K, SING); LUGAS HILL: *PEK 14* (K); MAMUT RIVER: 1200 m, *RSNB 1725* (K, SING); MARAI PARAI: 1500 m, *Clemens 32588* (BM, L), 1700 m, *32750* (BM); MARAI PARAI SPUR: 1500–1800 m, *Clemens 32614* (BM); MESILAU CAVE: 1800 m, *RSNB 4654* (K); MINITINDUK GORGE: 900–1200 m, *Clemens 28399* (BM, K, L, SING); PINOSUK PLATEAU: 1400 m, *Beaman 10743* (K); PORING: 900–1200 m, *Meijer SAN 24064* (K); POWER STATION: 2000 m, *Dransfield, J. et al. JD 5676* (K, L); SAYAP: *Jatin Tungking 2* (K); TEKUTAN: *Dius Tadong 62* (K); TENOMPOK: 1500 m, *Clemens 26643* (BM, K, L, SING), 1600 m, *29355* (BM, K, L, SING), 1500 m, *29547* (BM, K, SING).

40.16.5. Pinanga keahii Furtado, *Repert. Spec. Nov. Regni Veg.* 35: 280 (1934). Type: TENOMPOK, 1500 m, *Clemens 28670* (holotype SING!; isotypes BM!, K!, L!).

Slender, clustering, undergrowth palm to 2 m tall. Hill forest, lower montane forest. Elevation: 1300–1500 m.

Additional material examined. HIMBAAN: *Doinis Soibeh 259* (K), *682* (K); LUGAS HILL: 1300 m, *Beaman 10567* (K); MESILAU CAMP: 1500 m, *RSNB 6036* (K).

40.16.6. Pinanga lepidota Rendle, *J. Bot.*: 177 (1901).

Very slender, clustering, undergrowth palmlet to 1 m tall. Lowlands, hill forest, on ultramafic substrate. Elevation: 600–1000 m.

Material examined. HEMPUEN HILL: 700 m, *Dransfield, J. et al. JD 5581* (K, L), 600 m, *Meijer SAN 20955* (K); KIAU: *Jusimin Duaneh 125* (K); LUGAS HILL: *PEK 17* (K); PENIBUKAN: 1000 m, *Nooteboom & Aban 1571* (L).

40.16.7. Pinanga pilosa (Burret) J. Dransf., *Kew Bull.* 34: 775 (1980).

Pseudopinanga pilosa Burret, *Notizbl. Bot. Gart. Mus. Berlin-Dahlem* 13: 186 (1936). Type: TENOMPOK, 1500 m, *Clemens 29205* (holotype B†; isotypes BM!, K!, L!, SING).

Pseudopinanga pilosa Burret var. *gracilior* Burrett, *Notizbl. Bot. Gart. Mus. Berlin-Dahlem* 13: 186 (1936). Type: TENOMPOK, 1400 m, *Clemens 29204* (holotype B†; isotypes BM!, K!, L!, SING!).

Very slender, clustering, undergrowth palmlet to 1.5 m tall. Hill forest, lower montane forest. Elevation: 1200–2100 m.

Additional material examined. EASTERN SHOULDER: 2000 m, *RSNB 169* (K, SING); GOLF COURSE SITE: 1700–1800 m, *Beaman 7206* (K, MSC), 1700 m, *9039* (K), 1700–1800 m, *10660* (K); HIMBAAN: *Doinis Soibeh 258* (K); KILEMBUN RIVER HEAD: 1500 m, *Clemens 32451* (BM), 1400 m, *32532* (BM); KINATEKI RIVER: *Clemens 31394* (SING), 1200–1500 m, *31400* (SING); KINATEKI RIVER HEAD: 1500 m, *Clemens 32842* (BM); KUNDASANG: 1200 m, *Clemens 51381* (BM, SING); LIWAGU RIVER: 1500 m, *Darnton 560* (BM); LIWAGU RIVER TRAIL: *Benedict Busin 3* (K), 1700 m, *Dransfield, J. et al. JD 5692* (K, L); LIWAGU/MESILAU RIVERS: 1200 m, *RSNB 2651* (K, SING); LUGAS HILL: *PEK 23* (K); MENTEKI RIVER: 1600 m, *Beaman 10759* (K); MESILAU CAMP: 1500 m, *RSNB 4000* (K, SING); MESILAU CAVE: 1800 m, *RSNB 4655* (K, SING), *Collenette 21585* (K, L); MESILAU CAVE TRAIL: 1700–1900 m, *Beaman 8007* (K, MSC), 1700–1900 m, *9115* (K), 2100 m, *Meijer SAN 48092* (K); MESILAU RIVER: 1500 m, *RSNB 4035* (K, SING), 2100 m, *Clemens 51677* (BM); MESILAU TRAIL: *Aban Gibot SAN 91652* (K, L); MT. NUNGKEK: 1200 m, *Clemens 32727* (BM); MT. TEMBUYUKEN: 1500 m, *Meijer SAN 28709* (K, L); NALUMAD: *Daim Andau 680* (K); PAHU: *Doinis Soibeh 727* (K); PINOSUK PLATEAU: 1400 m, *Beaman 10738* (K, L, MSC), 1500 m, *Dransfield, J. et al. JD 5553* (K); POWER STATION: 2000 m, *Dransfield, J. et al. JD 5675* (K, L); SUMMIT TRAIL: 1800–2000 m, *Jacobs 5791* (K, L); TENOMPOK: 1600 m, *Aban Gibot SAN 62249* (K), 1400 m, *Clemens 29204a* (SING), *29205a* (SING); TINEKEK FALLS: 2000 m, *Clemens 40881* (BM, SING), 1800 m, *50026* (BM, SING); WEST MESILAU RIVER: 1600–1700 m, *Beaman 7445* (K, MSC), 1600–1700 m, *8694* (K), 1600 m, *9016* (K, MSC), 1600 m, *9186* (K).

40.16.8. Pinanga salicifolia Blume, *Rumphia* 2: 93 (1843).

Slender clustering palm to 3 m tall; juvenile foliage usually markedly different from the adult. Primary hill forest on ultramafic substrate; usually on podsolised soils on ridgetops and kerangas.

Material examined. MELANGKAP TOMIS: *Lorence Lugas 1436* (K).

40.16.9. Pinanga tenella (H. Wendl.) Scheff., *Bot. Zeit.*: 63 (1859). Plate 12E. Type: BUNGOL, *Lobb s.n.* (holotype K!).

Slender, densely clustering rheophytic palm. Lowlands or hill forest along fast-flowing rocky rivers. On Mount Kinabalu known only from the type collected in 1856, the first palm collected on the mountain.

40.16.10. Pinanga variegata Becc.

a. var. **hallieriana** Becc. ex Martelli, *Atti. Soc. Tosc. Sci. Nat. Pisa Mem.* 44: 128 (1934).

Slender, clustering, undergrowth palmlet. Lowlands, hill forest, along streams. Elevation: 1100 m.

Material examined. EASTERN SHOULDER: 1100 m, *RSNB 565* (K, L, SING); TEKUTAN: *Dius Tadong 294* (K).

40.17. PLECTOCOMIA Mart. ex Blume

Dransfield, J. (1984). *The rattans of Sabah.* Sabah Forest Records, No. 13. Forest Dept., Sabah. Madulid, D. A. (1981). A monograph of *Plectocomia* (Palmae: Lepidocaryoideae). *Kalikasan* 10: 11–94.

40.17.1. Plectocomia elongata Mart. ex Schult. & Schult. f., *Syst. Veg.* 7: 1333 (1830).

a. var. elongata

Robust, clustering, cirrate rattan, climbing to 25 m. Disturbed sites in lowlands to lower montane forest. Elevation: 1200–1700 m.

Material examined. MELANGKAP TOMIS: *Lorence Lugas 409* (K); MESILAU RIVER: 1500 m, *RSNB 4194* (K, L); MT. NUNGKEK: 1700 m, *Clemens 32007* (BM, L); PENIBUKAN: 1200 m, *Clemens 31039* (BM, K, L).

40.17.2. Plectocomia cf. elongata Mart. ex Schult. & Schult. f., *Syst. Veg.* 7: 1333 (1830).

Massive hapaxanthic rattan with plumose leaves. Lower montane forest, apparently restricted to poor soils. Elevation: 1500 m.

Material examined. GOLF COURSE: 1500 m, *Robert Nasi s.n.* (K).

40.17.3. Plectocomia mulleri Blume, *Rumphia* 3: 71 (1847).

Massive hapaxanthic rattan with plumose leaves. Lowlands, hill forest on poor sandy soils.

Material examined. KIAU: *Jusimin Duaneh 183* (K); LUGAS HILL: *PEK 9* (K), *45* (K); MELANGKAP TOMIS: *Lorence Lugas 1570* (K); NARAWANG: *Suni Sambuling 46* (K); PORING: *Meliden Giking 24* (K); TEKUTAN: *Dius Tadong 479* (K).

40.18. PLECTOCOMIOPSIS Becc. in Hook. f.

Dransfield, J. (1982). A reassessment of the genera *Plectocomiopsis, Myrialepis* and *Bejaudia* (Palmae: Lepidocaryoideae). *Kew Bull.* 37: 455–457. Dransfield, J. (1984). *The rattans of Sabah.* Sabah Forest Records, No. 13. Forest Dept., Sabah.

40.18.1. Plectocomiopsis geminiflora (Griff.) Becc. in Hook. f., *Fl. Brit. India* 6: 479 (1893).

Robust, clustering, cirrate rattan, climbing to 40 m. Lowlands, hill forest. Elevation: 600 m.

Material examined. PORING: 600 m, *Dransfield, J. et al. JD 5573* (K), *Meliden Giking 20* (K), *Sani Sambuling 23* (K).

40.18.2. Plectocomiopsis mira J. Dransf., *Kew Bull.* 37: 247 (1982). Type: HEMPUEN HILL, 700 m, *Dransfield JD 5583* (holotype K!).

Robust, clustering, high-climbing, cirrate rattan. Lowlands, hill forest, sometimes on ultramafic substrate. Elevation: 700 m.

Additional material examined. PORING: *Sani Sambuling 139* (K).

40.19. SALACCA Reinw.

Dransfield, J., and Mogea, J. P. (1981). A reassessment of the genus *Lophospatha* Burret. *Principes* 25: 178–180.

40.19.1. Salacca clemensiana Becc., *Philipp. J. Sci.* 4: 618 (1909). Plate 12F.

Massive clustering "stemless" palm with spiny petioles; fruit borne among the leaf sheaths. Valley bottoms in hill forest, 800–1000 m, rarely in the lowlands.

Material examined. HIMBAAN: *Doinis Soibeh 613* (K).

40.19.2. Salacca dolicholepis Burret, *Notizbl. Bot. Gart. Mus. Berlin-Dahlem* 15: 731 (1942). Plate 13A. Type: TENOMPOK, 1400 m, *Clemens 28819* (holotype B†; isotypes BM!, K!, L!, SING!).

"Stemless," clustering, very spiny palm. Hill forest, lower montane forest. Elevation: 900–1700 m. Endemic to Mount Kinabalu.

Additional material examined. BAMBANGAN RIVER: 1500 m, *RSNB 4438* (K, L); BUNDU TUHAN: *Doinis Soibeh 352* (K); DALLAS: 900 m, *Clemens 26659* (BM, K); KIAU: *Jusimin Duaneh 124* (K); LIWAGU RIVER TRAIL: 1700 m, *Dransfield, J. et al. JD 5704* (K); SERINSIM: *Jibrin Sibil 61* (K); TEKUTAN: *Dius Tadong 212* (K); TENOMPOK: 1500 m, *Clemens 28769* (BM), 1400 m, *29190* (BM, K, L); TINEKEK FALLS: 1700 m, *Clemens 40834* (BM, K), 1700 m, *40884* (BM).

40.19.3. Salacca lophospatha J. Dransf. & Mogea, *Principes* 25: 180 (1981).

Lophospatha borneensis Burret, *Notizbl. Bot. Gart. Mus. Berlin-Dahlem* 15: 753 (1942). Type: DALLAS, 900 m, *Clemens 26380* (holotype B†; isotype K!).

"Stemless," clustering, very spiny palm. Hill forest. Elevation: 900 m.

40.19.4. Salacca aff. vermicularis Becc., *Malesia* 3: 66 (1886).

Massive clustering "stemless" palm with fiercely spiny petioles; fruit borne at ground level among the sheaths. Lowlands, usually in valley bottoms.

Material examined. TEKUTAN: *Dius Tadong 60* (K).

40.19.5. Salacca indet.

Material examined. TEKUTAN: *Dius Tadong 418* (K).

41. BURMANNIACEAE

Beccari, O. (1878). Piante nuove o rare dell' arcipelago Malese e della Nuova Guinea, raccolte, descritte od illustrate da O. Beccari: Burmanniaceae. *Malesia* 1: 240–254, Pl. 10–15. Jonker, F. P. (1938). *A monograph of the Burmanniaceae.* Doctoral thesis, Univ. Utrecht. Jonker, F. P. (1948). Burmanniaceae. *Fl. Males.* I, 4: 13-26.

41.1. BURMANNIA L.

41.1.1. Burmannia coelestis D. Don, *Prodr. Fl. Nepal.*: 44 (1825).

Autotrophic green annual herb to 30 cm high, with simple or branched stems; leaves linear or lanceolate. Hill forest in disturbed situations. Elevation: 500 m.

Material examined. KAUNG/SINGGAREN: 500 m, *Darnton 503* (BM).

41.1.2. Burmannia longifolia Becc., *Malesia* 1: 244 (1878).

Green perennial herb. Lower montane forest, occasionally on ultramafic substrate. Elevation: 1100–2400 m.

Material examined. EAST MESILAU/MENTEKI RIVERS: 1700 m, *Beaman 8754* (MSC); GIGISSEN CREEK/MARAI PARAI: 1400 m, *Clemens 32389* (L); KINATEKI RIVER: 1100 m, *Collenette A 112* (BM); LUMU-LUMU: 1800 m, *Clemens s.n.* (BM, BM, BM); MAMUT COPPER MINE: 1600–1700 m, *Beaman 9945* (L, MSC); MARAI PARAI: 1500 m, *Clemens s.n.* (BM, BM); MARAI PARAI SPUR: 2100 m, *Gibbs 4033* (BM); MESILAU BASIN: 2100–2400 m, *Clemens 29713* (BM, BO n.v.); MESILAU CAVE TRAIL: 1800 m, *Beaman 7462* (MSC); MESILAU RIVER: 1500 m, *RSNB 4863* (K); MOUNT KINABALU: *Low s.n.* (K); MT. NUNGKEK: 1600 m, *Darnton s.n.* (BM); PARK HEADQUARTERS: 1400 m, *Price 132* (K); PENIBUKAN: *Clemens s.n.* (BM); WEST MESILAU RIVER: 1600 m, *Beaman 9036* (MSC).

41.1.3. Burmannia lutescens Becc., *Malesia* 1: 246 (1878).

Burmannia papillosa Stapf, *Trans. Linn. Soc. London, Bot.* 4: 232 (1894). Type: TINEKEK RIVER, 1000 m, *Haviland 1329* (holotype K!).

Saprophytic echlorophyllous herb. Hill forest. Elevation: 700–1200 m.

Additional material examined. MELANGKAP KAPA: 700–1000 m, *Beaman 8777* (L, MSC); PENIBUKAN: 1200 m, *Clemens 32192* (BM), 1200 m, *51680* (BM); SINGH'S PLATEAU: *RSNB s.n.* (K).

41.2. GYMNOSIPHON Blume

41.2.1. Gymnosiphon aphyllus Blume, *Enum. Pl. Javae* 1: 29 (1827).

Saprophytic echlorophyllous herb. Hill forest, possibly also lower montane forest, mostly on ultramafic substrate. Elevation: 800–1200 m.

Material examined. MARAI PARAI SPUR: *Clemens, M. S. 11029* (L); MT. NUNGKEK: 800–1100 m, *Clemens 32757* (BM, L); PENIBUKAN: 1200 m, *Clemens 50101* (BM, K).

41.3. THISMIA Griff.

41.3.1. Thismia episcopalis (Becc.) F. Muell., *Pap. & Proc. Roy. Soc. Tasmania* for 1890: 235 (1891). Plate 13B.

Echlorophyllous saprophytic small herb. Lower montane forest. Elevation: 1200–1700 m.

Material examined. DALLAS/TENOMPOK: 1200 m, *Clemens 27566* (BM); PINOSUK PLATEAU: 1700 m, *RSNB 1909* (K); TENOMPOK: 1500 m, *Clemens 29905* (BM); WEST MESILAU RIVER: 1600 m, *Collenette 638* (K).

41.3.2. Thismia ophiuris Becc., *Malesia* 1: 252 (1878).

Echlorophyllous saprophytic small herb. Hill forest or lower montane forest on ultramafic substrate. First collected in western Sarawak.

Material examined. MESILAU BASIN: *Clemens s.n.* (BM); PENIBUKAN: *Clemens s.n.* (BO n.v.).

42. CANNACEAE

42.1. CANNA L.

42.1.1. Canna indica L., *Sp. Pl.*: 1 (1753).

Large perennial herb with creeping rhizome. Cultivated as an ornamental, and escaped?

Material examined. MELANGKAP TOMIS: *Lorence Lugas 82* (K); MENGGIS: *Matamin Rumutom 333* (K); SERINSIM: *Kinsun Bakia 424* (K).

43. CENTROLEPIDACEAE

Hou, D. (1957). Centrolepidaceae. *Fl. Males.* I, 5: 421–428.

43.1. CENTROLEPIS Labill.

43.1.1. Centrolepis philippinensis Merr., *Philipp. J. Sci. C.* 2: 264 (1907). Plate 13C.

Centrolepis kinabaluensis Gibbs, *Fl. Arfak Mts.*: 99 (1917). Type: SUMMIT AREA, 4000 m, *Gibbs 4207* (holotype BM!; isotypes K!, L!).

Perennial mat-forming cushion plant. Summit area in cracks in granitic substrate. Elevation: 3000–4000 m.

Additional material examined. Marai Parai Spur: 3000–3400 m, *Clemens 32415* (BM, L); Paka-paka Cave: 3100 m, *Sleumer 4706* (L); Paka-paka Cave/Kadamaian River: 3200 m, *Clemens 50883* (BM); Paka-paka Cave/Low's Peak: *Clemens, M. S. 10625* (BM, K, L); Paka-paka Cave/Panar Laban: 3200 m, *Sinclair et al. 9136* (K, L); Sayat-sayat: 3800 m, *Meijer SAN 28570* (K); Summit Area: 3000–4000 m, *Clemens 27776* (BM, L), 3400–4000 m, *28940* (BM, K), 3900 m, *Gibbs 4209* (BM), 3600 m, *Hou 255* (K), 3600–3800 m, *Sleumer 4718* (L); Victoria Peak: 3800 m, *Clemens s.n.* (BM), *51404* (K, L).

44. COMMELINACEAE

In collaboration with E. J. Cowley (K) and R. B. Faden (US)

44.1. AMISCHOTOLYPE Hassk.

44.1.1. Amischotolype glabrata (Hassk.) Hassk., *Flora* 46: 391 (1863).

Prostrate herb, stems to 2 m. Hill forest, lower montane forest in wet areas. Elevation: 800–1800 m.

Material examined. Bambangan River: *Amin et al. SAN 123564* (K); Bundu Tuhan: *Doinis Soibeh 24* (K); Dallas: *Clemens 26396* (BM, K); Eastern Shoulder, Camp 1: 1200 m, *RSNB 1197* (K); Golf Course Site: 1700–1800 m, *Beaman 7205* (K, MSC, US); Lugas Hill: 1300 m, *Beaman 9513* (MSC); Melangkap Tomis: *Lorence Lugas 1192* (K), *1536* (K); Menteki River: 1600 m, *Beaman 10785* (K, MSC, US); Mesilau Camp: 1500 m, *RSNB 4001* (K, L); Mesilau Trail: 1500–1800 m, *Chow & Leopold 76433* (K, L); Nalumad: *Daim Andau 953* (K); Penataran Basin: 1200 m, *Clemens 34200* (BM, L); Penataran River: 900–1200 m, *Clemens 34272* (BM); Pinosuk Plateau: *Poore 3863* (K); Sayap: 800–1000 m, *Beaman 9759* (MSC); Tahubang Falls: 1200 m, *Clemens 30681* (BM, K, L); Tahubang River: 1100 m, *Kanis SAN 51478* (K, L); West Mesilau River: 1600–1700 m, *Beaman 7452* (K, MSC, US), 1600–1700 m, *8637* (K, MSC, US).

44.1.2. Amischotolype griffithii (C. B. Clarke) I. M. Turner, *Novon* 6: 221 (1996).

Erect herb to 1.5 m. Hill forest. Elevation: 500–1200 m.

Material examined. Dallas: 900 m, *Clemens 26039* (BM), 1200 m, *26476* (BM, K, L); Pinawantai: 500 m, *Shea & Aban SAN 76938* (K).

44.1.3. Amischotolype irritans (Ridl.) I. M. Turner, *Novon* 6: 221 (1996).

Erect herb to 1.5 m. Hill forest. Elevation: 900–1200 m.

Material examined. Dallas: 900–1200 m, *Clemens 26141* (BM, K, L).

44.1.4. Amischotolype marginata (Blume) Hassk., *Flora* 46: 392 (1863).

Erect or prostrate herb to 1.5 m. Hill forest in wet areas. Elevation: 700–1300 m.

Material examined. LUGAS HILL: 1300 m, *Beaman 8451* (US); MELANGKAP TOMIS: *Lorence Lugas 577* (K); NALUMAD: *Daim Andau 567* (K); PENIBUKAN: 1200 m, *Clemens 30889* (BM); PORING: 700–800 m, *Beaman 10923* (K); SAYAP: 800–1000 m, *Beaman 9754* (K, MSC, US); SERINSIM: *Kinsun Bakia 61* (K); TEKUTAN: *Lomudin Tadong 221* (K).

44.1.5. Amischotolype mollissima (Blume) Hassk., *Flora* 46: 392 (1863). Plate 13D.

Perennial herb. Lowlands in primary forest.

Material examined. NALUMAD: *Daim Andau 566* (K).

44.1.6. Amischotolype indet.

Material examined. DALLAS: 900 m, *Clemens s.n.* (BM); KINATEKI RIVER: 1500 m, *Clemens 50405a* (BM); KUNDASANG: 1400 m, *Cox 951* (L); MENGGIS: *Matamin Rumutom 287* (K); PENIBUKAN: 1200–1500 m, *Clemens s.n.* (BM), 1200–1500 m, *31295* (BM), 1200 m, *32085* (BM, L); PINOSUK PLATEAU: *Poore 3863* (L); SAYAP: *Yalin Surunda 101* (K).

44.2. BELOSYNAPSIS Hassk.

44.2.1. Belosynapsis ciliata (Blume) R. S. Rao, *Notes Roy. Bot. Gard. Edinburgh* 25: 187 (1964).

Prostrate creeping herb, rooting at the nodes. Lowlands, hill forest; rocky areas near water. Elevation: 500–900 m.

Material examined. BUNDU TUHAN: *Doinis Soibeh 29* (K); DALLAS: 900 m, *Clemens 28163* (L); DALLAS/TENOMPOK: 900 m, *Clemens 28163* (BM, K); NALUMAD: 500 m, *Shea & Aban SAN 77324* (K).

44.3. COMMELINA L.

44.3.1. Commelina diffusa N. L. Burm., *Fl. Indica*: 18, t. 7, f. 2 (1768).

Trailing herb. Lowlands or secondary hill forest. Elevation: 500 m.

Material examined. MELANGKAP TOMIS: *Lorence Lugas 592* (K); RANAU: 500 m, *Darnton 147* (BM); TEKUTAN: *Lomudin Tadong 106* (K).

44.3.2. Commelina paludosa Blume, *Enum. Pl. Javae* 1: 2 (1827).

Creeping herb. Hill forest, near rivers. Elevation: 1200 m.

Material examined. KUNDASANG: 1200 m, *RSNB 1471* (K).

44.4. FLOSCOPA Lour.

44.4.1. Floscopa scandens Lour., *Fl. Cochinch.* 1: 193 (1790).

Trailing perennial herb, rooting at the nodes. Lowlands or hill forest in thickets.

Material examined. MELANGKAP TOMIS: *Lorence Lugas 620* (K), *1330* (K).

44.5. MURDANNIA Royle

44.5.1. Murdannia japonica (Thunb.) Faden, *Taxon* 26: 142 (1977).

Herb to 0.5 m. Lowlands in secondary forest. Elevation: 500–800 m.

Material examined. HEMPUEN HILL: 800 m, *Madani SAN 89508* (K, L); MENGGIS: *Matamin Rumutom 128* (K); TEKUTAN: *Lomudin Tadong 33* (K), *268* (K), 500 m, *Shea & Aban SAN 77195* (K, L).

44.5.2. Murdannia loriformis (Hassk.) R. S. Rao & Kammathy, *Bull. Bot. Surv. India* 3: 393 (1962).

More or less prostrate herb with runners. Hill forest in secondary situations. Elevation: 900–1200 m.

Material examined. BUNDU TUHAN: 900 m, *Darnton 535* (BM); KIAU: 900 m, *Clemens 33059* (BM, K, L); LIWAGU/MESILAU RIVERS: 1200 m, *RSNB 2613a* (K, L).

44.6. POLLIA Thunb.

44.6.1. Pollia secundiflora (Blume) Bakh. f., *Blumea* 6: 399 (1950).

Stoloniferous herb to 1 m. Hill forest, lower montane forest. Elevation: 700–2100 m.

Material examined. KIAS: *Jibrin Sibil 296* (K); LIWAGU/MESILAU RIVERS: 1500 m, *RSNB 2824* (K, L); MELANGKAP KAPA: 700–1000 m, *Beaman 8810* (MSC); MESILAU BASIN: 1500–2100 m, *Clemens 29726* (BM, K); NAPONG: *Sani Sambuling 499* (K); PENATARAN BASIN: 900 m, *Clemens 34216* (L); PENATARAN RIVER: 900 m, *Clemens 34216* (BM, K); PORING: *Sani Sambuling 339* (K); SAYAP: 700 m, *Beaman 9461* (K, MSC, US); SERINSIM: *Kinsun Bakia 71* (K); TAHUBANG RIVER: *Haviland 1330* (K).

44.6.2. Pollia thyrsiflora (Blume) Steud., *Nomencl. Bot.,* ed. 2, 2: 368 (1841).

Erect perennial herb to 0.5 m high. Primary hill forest, sometimes in open situations. Elevation: 600–1500 m.

Material examined. LANGANAN FALLS: 1200 m, *Beaman 10963* (K); LOHAN/MAMUT COPPER MINE: 900 m, *Beaman 10604* (US); MELANGKAP KAPA: 600–700 m, *Beaman 8580* (US); MELANGKAP TOMIS: *Lorence Lugas 604* (K); NALUMAD: *Daim Andau 167* (K), *769* (K); TAHUBANG RIVER: 1200–1500 m, *Clemens 31518* (BM).

44.6.3. Pollia sp.

Erect perennial herb. Hill forest, probably on ultramafic substrate. Elevation: 1200 m.

Material examined. PENIBUKAN: 1200 m, *Clemens s.n.* (BM); TAHUBANG RIVER: 1200 m, *Clemens 40303* (BM).

44.7. RHOPALEPHORA Hassk.

44.7.1. Rhopalephora scaberrima (Blume) Faden, *Phytologia* 37: 480 (1977).

Creeping herb to 1 m. Hill forest in shade. Elevation: 500–900 m.

Material examined. DALLAS: 900 m, *Clemens 29682* (K, L); KAUNG/DALLAS: 500 m, *Collenette A 131* (BM).

44.7.2. Rhopalephora vitiensis (Seem.) Faden, *Phytologia* 37: 480 (1977).

Decumbent perennial herb to 0.4 m high, rooting at the lower nodes. Lowlands or hill forest in thickets, probably on ultramafic substrate.

Material examined. MENGGIS: *Matamin Rumutom 247* (K).

44.8. TRICARPELEMA J. K. Morton

44.8.1. Tricarpelema philippense (Panigrahi) Faden, *Smithsonian Contr. Bot.* 76: 153 (1991).

Erect herb. Hill forest on ultramafic substrate. Elevation: 700–900 m.

Material examined. LOHAN RIVER: 700–900 m, *Beaman 9227* (MSC); LOHAN/MAMUT COPPER MINE: 900 m, *Beaman 10627* (K, MSC, US); TEKUTAN: 800 m, *Shea & Aban SAN 77245* (K).

45. CONVALLARIACEAE

In collaboration with E. J. Cowley (K) and B. Mathew (K)

45.1. OPHIOPOGON Ker Gawl.

Jessop, J. P. (1979). Liliaceae–I. *Fl. Males.* I, 9: 225–227.

45.1.1. Ophiopogon caulescens (Blume) Backer, *Handb. Fl. Java* 3: 74 (1924).

Rhizomatous perennial with tufts of grassy leaves on short, erect stems. Hill forest, lower montane forest. Elevation: 600–1700 m.

Material examined. DALLAS: 900 m, *Clemens 26034* (BM, L); EASTERN SHOULDER: 1100 m, *RSNB 639* (K, L); EASTERN SHOULDER, CAMP 1: 1200 m, *RSNB 1209* (K, L); GURULAU SPUR: 1500 m, *Clemens 50449* (BM, K), 1500 m, *Gibbs 3996* (BM); HIMBAAN: *Doinis Soibeh 189* (K); KADAMAIAN RIVER: 1400 m, *Gibbs 4103* (BM); LIWAGU/MESILAU RIVERS: 1500 m, *RSNB 2877a* (K); LUGAS HILL: 1300 m, *Beaman 9504* (K, L, MSC); MELANGKAP KAPA: 600–700 m, *Beaman 8593* (K, MSC), 700–1000 m, *8797* (MSC); MESILAU RIVER: 1500 m, *RSNB 4926* (K); MESILAU TRAIL: 1400 m, *Sinanggul SAN 38368* (K, L); NALUMAD: *Daim Andau 770* (K), *934* (K), *Sigin et al. SAN 110647* (K); PARK HEADQUARTERS: 1500–1600 m, *Kokawa & Hotta 2854* (L), 1400 m, *Lajangah SAN 44765* (K); PENIBUKAN: 1200 m, *Clemens s.n.* (BM); SINGH'S PLATEAU: 1100 m, *RSNB s.n.* (K); TAHUBANG RIVER: 1100 m, *Phillipps, A. SNP 1012* (L); TAMOR RIVER (TENOMPOK AREA): 1300 m, *Sario SAN 28509* (K); WEST MESILAU RIVER: 1600–1700 m, *Beaman 7428* (K, L, MSC).

45.2. PELIOSANTHES Andr.

Jessop, J. P. (1976). A revision of *Peliosanthes* (Liliaceae). *Blumea* 23: 141–159. Jessop, J. P. (1979). Liliaceae–1. Fl. Males. I, 9: 227–229.

45.2.1. Peliosanthes teta Andrews

a. subsp. **humilis** (Andrews) Jessop, *Blumea* 23: 155 (1976).

Erect, stemless perennial herb with long-petioled elliptic leaves. Hill forest, lower montane forest. Elevation: 1100–1500 m.

Material examined. KILEMBUN RIVER HEAD: 1500 m, *Clemens 33056* (BM); LIWAGU/MESILAU RIVERS: 1500 m, *RSNB 2814* (K); MARAI PARAI: 1200 m, *Clemens 34473* (BM); PARK HEADQUARTERS: 1400 m, *Price 169* (K); PENIBUKAN: 1200 m, *Clemens s.n.* (BM); TAHUBANG RIVER: 1100–1200 m, *Clemens 40298* (BM, K); WEST MESILAU RIVER: 1500 m, *Collenette 588* (K).

46. COSTACEAE

In collaboration with E. J. Cowley (K) and R. M. Smith (E)

46.1. COSTUS L.

Maas, P. J. M. (1979). Notes on Asiatic and Australian Costoideae (Zingiberaceae). *Blumea* 25: 543–549. Schumann, K. (1904). Zingiberaceae. *Pflanzenr.* IV. 46: 1–458 [subfamily Costoideae on pp. 377–437].

46.1.1. Costus globosus Blume, *Enum. Pl. Javae*: 62 (1827). Plate 13E.

Tall perennial herb to 3 m, inflorescence basal; flowers orange. Hill forest. Elevation: 600–1200 m.

Material examined. EASTERN SHOULDER: 900 m, *RSNB 662* (K); KINATEKI RIVER: 600 m, *Collenette A 125* (BM); MINITINDUK GORGE: 900–1200 m, *Clemens 29620* (BM, K).

46.1.2. Costus aff. **globosus** Blume, *Enum. Pl. Javae*: 62 (1827).

Tall perennial herb to 3 m, inflorescence basal, red. Hill forest by rivers. Elevation: 1100–1500 m.

Material examined. EASTERN SHOULDER: 1100 m, *RSNB 635* (K); PENIBUKAN: 1200–1500 m, *Clemens 31343* (BM, K).

46.1.3. Costus aff. **microcephalus** K. Schum., *Pflanzenr.* IV. 46 (Heft 20): 412 (1904).

Small perennial herb; inflorescence less than 3 cm in diameter. Lowland primary forest.

Material examined. TEKUTAN: *Lomudin Tadong 337* (K).

46.1.4. Costus speciosus (J. König) Sm., *Trans. Linn. Soc. London* 1: 249 (1791).

Tall perennial herb to 4 m, inflorescence terminal. Lowlands, hill forest. Elevation: 600–900 m.

Material examined. DALLAS: 900 m, *Clemens 26379* (BM); MAMUT: *Amin & Peter 129389* (K); MELANGKAP TOMIS: *Lorence Lugas 783* (K), *1729* (K); NALUMAD: *Daim Andau 63* (K); PORING: *Sani Sambuling 9* (K), *376* (K); PORING HOT SPRINGS: 600 m, *Beaman 7550* (E, MSC); SERINSIM: *Jibrin Sibil 155* (K), *275* (K); TEKUTAN: *Dius Tadong 50* (K).

46.1.5. Costus indet.

Material examined. TEKUTAN: *Lomudin Tadong 356* (K).

47. CYPERACEAE

In collaboration with D. A. Simpson (K)

Kern, J. H. (1974). Cyperaceae. *Fl. Males.* I, 7: 435–753.

47.1. CAREX L.

Kern, J. H. & Nooteboom, H. P. (1979). Cyperaceae–II. *Fl. Males.* I, 9: 107–187. Kükenthal, G. (1909). Cyperaceae-Caricoideae. *Pflanzenr.* IV. 20 (Heft 38): 1–824. Nelmes, E. (1951). The genus *Carex* in Malaysia. *Reinwardtia* 1: 221–450.

47.1.1. Carex breviculmis R. Br., *Prodr.*: 242 (1810).

a. var. **breviculmis**

Tufted perennial sedge with short rhizome. Summit area in open places on granitic substrate above treeline. Elevation: 3700–4100 m.

Material examined. SUMMIT AREA: 4000 m, *Clemens 28028* (BM), 3700–4000 m, *51180* (K); UPPER KINABALU: 4100 m, *Clemens 28023* (K); VICTORIA PEAK: 3800 m, *Clemens 51349* (BM, K).

b. var. **perciliata** Kük., *Pflanzenr.* IV. 20 (Heft 38): 469 (1909).

Tufted perennial sedge with short rhizome. Open areas, rocky crevices, upper montane forest to above treeline. Elevation: 3000–4000 m.

Material examined. MARAI PARAI: 3200 m, *Clemens 32344* (BM, L); PAKA-PAKA CAVE/PANAR LABAN: 3400–4000 m, *Clemens 29005* (BM, K, L); PAKA-PAKA CAVE/SUMMIT AREA: 3000–4000 m, *Clemens s.n.* (BM); PANAR LABAN/UPPER MESILAU VALLEY: 3600 m, *Smith 537* (L); SUMMIT AREA: 3700–4000 m, *Clemens 51180* (BM, K).

47.1.2. Carex capillacea Boott, *Ills. Carex* 1: 44 (1858).

a. var. **capillacea**

Densely tufted perennial sedge with short rhizome. Upper montane mossy forest. Elevation: 2400–3500 m.

Material examined. EASTERN SHOULDER: 2700 m, *RSNB 718* (K, L), 3000 m, *RSNB 734* (K, L); KEMBURONGOH: 2400 m, *Clemens 28021* (BM, K, L); PAKA-PAKA CAVE: 3000 m, *Meijer SAN 22017* (K), 3000 m, *Smith s.n.* (K); PANAR LABAN: 3300 m, *Smith 465* (L); SUMMIT TRAIL: 3000–3500 m, *Jacobs 5759* (K, L).

b. var. **sachalinensis** (F. Schmidt) Ohwi, *Mem. Coll. Sci. Kyoto Imp. Univ.*, Ser. B, 11: 442 (1936).

Densely tufted perennial sedge with short rhizome. Upper montane forest and summit area in granitic crevices. Elevation: 3000–4100 m.

Material examined. MOUNT KINABALU: 3400 m, *Haviland 1393* (K); PAKA-PAKA CAVE: 3000 m, *Sinclair et al. 9188* (K, L); SUMMIT AREA: 4100 m, *Gibbs 4196* (K).

47.1.3. Carex cruciata Wahlenb., *Kongl. Vetensk. Acad. Handl.* 24: 149 (1803).

a. var. **cruciata**

Tufted perennial sedge with short rhizome. Hill forest, lower montane forest, in open places, sometimes on ultramafic substrate. Elevation: 700–1700 m.

Material examined. BAMBANGAN RIVER: 1500 m, *RSNB 4407* (K); BIDAU-BIDAU FALLS: 1500 m, *Clemens 34123* (BM, L); DALLAS: 900 m, *Clemens 30068* (K); GURULAU SPUR: 1500 m, *Gibbs 4004* (K), 1400–1500 m, *Kanis & Kuripin SAN 53954* (K, L); KIAU: *Amin G. & Francis SAN 121603* (K), *Jusimin Duaneh 385* (K); LIWAGU/MESILAU RIVERS: 1200 m, *RSNB 2593* (K, L); LOHAN RIVER: 700–900 m, *Beaman 9224* (L, MSC, NY); MAMUT RIVER: 1200 m, *RSNB 1706* (K, L); MARAI PARAI SPUR: *Clemens, M. S. 10885* (K); MT. NUNGKEK: 900–1200 m, *Clemens 32763* (BM); PARK HEADQUARTERS/POWER STATION: 1700 m, *Simpson 89/200* (K); PENATARAN RIVER: 900 m, *Clemens 32581* (BM), 900 m, *34045* (K, L), 1500 m, *34123* (K); TAHUBANG FALLS: 1200 m, *Clemens 30691* (BM); TEKUTAN: *Dius Tadong 183* (K); TENOMPOK/KUNDASANG: 1500 m, *Sinclair et al. 9230* (K).

b. var. **rafflesiana** (Boott) Noot., *Fl. Males.* I, 9: 122 (1979).

Carex spongoneura Nelms, *Kew Bull.* 1946: 18 (1946). Type: PENIBUKAN, 1500 m, *Gibbs 4093* (holotype K!; isotype BM!).

Tufted perennial sedge with short rhizome. Lower montane forest, probably in open places. Elevation: 900–2100 m.

Additional material examined. MARAI PARAI: *Clemens s.n.* (BM), 1500 m, *33157* (BM, K, L); MESILAU RIVER: 2100 m, *Collenette 21619* (L); PENATARAN RIVER: 900 m, *Clemens 34045* (BM).

47.1.4. Carex cryptostachys Brongn. in Duperrey, *Voy. Monde*: 152 (1829).

Perennial sedge with elongate rhizome. Lower montane forest. Elevation: 1400–1800 m.

Material examined. MESILAU CAVE TRAIL: 1800 m, *Beaman 7490* (MSC); PINOSUK PLATEAU: 1400–1600 m, *Kokawa & Hotta 4506* (L).

47.1.5. Carex filicina Nees in Wight, *Contr. Bot. India*: 123 (1834). Plate 14A.

Carex saturata C. B. Clarke, *J. Linn. Soc., Bot.* 37: 12 (1904). Type: MOUNT KINABALU, 3200 m, *Haviland 1402* (holotype K!).
Carex havilandii C. B. Clarke, *J. Linn. Soc., Bot.* 37: 13 (1904). Type: MOUNT KINABALU, 3000 m, *Haviland 1403* (holotype K!).
Carex verticillata Zoll.& Moritzi in Moritzi var. *havilandii* (C. B. Clarke) Nelmes, *Kew Bull.* 5: 195 (1950).
Carex hypsophila Miq. var. *havilandii* (C. B. Clarke) Kük., *Pflanzenr.* IV. 20 (Heft 38): 546 (1909).

Loosely tufted perennial sedge with short, thick rhizome. Lower montane forest, upper montane forest, summit area, often on ultramafic substrate. Elevation: 900–4100 m.

Additional material examined. EASTERN SHOULDER: 2400 m, *RSNB 192* (K, L), 3000 m, *RSNB 732* (K, L), 2700 m, *RSNB 912* (K, L); EASTERN SHOULDER, CAMP 4: 2900 m, *RSNB 1123* (K), 2900 m, *RSNB*

1123b (L); GURULAU SPUR: 2400–2700 m, *Clemens 50644* (BM), 2400 m, *51183* (BM, K), 3000 m, *51455* (K, L); KEMBURONGOH: 2400 m, *Clemens s.n.* (BM), 2100 m, *Gibbs 4230* (K), 2100–2400 m, *4243* (K); KINATEKI RIVER: 1500 m, *Clemens 32961* (BM); KINATEKI RIVER HEAD: 2400 m, *Clemens 31749* (BM); LAYANG-LAYANG/PAKA-PAKA CAVE: 2700–3200 m, *Kokawa & Hotta 3515* (L); MARAI PARAI: 1500–2700 m, *Clemens s.n.* (BM), 2100 m, *32342A* (BM), 1800 m, *32844* (BM), 1500 m, *32961* (K, L); MESILAU CAMP: *Poore H 509* (K); MESILAU CAVE: 1800 m, *RSNB 4696* (K); MESILAU RIVER: 1500 m, *RSNB 4136* (K), 2100 m, *Clemens 51341* (BM); MOUNT KINABALU: 2700–3000 m, *Meijer SAN 22042* (K); PAKA-PAKA CAVE: 3000 m, *Clemens 29007* (BM, K, L), 3000–3400 m, *Gibbs 4193* (K), 3000 m, *Sinclair et al. 9124* (K, L), 3000 m, *9187* (K), 3000 m, *Smith s.n.* (K); PENATARAN RIVER: 900 m, *Clemens 32580* (BM, L); SAYAT-SAYAT: 3800 m, *Smith 501* (L); SHANGRI LA VALLEY: 3400 m, *Collenette 21512* (K, L); SUMMIT AREA: 4100 m, *Gibbs 4191* (K); TAHUBANG RIVER HEAD: 2100 m, *Clemens 32929* (BM, K, L); TINEKEK FALLS: 2100 m, *Clemens 50037* (BM); WASAI RIVER: 1100 m, *Clemens 32573* (BM).

47.1.6. Carex graeffeana Boeck., *Flora* 58: 123 (1875).

Carex exploratorum Nelmes, *Kew Bull.* 1938: 108 (1938). Type: PENATARAN RIVER, 1200 m, *Clemens 34297* (holotype K!; isotypes BM!, L!).

Densely tufted rhizomatous perennial sedge forming large clumps. Hill forest on ultramafic substrate. Elevation: 1200 m.

Additional material examined. MAMUT RIVER: 1200 m, *RSNB 1708* (K).

47.1.7. Carex hypolytroides Ridl., *J. Fed. Malay States Mus.* 8: 124 (1917).

Tall rhizomatous perennial sedge with strong stolons. Lower montane forest in wet places. Elevation: 1600–1900 m. A remarkable species that could initially be confused with *Hypolytrum*.

Material examined. LUMU-LUMU: 1900 m, *Sinclair et al. 9220* (K); MESILAU CAVE: 1800 m, *RSNB 4691* (K); PARK HEADQUARTERS/POWER STATION: 1700 m, *Simpson 89/185* (K); WEST MESILAU RIVER: 1600–1700 m, *Beaman 8633* (MSC, NY).

47.1.8. Carex indica L., *Mant. Pl.* 2: 574 (1771).

Loosely tufted perennial sedge with short creeping rhizome. Hill forest, wet places. Elevation: 900 m.

Material examined. MT. NUNGKEK: 900 m, *Meijer SAN 24098* (L). The locality stated on the label is "Ladang Nungkop," which might be neither Mt. Nungkek nor a Kinabalu locality.

47.1.9. Carex perakensis C. B. Clarke, *Fl. Brit. India* 6: 720 (1894).

a. var. perakensis

Tufted perennial sedge with short, thick rhizome. Hill forest, lower montane forest, upper montane forest, in wet places, sometimes on ultramafic substrate. Elevation: 900–2700 m.

Material examined. EASTERN SHOULDER: 2700 m, *RSNB 720* (K, L), 1100 m, *Meijer SAN 24085* (K); PENATARAN RIVER: 2600 m, *Clemens 33652* (BM, K, L); SINGH'S PLATEAU: 900 m, *RSNB 1009* (K, L).

b. var. borneensis (C. B. Clarke) Noot., *Fl. Males.* I, 9: 141 (1979).

Carex borneensis C. B. Clarke, *J. Linn. Soc., Bot.* 37: 14 (1904). Type: MOUNT KINABALU, 3400 m, *Haviland 1404* (holotype K!).
Carex kinabaluensis Stapf in Gibbs, *J. Linn. Soc., Bot.* 42: 183 (1914). Type: PAKA-PAKA CAVE, 2700 m, *Gibbs 4240* (holotype BM!; holotype fragment K!).

Tufted perennial sedge with short, thick rhizome. Hill forest, lower montane forest, upper montane forest, sometimes on ultramafic substrate, in wet places. Elevation: 600–3400 m. Distinguished from the typical variety by having compound or fasciculate spikelets.

Additional material examined. BAMBANGAN RIVER: 1700 m, *RSNB 1316* (K); DALLAS: 900 m, *Clemens 30068* (L); EASTERN SHOULDER: 3000 m, *RSNB 852* (K); GOLF COURSE SITE: 1700 m, *Beaman 8551* (K, L, MSC); GURULAU SPUR: 2400 m, *Clemens 50881* (BM); KEMBURONGOH/PAKA-PAKA CAVE: 2400–2700 m, *Clemens 28923* (BM, K); KILEMBUN BASIN: 1700 m, *Clemens 40044* (BM); KINATEKI RIVER: 1800 m, *Clemens 31412* (BM), 1200–1500 m, *31413* (BM); KINATEKI RIVER HEAD: 2100 m, *Clemens 31778* (BM); LETENG RIVER (NEAR KUNDASANG): 1200 m, *Meijer SAN 24116* (K, L); MAMUT HILL: 1400–1700 m, *Kokawa & Hotta 5300* (L); MARAI PARAI: 2700 m, *Clemens 32352* (BM, K, L), 1500 m, *35063* (L), 1500 m, *35643* (BM); MARAI PARAI SPUR: 2100 m, *Gibbs 4097* (K); MESILAU CAMP: *Poore H 242* (K); MESILAU RIVER: 2100 m, *Clemens 51279* (BM); MINATUKAN SPUR: 1500 m, *Clemens 34431* (BM); MOUNT KINABALU: 900 m, *Forster 523* (K); MT. NUNGKEK: 600–900 m, *Clemens 32700* (BM, L); MT. TEMBUYUKEN: 2500 m, *Meijer SAN 28810* (K, L); PANAR LABAN: 3300 m, *Jermy & Rankin J 15115* (BM), 3300 m, *Smith 522* (L); PARK HEADQUARTERS: 1500 m, *Price 155* (K), 1500 m, *Simpson 89/186* (K); PENATARAN BASIN: 2700 m, *Clemens 33651* (BM, L); PENIBUKAN: 1200–1500 m, *Clemens s.n.* (BM, BM), 1500 m, *30986* (BM), 1200–1500 m, *31413* (K, L), 1200 m, *40426* (BM, K, L), 1200 m, *Gibbs 4070* (BM, K); PIG HILL: 2100 m, *RSNB 4373* (K); SUMMIT TRAIL: 2300 m, *Simpson 89/217* (K); TAHUBANG RIVER HEAD: 2100 m, *Clemens 32928* (BM, L).

47.1.10. Carex tristachya Thunb.

a. var. pocciliformis (Boott) Kük., *Pflanzenr.* IV. 20 (Heft 38): 473 (1909).

Densely tufted perennial sedge with short rhizome. Lower montane forest in open areas. Elevation: 1400 m.

Material examined. TAHUBANG FALLS: 1400 m, *Clemens 40281* (BM, K).

47.1.11. Carex verticillata Zoll. & Moritzi in Moritzi, *Syst. Verz.*: 98 (1846).

Perennial sedge with solitary or somewhat tufted stems and creeping rhizome. Upper montane forest, summit area, in damp open places. Elevation: 1800–4100 m.

Material examined. EASTERN SHOULDER: 2700 m, *RSNB 716* (K, L); EASTERN SHOULDER, CAMP 4: 3000 m, *RSNB 736* (K, L), 2900 m, *RSNB 1123* (L); GURULAU SPUR: 2400–3800 m, *Clemens 50940* (BM, K, L); JANET'S HALT/SHEILA'S PLATEAU: 2900 m, *Collenette 21529* (L); KEMBURONGOH: 2400 m, *Clemens 28925* (BM), 2100 m, *Gibbs 4230* (BM), 2300 m, *Meijer SAN 20370* (K, L); KILEMBUN RIVER: 2600 m, *Clemens 33808* (BM, L); KILEMBUN RIVER HEAD: 2700 m, *Clemens 35064* (BM); KING GEORGE PEAK: 4100 m, *RSNB 5971* (K); LOW'S PEAK: 4100 m, *Clemens 27080* (BM, K); MARAI PARAI: 2100 m, *Clemens 32342* (BM, K, L), 3200 m, *33231* (BM, L); MESILAU: 2700 m, *RSNB 5998* (K); MESILAU RIVER: 1800 m, *Clemens 51342* (BM); PAKA-PAKA CAVE: *Clemens, M. S. 10578* (BM), 3000–3400 m, *Gibbs 4194* (BM); PAKA-PAKA CAVE/LOW'S PEAK: *Clemens, M. S. 10615* (K); SAYAT-SAYAT: 3800 m, *Smith 521* (L); SUMMIT AREA: 3700 m, *Clemens 28022* (BM), 3800 m, *51347* (BM); SUMMIT TRAIL: 2300 m, *Simpson 89/220* (K).

47.2. COSTULARIA C. B. Clarke

47.2.1. Costularia pilisepala (Steud.) J. Kern, *Fl. Males.* I, 5: 420 (1957).

Perennial sedge with short woody rhizome. Lower montane forest, upper montane forest, sometimes on ultramafic substrate, in marshy or boggy areas. Elevation: 1500–2400 m. Marai Parai is named for the plant, which looks like hill rice blowing in the wind.

Material examined. BAMBANGAN RIVER: 1600 m, *RSNB 4966* (K); GURULAU SPUR: 2400 m, *Clemens 51062* (BM, K); MARAI PARAI: *Clemens 30806* (BM), 1500 m, *32627* (BM, K), 1600 m, *40232* (BM, K), 1600 m, *Collenette A 98* (BM), *Gibbs 4095* (K); MARAI PARAI SPUR: *Clemens, M. S. 10897* (K), 1700 m, *Gibbs 4094* (BM); MESILAU RIVER: 2100 m, *Collenette 21622* (L); PENATARAN BASIN: 2000 m, *Clemens 34214* (BM).

47.3. CYPERUS L.

Kükenthal, G. (1935 36). Cyperaceae-Scirpoideae-Cypereae. *Pflanzenr.* IV. 20 (Heft 101): 41–320.

47.3.1. Cyperus castaneus Willd., *Sp. Pl.* 1: 278 (1797).

Annual sedge. Lowlands in open areas. Elevation: 300 m.

Material examined. KEBAYAU/KAUNG: 300 m, *Clemens 27674* (L).

47.3.2. Cyperus compactus Retz., *Observ. Bot.* 5: 10 (1788). Plate 14B.

Tufted perennial sedge with short rhizome. Lowlands in open, secondary situations.

Material examined. KIAS: *Jibrin Sibil 303* (K).

47.3.3. Cyperus compressus L., *Sp. Pl.*: 46 (1753).

Glabrous annual sedge with tufted stems. Hill forest in open secondary situations.

Material examined. MELANGKAP TOMIS: *Lorence Lugas 641* (K).

47.3.4. Cyperus cyperoides (L.) Kuntze, *Revis. Gen. Pl.* 3(2): 333 (1898).

Tufted perennial sedge with short rhizome. Lowlands in open situations and thickets.

Material examined. SERINSIM: *Kinsun Bakia 519* (K); TEKUTAN: *Lomudin Tadong 176* (K).

47.3.5. Cyperus diffusus Vahl, *Enum. Pl.* 2: 321 (1806).

Perennial sedge with short rhizome. Lowlands in damp, semi-shaded situations. Elevation: 500 m.

Material examined. PORING HOT SPRINGS: 500 m, *Simpson 89/214* (K).

47.3.6. Cyperus haspan L., *Sp. Pl.*: 45 (1753).

Annual or perennial sedge with short rhizome and slender, weak, trailing stems. Hill forest, lower montane forest, in drainage ditches, margins of shallow standing water and paddy fields. Elevation: 900–1500 m. Cf. Wilson (1991) concerning spelling of the epithet *haspan*.

Material examined. HIMBAAN: *Doinis Soibeh 111* (K); KIAU: 900 m, *Gibbs 3967* (BM, K); LOHAN/MAMUT COPPER MINE: 900 m, *Beaman 10617* (MSC, NY); PARK HEADQUARTERS: 1500 m, *Simpson 89/204* (K).

47.3.7. Cyperus laxus Lam.

a. var. macrostachyus (Boeck.) Karth., *Fl. Ind. Enum., Monocot.*: 46 (1989).

Tufted perennial sedge with short rhizome. Lowland primary forest near streams, sometimes on ultramafic substrate.

Material examined. MENGGIS: *Matamin Rumutom 304* (K); TEKUTAN: *Lomudin Tadong 288* (K).

47.3.8. Cyperus pilosus Vahl, *Enum. Pl.* 2: 354 (1806).

Perennial sedge with rhizome producing slender stolons. Lowlands, hill forest, lower montane forest, in drainage ditches in open areas. Elevation: 500–1500 m.

Material examined. MELANGKAP TOMIS: *Lorence Lugas 640* (K), *1356* (K), *2560* (K); PARK HEADQUARTERS: 1500 m, *Simpson 89/203* (K); PORING HOT SPRINGS: 500 m, *Simpson 89/205* (K); TENOMPOK: 1500 m, *Clemens 28073* (BM, K).

47.3.9. Cyperus sphacelatus Rottb., *Descr. Icon. Rar. Pl.*: 26 (1773).

Tufted, fibrous-rooted annual sedge to 0.5 m. Lowlands or hill forest in disturbed situations.

Material examined. MELANGKAP TOMIS/BUNDU PAKA: *Lorence Lugas 2643* (K).

47.4. ELEOCHARIS R. Br.

47.4.1. Eleocharis congesta D. Don, *Prodr. Fl. Nepal.*: 41 (1825).

Annual or perennial tufted sedge. Lower montane forest, in the open in drainage ditches. Elevation: 1800 m.

Material examined. PARK HEADQUARTERS/POWER STATION: 1800 m, *Simpson 89/218* (K).

47.5. FIMBRISTYLIS Vahl

47.5.1. Fimbristylis complanata (Retz.) Link, *Hort. Berol.* 1: 292 (1827).

Densely tufted perennial sedge with short rhizome. Lowlands, hill forest, in open, grassy places. Elevation: 500–800 m.

Material examined. DALLAS/KAUNG: 800 m, *Clemens 51558* (BM); PORING HOT SPRINGS: 500 m, *Simpson 89/209* (K).

47.5.2. Fimbristylis dichotoma (L.) Vahl, *Enum. Pl.* 2: 287 (1806).

Annual or perennial sedge with short rhizome. Lowlands, lower montane forest, on sunny or semi-shaded grassy banks and roadsides. Elevation: 500–1700 m.

Material examined. KEBAYAU/MELANGKAP TOMIS: *Lorence Lugas 2644* (K); NALUMAD: *Daim Andau 60* (K); PARK HEADQUARTERS: 1500 m, *Simpson 89/188* (K), 1500 m, *89/197* (K); PARK HEADQUARTERS/POWER STATION: 1700 m, *Simpson 89/199* (K); PORING HOT SPRINGS: 500 m, *Simpson 89/210* (K), 500 m, *89/212* (K).

47.5.3. Fimbristylis dura (Zoll. & Moritzi) Merr., *Philipp. J. Sci., Bot.* 11: 53 (1916).

Perennial sedge with short, thick woody rhizome. Lowlands and hill forest in disturbed situations and thickets.

Material examined. NALUMAD: *Daim Andau 171* (K); TEKUTAN: *Lomudin Tadong 165* (K).

47.5.4. Fimbristylis littoralis Gaudich. in Freyc., *Voy. Uranie*: 413 (1829).

Annual or perennial densely tufted sedge. Lowlands, in drainage ditches. Elevation: 500 m.

Material examined. PORING HOT SPRINGS: 500 m, *Simpson 89/207* (K).

47.5.5. Fimbristylis miliacea (L.) Vahl, *Enum. Pl.* 2: 287 (1805).

Glabrous annual sedge with fibrous roots. Lowlands or hill forest in disturbed situations.

Material examined. NALUMAD: *Daim Andau 176* (K); TEKUTAN: *Lomudin Tadong 213* (K).

47.5.6. Fimbristylis obtusata (C. B. Clarke) Ridl., *Fl. Malay Penins.* 5: 157 (1925).

Small tufted annual sedge. Lowlands. Elevation: 300–600 m.

Material examined. KAUNG: 300–600 m, *Clemens 27681* (BM, K, L).

47.5.7. Fimbristylis umbellaris Vahl, *Enum. Pl.* 2: 291 (1805).

Perennial sedge with short creeping rhizome. Lowlands and hill forest in open, wet secondary situations. The name *F. globulosa* (Retz.) Kunth was used for this species in *Fl. Males.* I, 7: 551 (1974).

Material examined. MELANGKAP TOMIS: *Lorence Lugas 485* (K), *638* (K), *1844* (K); TEKUTAN: *Dius Tadong 343* (K), *746* (K), *Lomudin Tadong 70* (K).

47.6. FUIRENA Rottb.

47.6.1. Fuirena umbellata Rottb., *Descr. Icon. Rar. Pl.*: 70, t. 19, f. 3 (1773).

Perennial sedge with short, thick rhizome. Lowlands and hill forest in open, wet areas.

Material examined. MELANGKAP TOMIS: *Lorence Lugas 154* (K), *1366* (K); TEKUTAN: *Dius Tadong 349* (K), *Lomudin Tadong 269* (K).

47.7. GAHNIA J. R. Forst. & G. Forst.

Kern, J. H. (1957). The genus *Gahnia. Taxon* 6: 153–154. Kern, J. H. (1962). On the delimitation of the genus *Gahnia* (Cyperaceae). *Acta Bot. Neerl.* 11: 216–224.

47.7.1. Gahnia baniensis Benl, *Repert. Spec. Nov. Regni Veg.* 44: 197 (1938).

Large tufted perennial sedge. Lower montane forest. Elevation: 1700 m.

Material examined. MARAI PARAI SPUR: *Clemens, M. S. 11087* (BM); PARK HEADQUARTERS: 1700 m, *Jermy & Rankin J 15204* (BM).

47.7.2. Gahnia javanica Zoll. & Moritzi ex Moritzi, *Syst. Verz.*: 98 (1846). Plate 15A, B.

Large tussock-forming perennial sedge. Rarely lower montane forest, mostly upper montane forest, in open areas. Elevation: 1500–3300 m.

Material examined. GURULAU SPUR: 2400 m, *Clemens s.n.* (BM); KEMBURONGOH: 2400 m, *Clemens 29001* (BM, K, L), 2900 m, *Fuchs 21071* (L), 2100 m, *Gibbs 4160* (K); LUBANG/PAKA-PAKA CAVE: *Clemens, M. S. 10709* (K); MESILAU: *Poore H 502* (K); MOUNT KINABALU: 2400 m, *Whitehead s.n.* (BM); PAKA-PAKA CAVE: 2900 m, *Sinclair et al. 9194* (L); PANAR LABAN: 3300 m, *Jermy & Rankin J 15114* (BM); PARK HEADQUARTERS: 1500 m, *Simpson 89/198* (K); PENATARAN BASIN: 2000 m, *Clemens 34213* (BM); PIG HILL: 2300 m, *RSNB 4502* (K).

47.7.3. Gahnia tristis Nees ex Hook. & Arn., *Bot. Beechey Voy.*: 228 (1837).

Perennial sedge with short woody rhizome. Hill forest.

Material examined. MT. NUNGKEK: *Lorence Lugas 1513* (K).

47.8. HYPOLYTRUM Rich.

47.8.1. Hypolytrum nemorum (Vahl) Spreng., *Syst. Veg.* 1: 233 (1835). Plate 15C.

Large perennial sedge with thick rhizome. Hill forest, in swampy places. Elevation: 700–1200 m.

Material examined. BUNGOL: 800 m, *Clemens 51559* (BM); DALLAS/TENOMPOK: 1200 m, *Clemens 26830* (BM, K); MELANGKAP KAPA: 700–1000 m, *Beaman 8781* (MSC, NY); MELANGKAP TOMIS: 900–1000 m, *Beaman 8975* (MSC, NY), *Lorence Lugas 1552* (K).

47.9. ISOLEPIS R. Br.

47.9.1. Isolepis habra (Edgar) Sojàk, *Čas. Nár. Mus., Odd. Přr.* 148: 194 (1980).

Tufted perennial sedge up to 30 cm high, with descending rhizomes. Upper montane forest, summit area. Elevation: 2700–3400 m.

Material examined. EASTERN SHOULDER: 2700 m, *RSNB 717* (K); MOUNT KINABALU: 3400 m, *Haviland 1394* (K), *Haviland 1397* (K).

47.9.2. Isolepis subtilissima Boeck., *Flora* 41: 416 (1858).

Low tufted perennial sedge up to 15 m high, with horizontal rhizomes, forming large mats. Upper montane forest, summit area, in boggy areas with sphagnum. Elevation: 3000–3500 m.

Material examined. 3000 m, *Meijer SAN 22059* (K); Summit Trail: 3000–3500 m, *Jacobs 5760* (K).

47.10. KYLLINGA Rottb.

Kükenthal, G. (1936). Cyperaceae-Scirpoideae-Cypereae. *Pflanzenr.* IV. 20 (Heft 101): 1–671 [*Kyllinga* on pp. 566–614].

47.10.1. Kyllinga brevifolia Rottb., *Descr. Icon. Rar. Pl.*: 13 (1773).

Perennial sedge with horizontally creeping rhizome. Lower montane forest on roadside verges in sun and semi-shade. Elevation: 1500 m.

Material examined. Park Headquarters: 1500 m, *Simpson 191* (K).

47.10.2. Kyllinga melanosperma Nees in Wight, *Contr. Bot. India*: 91 (1834).

Perennial sedge with creeping rhizome. Lower montane forest, in open, damp places, often by roadsides. Elevation: 1500 m.

Material examined. Park Headquarters: 1500 m, *Simpson 89/190* (K); Tenompok: 1500 m, *Clemens 51233* (K).

47.10.3. Kyllinga nemoralis (J. R. Forst. & G. Forst.) Dandy ex Hutch. & Dalziel, *Fl. W. Trop. Afr.* 2: 487 (1936). Plate 15D.

Perennial sedge with long slender rhizome. Lowlands in secondary forest and open grassy places. Elevation: 500 m.

Material examined. Melangkap Tomis: *Lorence Lugas 1920* (K); Poring Hot Springs: 500 m, *Simpson 89/208* (K); Tekutan: *Dius Tadong 660* (K).

47.10.4. Kyllinga indet.

Material examined. Himbaan: *Doinis Soibeh 141* (K); Melangkap Tomis: *Lorence Lugas 1854* (K); Serinsim: *Kinsun Bakia 213* (K).

47.11. LEPIDOSPERMA Labill.

47.11.1. Lepidosperma chinense Nees & Meyen, *En.* 2: 320 (1837).

Cladium borneensis C. B. Clarke in Stapf, *Trans. Linn. Soc. London, Bot.* 4: 245 (1894). Type: MOUNT KINABALU, 1500 m, *Low s.n.* (holotype K!).
Mariscus borneensis (C. B. Clarke) Fern., *Rhodora* 25: 52 (1923).
Machaerina borneensis (C. B. Clarke) T. M. Koyama, *Bot. Mag. Tokyo* 69: 62 (1956).

Perennial sedge with short rhizome. Lower montane forest, upper montane forest, in rocky or heathy areas on ultramafic substrate. Elevation: 1500–2400 m.

Additional material examined. GURULAU SPUR: 2400 m, *Clemens 51059* (BM, L), 2400 m, *51181* (BM, K); MARAI PARAI: 1600 m, *Clemens 40233* (BM, L), 1500 m, *Collenette A 83* (BM); MARAI PARAI SPUR: 1500 m, *Gibbs 4037* (K).

47.12. LIPOCARPHA R. Br.

47.12.1. Lipocarpha chinensis (Osbeck) J. Kern, *Blumea*, Suppl. 4: 167 (1958).

Tufted annual or perennial sedge. Lower montane forest, drainage ditches, damp roadside verges in open areas. Elevation: 1500 m.

Material examined. MELANGKAP TOMIS: *Lorence Lugas 636* (K), *1146* (K), *1368* (K); PARK HEADQUARTERS: 1500 m, *Simpson 89/189* (K); TENOMPOK: 1500 m, *Clemens 28074* (BM, K, L), 1500 m, *51041* (BM, K).

47.13. MACHAERINA Vahl

Kern, J. H. (1959). Florae Malesianae Precursores XXII. *Cladium & Machaerina. Acta Bot. Neerl.* 8: 263–268.

47.13.1. Machaerina aspericaulis (Kük.) T. Koyama, *Bot. Mag. (Tokyo)* 69: 62 (1956).

Cladium samoense C. B. Clarke in Stapf, *Trans. Linn. Soc. London, Bot.* 4: 245 (1894) p.p. quoad *Haviland 1406*. Type: MARAI PARAI, 1700 m, *Haviland 1406* (syntype K!).
Cladium aspericaule Kük., *Bull. Jard. Bot. Buitenzorg,* ser. 3, 16: 309 (1940). Type: MARAI PARAI SPUR, 1500 m, *Clemens 32404* (holotype B n.v.; isotypes BM!, L!).

Perennial sedge with short rhizome. Lower montane forest, in open areas on ultramafic substrate. Elevation: 1500–1700 m. Endemic to Mount Kinabalu.

Additional material examined. MARAI PARAI: 1500 m, *Clemens 32628* (K, L); MARAI PARAI SPUR: *Clemens, M. S. 11088* (K).

47.13.2. Machaerina disticha (C. B. Clarke) T. Koyama, *Bot. Mag. (Tokyo)* 69: 63 (1956).

Perennial sedge with short, creeping rhizome. Lower montane forest, in open grassy areas, mostly on ultramafic substrate. Elevation: 900–1500 m.

Material examined. PENATARAN RIVER: 900 m, *Clemens 34122* (BM); PENIBUKAN: 1200–1500 m, *Clemens 30752* (BM, K, L); PINOSUK PLATEAU: 1500 m, *RSNB 1292* (K, L).

47.13.3. Machaerina falcata (Nees) T. Koyama, *Bot. Mag. (Tokyo)* 69: 63 (1956). Plate 15E.

Cladium samoense C. B. Clarke in Stapf, *Trans. Linn. Soc. London, Bot.* 4: 245 (1894) p.p. quoad *Haviland 1405*. Type: MOUNT KINABALU, 3500 m, *Haviland 1405* (syntype K!).
Vincentia malesiaca Stapf in Gibbs, *J. Linn. Soc., Bot.* 42: 178 (1914). Type: PAKA-PAKA CAVE, 3200 m, *Gibbs 4278* (holotype K!).

Tufted perennial sedge with short rhizome. Upper montane forest, *Leptospermum* forest and scrub, sometimes on ultramafic substrate. Elevation: 1500–3700 m.

Additional material examined. EASTERN SHOULDER: 2900 m, *RSNB 993* (K); GURULAU SPUR: 2400–2700 m, *Clemens 50604* (BM), 2400–2700 m, *50636* (BM, K, L), 2400 m, *51182* (BM, K); KILEMBUN BASIN: 2700 m, *Clemens 33853* (BM, L); MARAI PARAI: 2700 m, *Clemens 31674* (L), 1500 m, *32806* (BM); MEKEDEU VALLEY: 3700 m, *RSNB 5964* (K); MESILAU BASIN: 2400–3000 m, *Clemens s.n.* (BM), 2400–3000 m, *30063* (K); MOUNT KINABALU: 3200 m, *Rickards 163* (K); PAKA-PAKA CAVE: 2600 m, *Meijer SAN 22050* (K, L); PAKA-PAKA CAVE/LOW'S PEAK: *Clemens, M. S. 10608* (K); PAKA-PAKA CAVE/PANAR LABAN: 3500 m, *Clemens 28024* (BM, K, L), 3200 m, *Sinclair et al. 9140* (L); PANAR LABAN: 3400 m, *Smith 540* (L).

47.13.4. Machaerina glomerata (Gaudich.) T. Koyama, *Bot. Mag. (Tokyo)* 69: 63 (1956). Plate 15F.

Perennial sedge with short rhizome. Hill forest, lower montane forest, on ultramafic substrate in open areas. Elevation: 600–1800 m.

Material examined. MARAI PARAI: 600–900 m, *Clemens 32736* (K, L); MELANGKAP TOMIS: *Lorence Lugas 1982* (K); PENATARAN BASIN: 1500–1800 m, *Clemens 40033* (L); PENATARAN RIVER: 900 m, *Clemens 34274* (L); PENIBUKAN: 1200–1500 m, *Clemens 30900* (K, L), 1400 m, *50309* (K, L).

47.14. MAPANIA Aubl.

Simpson, D. A. (1992). *A revision of the genus* Mapania *(Cyperaceae)*. Royal Botanic Gardens, Kew.
Uittien, H. (1936). Studies in Cyperaceae-Mapanieae I–V. *Rec. Trav. Bot. Néerl.* 33: 133–155.

47.14.1. Mapania borneensis Merr., *J. Straits Branch Roy. Asiat. Soc.* 76: 78 (1917).

Large rhizomatous perennial sedge. Lower montane forest, in damp shady areas. Elevation: 1400–1500 m.

Material examined. KIAU VIEW TRAIL: 1500 m, *Simpson 89/202* (K); MT. NUNGKEK: 1400 m, *Darnton 579* (BM, L); TENOMPOK: *Clemens s.n.* (BM, K).

47.14.2. Mapania caudata Kük., *Repert. Spec. Nov. Regni Veg.* 29: 201 (1931).

Moderate-sized, rhizomatous perennial sedge. Hill forest, in damp shady areas. Elevation: 900 m.

Material examined. DALLAS: 900 m, *Clemens 27028* (BM, L), 900 m, *30064* (K, L).

47.14.3. Mapania cuspidata (Miq.) Uittien

a. var. petiolata (C. B. Clarke) Uittien, *J. Arnold Arbor.* 20: 213 (1939).

Large rhizomatous perennial sedge. Hill forest, in damp shady areas. Elevation: 900 m.

Material examined. DALLAS: 900 m, *Clemens 30065* (K); MENGGIS: *Matamin Rumutom 212* (K); SINGH'S PLATEAU: *RSNB s.n.* (K).

47.14.4. Mapania latifolia Uittien, *Recueil Trav. Bot. Néerl.* 32: 199 (1935).

Large rhizomatous, stoloniferous perennial sedge. Hill forest, lower montane forest, in damp shady areas. Elevation: 900–1500 m. Distinguished from *M. cuspidata* var. *petiolata* by its one central culm.

Material examined. DALLAS: 900–1200 m, *Clemens s.n.* (BM, BM), 900 m, *26886* (BM, K), 900 m, *30065* (K, L); DALLAS/TENOMPOK: 1200 m, *Clemens 27573* (BM, K, L); MARAI PARAI: *Clemens, M. S. 11092* (K); MARAI PARAI/MT. NUNGKEK: 1200 m, *Clemens 35016* (BM); MT. NUNGKEK: 1100 m, *Darnton 601* (BM); PENIBUKAN: 1200 m, *Clemens 30654* (BM), 1200–1500 m, *31307* (L), 1400 m, *50504* (BM, K).

47.14.5. Mapania palustris (Hassk. ex Steud.) Fern.-Vill., *Nov. App.* 309 (1882). Plate 16A, B.

Large rhizomatous perennial sedge, pandan-like. Hill forest (rarely), lower montane forest, sometimes on ultramafic substrate. Elevation: 900–1700 m.

Material examined. KILEMBUN BASIN: 1200 m, *Clemens 40041* (BM); LIWAGU RIVER TRAIL: 1700 m, *Simpson 89/195* (K); MAMUT RIVER: 1300 m, *RSNB 1630* (K), 1200 m, *RSNB 1745* (K, L); MARAI PARAI: 1500 m, *Clemens 32333* (BM, L); MESILAU: 1500–1700 m, *RSNB 8444* (K); PENIBUKAN: 1200 m, *Meijer SAN 54008* (K, L); PINOSUK PLATEAU: 1400 m, *Beaman 10721* (K, MSC, NY); SINGH'S PLATEAU: 900 m, *RSNB 1021* (K); TENOMPOK: 1500 m, *Clemens 28209* (BM, K, L).

47.14.6. Mapania squamata (Kurz) C. B. Clarke, *Kew Bull. add. ser.* 8: 53 (1908).

Large rhizomatous perennial sedge. Hill forest, in damp shady areas. Elevation: 1200 m.

Material examined. PENIBUKAN: 1200 m, *Clemens 32096* (BM, K, L).

47.14.7. Mapania indet.

Material examined. MEKEDEU RIVER: 400–500 m, *Beaman 9669* (NY).

47.15. OREOBOLUS R. Br.

Seberg, O. (1988). Taxonomy, phylogeny, and biogeography of the genus *Oreobolus* R. Br. (Cyperaceae), with comments on the biogeography of the South Pacific continents. *Bot. J. Linn. Soc.* 96: 119–195.

47.15.1. Oreobolus ambiguus Kük. & Steenis in Kük., *Bull. Jard. Bot. Buitenzorg,* ser. 3, 14: 48 (1936). Type: SUMMIT AREA, 4000 m, *Clemens 29006* (lectotype of Seberg [1988, p. 151] BM!; isolectotypes K!, L!).

Low, cushion-forming perennial sedge. Summit area, around shrubs and on wet rocks sheltered by boulders, and in damp hollows on granitic rocks. Elevation: 3600–4000 m. In Borneo only recorded from Mount Kinabalu.

Additional material examined. SAYAT-SAYAT: 3800 m, *Meijer SAN 28566* (K); SUMMIT AREA: 4000 m, *Clemens s.n.* (BM), 3600–3900 m, *Sleumer 4717* (K, L), 3800 m, *4722* (L); VICTORIA PEAK: 3800 m, *Clemens 51348* (BM, K, L).

47.16. PYCREUS P. Beauv.

Kükenthal, G. (1936). Cyperaceae-Scirpoideae-Cypereae. *Pflanzenr.* IV. 20: (Heft 101) 1–671 [*Pycreus* on pp. 326–402].

47.16.1. Pycreus pelophilus (Ridl.) C. B. Clarke in T. Durand & Schinz, *Consp. Fl. Afric.* 5: 540 (1894).

Tufted perennial sedge with slender rhizome. Secondary hill forest near rivers.

Material examined. MELANGKAP TOMIS: *Lorence Lugas 595* (K).

47.16.2. Pycreus sanguinolentus (Vahl) Nees

a. subsp. **cyrtostachys** (Miq.) D. A. Simpson, comb. nov.

Cyperus sanguinolentus Vahl subsp. *cyrtostachys* (Miq.) J. Kern, *Reinwardtia* 3: 57 (1954).

Annual or perennial sedge with short rhizome. Lowlands in drainage ditches in open places. Elevation: 500 m.

Material examined. PORING HOT SPRINGS: 500 m, *Simpson 89/206* (K).

47.17. RHYNCHOSPORA Vahl

47.17.1. Rhynchospora corymbosa (L.) Britton, *Trans. New York Acad. Sci.* 11: 84 (1892).

Perennial sedge with a short rhizome. Lowlands and hill forest in open, secondary situations, infrequently in primary forest.

Material examined. MELANGKAP TOMIS: *Lorence Lugas 49* (K), *622* (K), *1150* (K); TEKUTAN: *Dius Tadong 574* (K), *635* (K), *Lomudin Tadong 23* (K).

47.17.2. Rhynchospora rubra (Lour.) Makino, *Bot. Mag. (Tokyo)* 17: 180 (1903).

Perennial sedge with short rhizome. Lowlands in open situations. Elevation: 300 m.

Material examined. KEBAYAU/KAUNG: 300 m, *Clemens 27679* (BM, K).

47.17.3. Rhynchospora rugosa (Vahl) Gale, *Rhodora* 46: 275 (1944).

Perennial sedge with short rhizome. Lower montane forest, in open places on banks of streams and drainage ditches. Elevation: 1500 m.

Material examined. PARK HEADQUARTERS: 1500 m, *Simpson 89/184* (K).

47.18. SCHOENOPLECTUS (Rchb.) Palla

47.18.1. Schoenoplectus juncoides (Roxb.) V. I. Krecz., *Fl. Uzbekist., ed. Schreder,* 1: 328 (1941).

Tufted annual sedge with slender, erect stems. Secondary hill forest.

Material examined. HIMBAAN: *Doinis Soibeh 138* (K).

47.18.2. Schoenoplectus mucronatus (L.) Palla ex A. Kern., *Sched. Fl. Austr.-Hung.* 5: 91 (1888).

Leafless perennial sedge with very short rhizome. Lower montane forest, in open wet or swampy areas. Elevation: 1500–1700 m.

Material examined. MELANGKAP TOMIS: *Lorence Lugas 1851* (K); PARK HEADQUARTERS/POWER STATION: 1700 m, *Simpson s.n.* (K); TENOMPOK: 1500 m, *Clemens 28239* (BM, K).

47.19. SCHOENUS L.

47.19.1. Schoenus curvulus F. Muell., *Trans. Roy. Soc. Victoria* 1(2): 36 (1889).

Schoenus kinabaluensis Stapf in Gibbs, *J. Linn. Soc., Bot.* 42: 176 (1914). Type: MOUNT KINABALU, 2700 m, *Low s.n.* (syntype K!); PAKA-PAKA CAVE, 2700 m, *Gibbs 4220* (syntype K!; isosyntype BM!); SUMMIT AREA, *Gibbs 4188* (syntype K!), *4198* (syntype K!).

Tufted perennial sedge with short rhizome. Upper montane forest, summit area, in wet places in the open or under shrubs, sometimes on ultramafics. Elevation: 1800–4100 m.

Additional material examined. EASTERN SHOULDER: 3200 m, *RSNB 856a* (K); EASTERN SHOULDER, CAMP 4: 3200 m, *RSNB 1130* (K); GURULAU SPUR: 2400 m, *Clemens 51060* (BM, L), 2400–2700 m, *51228* (K); KADAMAIAN RIVER HEAD: 3000 m, *Clemens 50983* (BM, K, L); KEMBURONGOH: 2800 m, *Meijer SAN 20391* (K); KING GEORGE PEAK: 4000 m, *RSNB 5975* (K); LOW'S PEAK: 4100 m, *Clemens 27079* (BM, K, L); MARAI PARAI: 3400 m, *Clemens 33232* (BM, L); MESILAU RIVER: 1800 m, *Clemens 51617* (BM); MOUNT KINABALU: 2700 m, *Forster 519* (K), 2800 m, *Gibbs 4270* (BM), 4000 m, *Haviland 1395* (K), 3400 m, *1396* (K), 3000 m, *Meijer SAN 22035* (K), 3000 m, *SAN 22053A* (K); PAKA-PAKA CAVE: 3200 m, *Sinclair et al. 9135* (K); PANAR LABAN: 3500 m, *Beaman 8300* (K, MSC, NY), 3300 m, *Jermy & Rankin J 15082* (BM); SUMMIT AREA: 3800 m, *Clemens 30067* (K, L), 3600–3800 m, *Sleumer 4714* (L).

47.19.2. Schoenus delicatulus (Fernald) J. Kern, *Blumea,* Suppl. 4: 167 (1958).

Perennial sedge with short rhizome. Lower montane forest on ultramafic substrate. Elevation: 1500 m.

Material examined. MARAI PARAI: 1500 m, *Clemens 32382* (K, L).

47.19.3. Schoenus longibracteatus Kük., *Repert. Spec. Nov. Regni Veg.* 44: 70 (1938). Type: MARAI PARAI, 1500 m, *Clemens 32359* (syntype B n.v.; iso-syntypes BM!, K!, L!); PAKA-PAKA CAVE/LUBANG, *Clemens, M. S. 10708* (syntype BO n.v.).

Perennial sedge with short rhizome. Lower montane forest, upper montane forest, on ultramafic substrate, in open areas. Elevation: 1500–2900 m. Known only from Mount Kinabalu and Mt. Doorman in Irian Jaya.

Additional material examined. SUMMIT TRAIL: 2500–2900 m, *Jacobs 5719* (K, L).

47.19.4. Schoenus melanostachys R. Br., *Prodr.*: 231 (1810).

Densely tufted perennial sedge with short rhizome. Lower montane forest, upper montane forest, on ultramafic substrate. Elevation: 1500–2900 m. In Borneo only recorded from Mount Kinabalu.

Material examined. KEMBURONGOH: 2600 m, *Meijer SAN 20391* (L); MARAI PARAI: *Clemens 30805* (BM, K), 1700 m, *Haviland 1407* (K); MARAI PARAI SPUR: 1500 m, *Gibbs 4038* (K); MESILAU: 1900 m, *Meijer SAN 38570* (K, L); MESILAU CAMP: *Poore 504* (K); MESILAU CAVE: 1900–2200 m, *Beaman 9542* (K, MSC, NY), 2000 m, *Sleumer 4736* (K); MOUNT KINABALU: 2900 m, *Meijer SAN 29276* (K, L); PIG HILL: 2100 m, *RSNB 4374* (K).

47.20. SCIRPUS L.

47.20.1. Scirpus subcapitatus Thwaites, *Enum. Pl. Zeyl.*: 351 (1864).

Scirpus clarkei Stapf, *Trans. Linn. Soc. London, Bot.* 4: 244 (1894). Type: MOUNT KINABALU, 3400 m, *Haviland 1398* (holotype K!).
Scirpus pakapakensis Stapf in Gibbs, *J. Linn. Soc., Bot.* 42: 174 (1914). Type: PAKA-PAKA CAVE, 2700–3000 m, *Gibbs 4277* (holotype K!; isotype BM!).

Densely tufted perennial sedge with short creeping rhizome. Lower montane forest (rarely), upper montane forest, summit area, sometimes on ultramafic substrate, in open boggy places. Elevation: 1200–4100 m.

Additional material examined. DACHANG: 3000 m, *Clemens 29090* (BM, K); EAST MESILAU RIVER: 2000 m, *Collenette 21600* (K, L); EASTERN SHOULDER: 3200 m, *RSNB 856* (L); GURULAU SPUR: 1500 m, *Clemens 51229* (L), 1500 m, *51229A* (BM); KEMBURONGOH: 2400 m, *Clemens 27816* (BM), 2600 m, *Meijer SAN 20391a* (L); KEMBURONGOH/PAKA-PAKA CAVE: *Clemens 29002* (BM, L), 2700 m, *Sinclair et al. 9199* (L); KILEMBUN BASIN: 2700 m, *Clemens 33730* (BM); KILEMBUN RIVER: 2900 m, *Clemens 33842* (BM, L); KING GEORGE PEAK: 4000 m, *RSNB 5977* (K); LOW'S PEAK: 4100 m, *Clemens 27078* (BM, L); MARAI PARAI: 3200 m, *Clemens 32329* (BM, L); MESILAU CAMP: *Poore H 244* (K); MESILAU RIVER: 1500 m, *Clemens 51618* (BM), 2100 m, *51657* (BM, K); MOUNT KINABALU: 2700 m, *Meijer SAN 22053* (L), 4000 m, *Whitehead s.n.* (BM); PAKA-PAKA CAVE: 3000 m, *Sinclair et al. 9120* (L); PAKA-PAKA CAVE/LOW'S PEAK: *Clemens, M. S. 10609* (BM, K); PANAR LABAN: 3500 m, *Beaman 8299* (K, MSC, NY), 3300 m, *Jermy & Rankin J 15123* (BM), 3300 m, *Smith 466* (L); PENATARAN RIVER: 1200 m, *Clemens 34169* (BM, L); SUMMIT AREA: 3400–4000 m, *Clemens s.n.* (BM, BM), 3000–4000 m, *30062* (L), 4100 m, *Forster S 22* (K), 4100 m, *Gibbs 4190* (K), 3400–4000 m, *Holttum 25492* (K), 3600–3800 m, *Sleumer 4715* (K); SUMMIT TRAIL: 2500–2900 m, *Jacobs 5720* (K), 3000 m, *Jermy & Rankin J 15203* (BM); UPPER KINABALU: 2400 m, *Clemens 51061* (BM).

47.20.2. Scirpus ternatanus Reinw. ex Miq., *Fl. Ned. Ind.* 3: 307 (1856).

Perennial sedge. Lower montane forest, in open, wet situations and drainage ditches. Elevation: 1200–1800 m.

Material examined. GOLF COURSE SITE: 1700 m, *Beaman 8550* (K, MSC, NY); MAMUT RIVER: 1200 m, *RSNB 1707* (K); PARK HEADQUARTERS/POWER STATION: 1800 m, *Simpson 89/219* (K).

47.21. SCLERIA Bergius

47.21.1. Scleria ciliaris Nees in Wight, *Contr. Bot. India*: 117 (1834).

Stout perennial sedge. Lowlands and hill forest in disturbed situations.

Material examined. MELANGKAP TOMIS: *Lorence Lugas 95* (K), *818* (K), *1524* (K), *2426* (K); PORING: *Meliden Giking 53* (K), *Sani Sambuling 401* (K).

47.21.2. Scleria motleyi C. B. Clarke, *Philipp. J. Sci. C.* 2: 104 (1907).

Large perennial sedge. Hill forest, lower montane forest. Elevation: 500–1700 m.

Material examined. DALLAS: 900 m, *Clemens 27303* (BM, K); PORING HOT SPRINGS: 1100 m, *Meijer SAN 26433* (K), 500 m, *Simpson 89/216* (K); SINGH'S PLATEAU: 1400 m, *RSNB 1035* (K); WEST MESILAU RIVER: 1600–1700 m, *Beaman 8653* (MSC, NY).

47.21.3. Scleria purpurascens Steud., *Syn. Pl. Glumac.* 2: 169 (1855).

a. var. **purpurascens**

Large perennial sedge. Lowlands, hill forest, on roadsides in sunny and semi-shaded areas. Elevation: 500–900 m.

Material examined. DALLAS: 900 m, *Clemens s.n.* (BM), 900 m, *28269* (K); MELANGKAP TOMIS: *Lorence Lugas 483* (K), *1041* (K); MENGGIS: *Matamin Rumutom 328* (K); PORING HOT SPRINGS: 500 m, *Simpson 89/213* (K); SERINSIM: *Jibrin Sibil 76* (K), *280* (K), *Kinsun Bakia 415* (K); TEKUTAN: *Dius Tadong 384* (K), *471* (K), *551* (K), *646* (K), *803* (K).

47.21.4. Scleria scrobiculata Nees & Meyen ex Nees in Wight, *Contr. Bot. India*: 117 (1834).

Perennial sedge up to 2 m or more tall. Lowlands in damp, semi-shaded places. Elevation: 500 m.

Material examined. NALUMAD: *Daim Andau 16* (K); PORING HOT SPRINGS: 500 m, *Simpson 89/194* (K); TEKUTAN: *Dius Tadong 573* (K), *642* (K), *701* (K).

47.21.5. Scleria terrestris (L.) Fassett, *Rhodora* 26: 159 (1924).

Perennial sedge. Hill forest, lower montane forest, in semi-shaded areas. Elevation: 1100–2000 m.

Material examined. BUNDU TUHAN: *Doinis Soibeh 70* (K); EASTERN SHOULDER: 1200 m, *RSNB 289* (K); MARAI PARAI: 1500 m, *Clemens 32599* (BM); MT. NUNGKEK: 1200 m, *Clemens 32795* (BM); PARK

Cyperaceae

HEADQUARTERS: 1500 m, *Simpson 89/194* (K); SINGH'S PLATEAU: 1400 m, *RSNB 1032* (K); TENOMPOK: 1500 m, *Clemens s.n.* (BM); TINEKEK FALLS: 2000 m, *Clemens 50038* (BM); WASAI RIVER: 1100 m, *Clemens 32579* (BM).

47.21.6. Scleria cf. **terrestris** (L.) Fassett, *Rhodora* 26: 159 (1924).

Erect or scrambling perennial sedge. Hill forest thickets.

Material examined. KIAU: *Jusimin Duaneh 274* (K).

47.22. UNCINIA Pers.

Kern, J. H. & Nooteboom, H. P. (1979). Cyperaceae–II. *Fl. Males.* I, 9: 107–187. Nelmes, E. (1949). Notes on Cyperaceae: XX. The genus *Uncinia* in Malaysia. *Kew Bull.* 4: 140–145.

47.22.1. Uncinia compacta R. Br., *Prodr.*: 241 (1810).

Uncinia riparia R. Br. var. *stolonifera* Steen. & Kük., *Bull. Jard. Bot. Buitenzorg,* ser. 3, 13: 201 (1934). Type: MARAI PARAI, 3400 m, *Clemens 32333* (syntype BO n.v.; isosyntype K!), 3200 m, *32341* (syntype BO n.v.; isosyntype K!); PAKA-PAKA CAVE, 3400 m, *Clemens 29004* (syntype BO n.v.).
Uncinia subtrigona Nelmes, *Kew Bull.* 4: 144 (1949). Type: PAKA-PAKA CAVE, 3400 m, *Clemens 29004* (holotype K!).

Tufted perennial sedge with short rhizome. Upper montane forest, in open places. Elevation: 2900–3400 m.

Additional material examined. EASTERN SHOULDER: 2900 m, *RSNB 867* (K).

48. DIOSCOREACEAE

In collaboration with E. J. Cowley (K) and P. Wilkin (K)

Burkill, I. H. (1951). Dioscoreaceae. *Fl. Males.* I, 4: 293–335. Knuth, R. (1924). Dioscoreaceae. *Pflanzenr.* IV. 43 (Heft 87): 1–387.

48.1. DIOSCOREA L.

Prain, D., and Burkill, I. H. (1936). An account of the genus *Dioscorea* in the East. Part I. The species which twine to the left. *Ann. Roy. Bot. Gard. (Calcutta)* 14: 1–210 + i–vi. Prain, D., and Burkill, I. H. (1936). An account of the genus *Dioscorea* in the East. Part I. The species which twine to the left. *Ann. Roy. Bot. Gard. (Calcutta)* 14: Plates 1–85. Prain, D., and Burkill, I. H. (1938). An account of the genus *Dioscorea* in the East. Part II. The species which twine to the right: with addenda to Part I, and a summary. *Ann. Roy. Bot. Gard. (Calcutta)* 14: 211–528 + i–xx. Prain, D., and Burkill, I. H. (1938). An account of the genus *Dioscorea* in the East. Part II. The species which twine to the right. *Ann. Roy. Bot. Gard. (Calcutta)* 14: Plates 86–150.

48.1.1. Dioscorea alata L., *Sp. Pl.*: 1033 (1753).

Square-stemmed with usually 4-wings, climbing herb. Hill forest. Elevation: 900 m.

Material examined. DALLAS: 900 m, *Clemens 26455* (BM, K, L).

48.1.2. Dioscorea bulbifera L., *Sp. Pl.*: 1033 (1753).

Climber bearing "tubers" in leaf axils. Hill forest. Elevation: 900–1200 m.

Material examined. DALLAS: 1100 m, *Clemens 26389* (BM), 900 m, *26433* (L); DALLAS/ TENOMPOK: 1200 m, *Clemens 26433* (BM, K); HIMBAAN: *Doinis Soibeh 378* (K), *855* (K); KIAU: *Jusimin Duaneh 92* (K); MELANGKAP TOMIS: *Lorence Lugas 775* (K), *1215* (K).

48.1.3. Dioscorea bullata Prain & Burkill, *Kew Bull.* 1925: 60 (1925).

Climber. Hill forest. Elevation: 900–1200 m.

Material examined. DALLAS: 900 m, *Clemens 26937* (BM, K, L); KIAU: 900–1200 m, *Clemens 33061* (BM).

48.1.4. Dioscorea hispida Dennst., *Schlüssel Hortus Malab.*: 15 (1818).

Climber with large trifoliate leaves. Lowland secondary forest.

Material examined. SERINSIM: *Akungsai Bakia 66* (K), *Kinsun Bakia 470* (K); TEKUTAN: *Dius Tadong 282* (K), *Lomudin Tadong 298* (K).

48.1.5. Dioscorea laurifolia Wall. ex Hook. f., *Fl. Brit. India* 6: 293 (1892).

Climber; male flowers negatively geotropic. Hill forest.

Material examined. MELANGKAP TOMIS: *Lorence Lugas 931* (K).

48.1.6. Dioscorea nummularia Lam., *Encycl. Meth.* 3: 231 (1789).

Climbing herb to 4 m. Hill forest, lower montane forest. Elevation: 900–1500 m.

Material examined. DALLAS: 900 m, *Clemens 30091* (K); KAUNG/GURULAU: *Gibbs 4297* (K); PENIBUKAN: 1200–1500 m, *Clemens 30925* (BM, L), 1200 m, *30973* (BM), 1200 m, *40500* (BM, K, L); TENOMPOK: 1500 m, *Clemens 28056* (K, L).

48.1.7. Dioscorea pentaphylla L., *Sp. Pl.*: 1032 (1753).

Climber to 3 m; leaves 3- or 5-foliate. Hill forest. Elevation: 900–1200 m. Tubers eaten by natives.

Material examined. DALLAS: 900 m, *Clemens 26333* (L), 900 m, *26728* (K, L); DALLAS/KIAU: 900 m, *Clemens 26333* (BM, K); KIAU: *Clemens, M. S. 10026* (BM, K), *Jusimin Duaneh 101* (K); MELANGKAP TOMIS: *Lorence Lugas 624* (K), *869* (K); MOUNT KINABALU: 900–1200 m, *Clemens 26730* (BM).

48.1.8. Dioscorea polyclados Hook. f., *Fl. Brit. India* 6: 294 (1892).

Climber. Hill forest. Elevation: 1200 m.

Material examined. LOHAN: 1200 m, *Kuripin L. SAN 28773* (K, L).

48.1.9. Dioscorea pyrifolia Kunth, *Enum. Pl.* 5: 384 (1850).

Climbing herb. Hill forest, rarely lower montane forest. Elevation: 300–1500 m. Several varieties were listed by Prain and Burkill (1938).

Material examined. DALLAS: 900 m, *Clemens 26143* (BM, K, L), 900–1200 m, *26442* (BM, K, L), 900 m, *26698* (K, L), 900 m, *26699* (BM), 900–1200 m, *26729* (BM), 900 m, *26741* (L), 900–1200 m, *27041* (BM, K, L), 900 m, *27724* (BM, K), 900 m, *30092* (K); KAUNG/GENAMBAR: 300 m, *Gibbs 4297* (BM); KIAU: *Jusimin Duaneh 82* (K), *252* (K), 1200 m, *Kanis SAN 56149* (K, L); LOHAN RIVER: 700–800 m, *Beaman 9072* (L, MSC); LUGAS HILL: *Doinis Soibeh 417* (K); MELANGKAP TOMIS: *Lorence Lugas 1259* (K), *1605* (K); SOSOPODON: 1100 m, *Henry Tai SAN 42527* (K); TENOMPOK: 1500 m, *Clemens 28056* (BM).

48.1.10. Dioscorea cf. stenomeriflora Prain & Burkill, *J. & Proc. Asiat. Soc. Bengal* 10: 40 (1914).

Vine to 3–4 m. Lower montane forest. Elevation: 1800 m. The collection is sterile.

Material examined. GURULAU SPUR: 1800 m, *Clemens 50570* (BM, K).

48.1.11. Dioscorea sumatrana Prain & Burkill, *Kew Bull.* 1931: 90 (1931).

Climber with very large fruits. Lowland forest. First record of the species for Borneo.

Material examined. NALUMAD RIVER: *Daim Andau 935* (K).

48.1.12. Dioscorea sp.

Climber with stem very spiny near base. Hill forest on ultramafic substrate. Elevation: 1200 m. This species apparently occurs also in peninsular Thailand.

Material examined. PENIBUKAN: 1200 m, *Clemens 32163* (BM).

48.1.13. Dioscorea indet.

Material examined. DALLAS: 800 m, *Clemens s.n.* (BM); DALLAS/TENOMPOK: *Clemens 27161* (BM); MELANGKAP TOMIS: *Lorence Lugas 1811* (K); NALUMAD: *Sigin et al. SAN 112257* (L); PORING/ NALUMAD: 500–600 m, *Kokawa & Hotta 5247* (L); SERINSIM: *Kinsun Bakia 408* (K).

49. DRACAENACEAE

49.1. DRACAENA Vand. ex L.

In collaboration with E. J. Cowley (K) and B. Mathew (K)

49.1.1. Dracaena angustifolia Roxb., *Hort. Beng.*: 24 (1814).

Small tree to 12 m. Lowlands, hill forest. Elevation: 400–1200 m.

Material examined. DALLAS: 800 m, *Clemens 27306* (BM); HEMPUEN HILL: 800 m, *Abbe et al. 9962* (K, L); KAUNG: 400 m, *Darnton 377* (BM); LOHAN RIVER: 500 m, *RSNB 2964* (L); MELANGKAP TOMIS: 900 m, *Beaman 8397* (MSC), *Lorence Lugas 599* (K), *2293* (K); MINITINDUK GORGE: 900–1200 m, *Clemens 29669* (BM, L); NALUMAD: *Daim Andau 906* (K); PENATARAN RIVER: 500 m, *Beaman 9299* (K, MSC); PERANCANGAN: *Amin et al. 115942* (K); PINAWANTAI: 500 m, *Shea & Aban SAN 76771* (K); PORING: *Sani Sambuling 10* (K); SERINSIM: *Jibrin Sibil 84* (K), *319* (K), *Kinsun Bakia 175* (K); TEKUTAN: *Dius Tadong 358* (K), *428* (K), *Lomudin Tadong 196* (K).

49.1.2. Dracaena elliptica Thunb., *Diss. Dracaena*: 6 (1808).

Small tree to c. 2 m. Hill forest, lower montane forest? Elevation: 900–1500 m.

Material examined. EASTERN SHOULDER: 1000 m, *RSNB 66* (L), 900 m, *RSNB 225* (L); HIMBAAN: *Doinis Soibeh 240* (K); KUNDASANG/RANAU: *Lajangah SAN 33055* (K); LUGAS HILL: 1300 m, *Beaman 10547* (K, L, MSC); MENGGIS: *Matamin Rumutom 267* (K); NALUMAD: *Daim Andau 980* (K); PARK HEADQUARTERS: 1200 m, *Tai SAN 42538* (K); PENIBUKAN: 1200–1500 m, *Clemens 31199* (BM); PORING: *Sani Sambuling 12* (K); SOSOPODON: 1400 m, *Mikil SAN 37710* (K); TEKUTAN: *Dius Tadong 149* (K).

49.1.3. Dracaena sp.

Shrub to 3 m high. Lower montane forest, probably on ultramafic substrate. Elevation: 1200–1500 m.

Material examined. MELANGKAP TOMIS: *Lorence Lugas 1921* (K); PENIBUKAN: 1200 m, *Clemens s.n.* (BM), 1200–1500 m, *35031* (BM, K); TENOMPOK/DALLAS: *Clemens 26220* (BM).

50. ERIOCAULACEAE

50.1. ERIOCAULON L.

Royen, P. van. (1960). New species in *Eriocaulon*. *Blumea* 10: 126–135. Ruhland, W. (1903). Erio-caulaceae. *Pflanzenr.* IV. 30 (Heft 13): 1–294 [*Eriocaulon* on pp. 30–117].

50.1.1. Eriocaulon hookerianum Stapf, *Trans. Linn. Soc. London, Bot.* 4: 243 (1894).

a. var. **hookerianum.** Type: KEMBURONGOH, 2300 m, *Haviland 1153 bis* (syntype K n.v.), *1204* (syntype K n.v.); Marai Parai, *Low s.n.* (syntype fragment L!); Marai Parai Spur, 1500–1700 m, *Burbidge s.n.* (syntype K n.v.), 1500–1700 m, *Low s.n.* (syntype K n.v.; syntype fragment L!). The Kew specimens apparently have been misplaced.

Eriocaulon beccarii Suess. & Heine, *Mitt. Bot. Staatssamml. München,* Heft 2: 57 (1950). Type: KEMBURONGOH, 2400 m, *Clemens 27813* (syntype M n.v.; isosyntypes BM!, K!); MARAI PARAI, 1500 m, *Clemens 32629* (syntype M n.v.; isosyntypes BM!, K!).

Densely tufted perennial herb with scapes to ca. 30 cm. In open areas in lower and upper montane forest on ultramafic substrate. Elevation: 1200–2700 m.

Additional material examined. EASTERN SHOULDER: 2700 m, *RSNB 1092* (L); EASTERN SHOULDER, CAMP 3: 2700 m, *RSNB 1092* (K); GURULAU SPUR: 2400–2700 m, *Clemens 50643* (BM, K, L); MARAI PARAI: 1500 m, *Collector unknown SNP 1042* (L); MARAI PARAI SPUR: *Clemens, M. S. 10874* (BM, K), *10885* (K), 1200–1700 m, *Gibbs 3125* (BM), *Topping 1885* (BM), *1887* (L); MOUNT KINABALU: 2200 m, *Whitehead s.n.* (BM, BM), MT. TEMBUYUKEN: *Phillipps, A. SNP 190* (L); PENATARAN BASIN: 1500 m, *Clemens 34090* (BM).

50.1.2. Eriocaulon kinabaluense P. Royen, *Blumea* 10: 133 (1960). Type: PAKA-PAKA CAVE/LOW'S PEAK, *Clemens, M. S. 10611* (holotype L!; isotype K!).

Densely matted low cushion plant with scapes barely exceeding the leaves, to 3 cm high. Open boggy, rocky areas in upper montane forest and the summit area. Elevation: 3100–4000 m. Endemic to Mount Kinabalu but cf. *E. brevipedunculatum* Merr., the type from Mt. Halcon, Philippines.

Additional material examined. MARAI PARAI: 3400 m, *Clemens 32336* (BM, L); PAKA-PAKA CAVE: 3100 m, *Sleumer 4705* (L); SAYAT-SAYAT: 3800 m, *Meijer SAN 28569* (K); SUMMIT AREA: 4000 m, *Clemens 27089* (K, L), 3400 m, *30059* (K), *Gibbs 4209* (BM, K); SUMMIT TRAIL: 3800 m, *Sleumer 4719* (L); VICTORIA PEAK: 4000 m, *Clemens 27777* (BM, K, L), 3800 m, *51120* (BM, L).

51. FLAGELLARIACEAE

Backer, C. A. (1951). Flagellariaceae. *Fl. Males.* I, 4: 245–250.

51.1. FLAGELLARIA L.

51.1.1. Flagellaria indica L., *Sp. Pl.*: 333 (1753).

Clambering wiry-stemmed vine, the leaves terminating in climbing tendrils. Lowlands, generally in disturbed situations. Elevation: 300 m.

Material examined. KEBAYAU: 300 m, *Clemens 27691* (BM, K, L); MELANGKAP TOMIS: *Lorence Lugas 578* (K), *1587* (K), *2635* (K); MENGGIS: *Matamin Rumutom 206* (K); TEKUTAN: *Dius Tadong 254* (K).

52. HANGUANACEAE

In collaboration with E. J. Cowley (K)

Backer, C. A. (1951). Flagellariaceae. *Fl. Males.* I, 4: 245–250.

52.1. HANGUANA Blume

52.1.1. Hanguana major Airy Shaw, *Kew Bull.* 35: 819 (1981). Plate 16C. Type: MESILAU RIVER, 1500 m, *RSNB 4233* (holotype K!).

Tufted herb to 1 m; fruits large. Hill forest, lower montane forest. Elevation: 700–1500 m.

Additional material examined. BAMBANGAN RIVER: *Amin & Jarius SAN 116545* (L); DALLAS: 900 m, *Clemens 26035* (BM, K); DALLAS/TENOMPOK: 1200 m, *Clemens 27575* (BM, K); EASTERN SHOULDER: 1200 m, *RSNB 299* (K); KULUNG HILL: 800 m, *Beaman 7861* (K, L, MSC); LANGANAN/ MAMUT RIVERS: 1200 m, *RSNB 1715* (K); LIWAGU/MESILAU RIVERS: 1500 m, *RSNB 2882* (K); LUGAS HILL: 1300 m, *Beaman 10570* (L, MSC); MARAI PARAI SPUR: *Clemens, M. S. 11108* (BM, K); MELANGKAP KAPA: 700–1000 m, *Beaman 8792* (MSC); MELANGKAP TOMIS: *Lorence Lugas 731* (K); PARK HEADQUARTERS: 1400 m, *Price 147* (K); PENATARAN RIVER: 1500 m, *Clemens 33995* (BM, L); PENIBUKAN: 1200 m, *Clemens 30687* (BM, K, L), 1200–1500 m, *31296* (BM); PINOSUK PLATEAU: 1400 m, *Beaman 10705* (L, MSC); SOSOPODON: *Ampon & Saikeh SAN 71825* (K, L); TEKUTAN: *Lomudin Tadong 518* (K); TENOMPOK: 1500 m, *Clemens 29815* (BM).

52.1.2. Hanguana malayana (Jack) Merr., *Philipp. J. Sci., Bot.* 10: 3 (1915).

Tufted herb to 1 m; fruits small. Primary hill forest.

Material examined. NALUMAD: *Daim Andau 607* (K).

53. HYPOXIDACEAE

In collaboration with B. Mathew (K)

53.1. CURCULIGO Gaertn.

Geerinck, D. J. L. (1993). Amaryllidaceae (including Hypoxidaceae). *Fl. Males.* I, 11: 366–373; (*Curculigo* on pp. 366–370). Hilliard, O. R. and Burtt, B. L. (1978). Notes on some plants from southern Africa chiefly from Natal: VII. Cf. pp. 72–76, The genera of Hypoxideae [includes key to genera]. *Notes Roy. Bot. Gard. Edinburgh* 36: 43–76.

Hilliard and Burtt (1978) provide considerable evidence for recognizing the genus *Molineria* in addition to *Curculigo.* The *Flora Malesiana* treatment, however, treats all of the species as *Curculigo,* and we have reluctantly adopted this concept.

53.1.1. Curculigo capitulata (Lour.) Kuntze, *Revis. Gen. Pl.* 1: 703 (1891).

Stemless clump-forming perennial herb; leaves glabrous or sparsely pubescent; inflorescence elevated above leaf bases. Hill forest, lower montane forest. Elevation: 900–2000 m.

Material examined. EASTERN SHOULDER, CAMP 2: 2000 m, *RSNB 1049* (K, L); KIBAMBANG RIVER: 1200 m, *Clemens 34273* (BM); TAHUBANG RIVER: 900–1200 m, *Clemens s.n.* (BM).

53.1.2. Curculigo latifolia Dryand. in W. T. Aiton, *Hort. Kew.* ed. 2, 2: 253 (1811). Plate 16D.

Stemless clump-forming rhizomatous perennial herb; leaves glabrous or sparsely pubescent; inflorescence at leaf bases. Lowlands, hill forest, lower montane forest. Elevation: 600–1600 m.

Material examined. BUNDU TUHAN/PAHU: *Doinis Soibeh 767* (K); MARAI PARAI: 1200 m, *Clemens 35009* (BM); PINOSUK PLATEAU: 1600 m, *de Vogel 8009* (L), 1600 m, *8010* (L); PORING: 600 m, *Darnton 158* (BM); TEKUTAN: *Lomudin Tadong 21* (K), *194* (K); WEST MESILAU RIVER: 1600 m, *Beaman 7506* (MSC).

53.1.3. Curculigo villosa Wall. ex Kurz in Miq., *Ann. Mus. Bot. Lugduno-Batavum* 4: 176 (1869).

Stemless clump-forming rhizomatous perennial herb; leaves densely pubescent. Hill forest, probably on ultramafic substrate. The species is not recognized in the *Flora Malesiana* treatment, but the Kinabalu specimens seem very distinct from *C. latifolia.*

Material examined. MELANGKAP TOMIS: *Lorence Lugas 2425* (K); SERINSIM: *Kinsun Bakia 270* (K).

54. IRIDACEAE

In collaboration with B. Mathew (K)

Geerinck, D. J. L. (1977). Iridaceae. *Fl. Males.* I, 8: 77–84.

54.1. PATERSONIA R. Br.

54.1.1. Patersonia lowii Stapf, *Trans. Linn. Soc. London, Bot.* 4: 241 (1894). Plate 17A. Type: MARAI PARAI, 1700 m, *Haviland 1259* (syntype K!); MOUNT KINABALU, 1500 m, *Low s.n.* (syntype K!).

Patersonia borneensis Stapf, *Trans. Linn. Soc. London, Bot.* 4: 242 (1894). Type: KEMBURONGOH, 2300 m, *Haviland 1179* (holotype K!).

Tufted perennial herb. Lower montane forest, upper montane forest, mostly on ultramafic substrate. Elevation: 1500–2700 m.

Additional material examined. EASTERN SHOULDER: 2700 m, *RSNB 996* (K, L); GURULAU SPUR: 2400–2700 m, *Clemens 50642* (BM), 2400 m, *51034* (BM, K, L); KEMBURONGOH: 2200 m, *Carr SFN 27565* (BM), 2300 m, *Gibbs 3124* (BM, K); MARAI PARAI: *Clemens s.n.* (BM), 1700 m, *32931* (BM, K, L), 1500 m, *Collenette A 36* (BM); MOUNT KINABALU: 2400 m, *Whitehead s.n.* (BM); UPPER KINABALU: *Clemens 30061* (K, L).

55. JOINVILLEACEAE

Backer, C. A. (1951). Flagellariaceae. *Fl. Males.* I, 4(3): 245–250.

55.1. JOINVILLEA Gaudich.

Newell, T. K. (1969). A study of the genus *Joinvillea* (Flagellariaceae). *J. Arnold Arbor.* 50: 527–555.

55.1.1. Joinvillea ascendens Gaudich. ex Brongn. & Griseb.

a. subsp. borneensis (Becc.) Newell, *J. Arnold Arbor.* 50: 549 (1969). Plate 17B.

Tall, tufted grass-like perennial. Hill forest, lower montane forest. Elevation: 500–1700 m.

Material examined. BAMBANGAN RIVER: 1500 m, *RSNB 4387* (K); DALLAS/BUNGOL: 500 m, *Clemens 26015* (BM, K); MELANGKAP TOMIS: *Lorence Lugas 1537* (K); NALUMAD: *Daim Andau 33* (K); PARK HEADQUARTERS: 1400 m, *Newell 144* (K, L), 1400 m, *145* (L), 1400 m, *149* (L), 1400 m, *151* (L); PINOSUK PLATEAU: 1700 m, *RSNB 1809* (K); SOSOPODON/KUNDASANG: 1300 m, *Newell 143* (K, L); TAHUBANG RIVER HEAD: 1500 m, *Clemens 40708* (BM).

56. JUNCACEAE

Backer, C. A. (1951). Juncaceae. *Fl. Males.* I, 4: 210–215. Buchenau, Fr. (1906). Juncaceae. *Pflanzenr.* IV. 36 (Heft 25): 1–284.

56.1. JUNCUS L.

56.1.1. Juncus bufonius L., *Sp. Pl.*: 328 (1753).

Tufted decumbent to erect annual rush to 20 cm. In disturbed situations at upper edge of upper montane forest. Elevation: 3200–3400 m.

Material examined. PANAR LABAN: 3200 m, *Salick et al. 9023* (K), 3400 m, *Smith s.n.* (K).

56.1.2. Juncus effusus L., Sp. Pl.: 326 (1753).

Erect rhizomatous perennial rush. Upper montane forest? Elevation: 2100–2400 m.

Material examined. MESILAU RIVER: 2100–2400 m, *Clemens 51674* (BM, K, L).

56.2. LUZULA DC.

56.2.1. Luzula effusa Buchenau, *Krit. Verz. Juncac.* 53: 88 (1879).

Tufted stoloniferous suberect perennial rush to 0.5 m. Upper montane forest. Elevation: 2700–3400 m.

Material examined. EASTERN SHOULDER: 3400 m, *Smith, J. M. B. s.n.* (K); MESILAU RIVER: 2700 m, *Clemens 51371* (BM, K, L); SHEILA'S PLATEAU/SHANGRI LA VALLEY: 3400 m, *Fuchs & Collenette 21439* (L).

57. LIMNOCHARITACEAE

57.1. LIMNOCHARIS Humb. & Bonpl.

Buchenau, Fr. (1903). Butomaceae. *Pflanzenr.* IV. 16 (Heft 16): 1–12. Steenis, C. G. G. van. (1954). Butomaceae. *Fl. Males.* I, 5: 118–120 [*Limnocharis* on pp. 119–120].

57.1.1. Limnocharis flava (L.) Buchenau, *Abh. Naturw. Ver. Bremen* 2: 2 (1868).

Erect perennial herb to 0.5 m. Open wet situations at low elevations; weed in old rice paddys.

Material examined. BAMBANGAN (SERINSIM AREA): *Kinsun Bakia 166* (K); MELANGKAP TOMIS: *Lorence Lugas 58* (K), *1548* (K); PORING: *Meliden Giking 98* (K).

58. LOWIACEAE

58.1. ORCHIDANTHA N. E. Br.

Holttum, R. E. (1970). The genus *Orchidantha* (Lowiaceae). *Gard. Bull. Singapore* 25: 239–246, 1 pl.
Schumann, K. (1900). Musaceae. *Pflanzenr.* IV. 45 (Heft 1): 1–45 [*Orchidantha* on pp. 40–42].

58.1.1. Orchidantha sp.

Large perennial terrestrial herb. Lowland primary forest, apparently on ultra-mafic substrate. The specimen is inadequate for determination, but may represent an undescribed taxon.

Material examined. SERINSIM: *Kinsun Bakia 200* (K).

59. MARANTACEAE

In collaboration with E. J. Cowley (K)

Schumann, K. (1902). Marantaceae. *Pflanzenr.* IV. 48 (Heft 11): 1–184.

59.1. DONAX Lour.

59.1.1. Donax canniformis (G. Forst.) K. Schum., *Bot. Jahrb. Syst.* 15: 440 (1893).

Large perennial herb to 3 m. Lowlands, hill forest. Elevation: 500–600 m.

Material examined. BAMBANGAN (SERINSIM AREA): *Kinsun Bakia 4* (K); HIMBAAN: *Doinis Soibeh 320* (K); MELANGKAP TOMIS: *Lorence Lugas 69* (K), *498* (K), *860* (K), *1003* (K), *1388* (K), *2590* (K); MENGGIS: *Matamin Rumutom 49* (K); NALUMAD: *Daim Andau 14* (K); NAPUNG: *Dius Tadong 482* (K), *Sani Sambuling 152* (K), *510* (K); PINAWANTAI: 500 m, *Shea & Aban 76834* (L); PORING: *Meliden Giking 76* (K), *Sani Sambuling 51* (K), *726* (K); PORING HOT SPRINGS: 600 m, *Amin, Francis & Jarius 117389* (K), 600 m, *Beaman 7856* (K, MSC); SERINSIM: *Akungsai Bakia 32* (K); TEKUTAN: *Dius Tadong 211* (K), *346* (K), *540* (K), *588* (K), *822* (K), *Lomudin Tadong 8* (K), *305* (K).

59.2. PHACELOPHRYNIUM K. Schum.

59.2.1. Phacelophrynium maximum (Blume) K. Schum., *Pflanzenr.* IV. 48 (Heft 11): 122 (1902).

Herb to 2 m. Hill forest, sometimes on ultramafic substrate. Elevation: 900–1200 m. Leaves used for thatching.

Material examined. DALLAS: 900–1200 m, *Clemens 26272* (K), 900–1200 m, *26369* (BM), 900–1200 m, *26443* (BM), 900 m, *27204* (BM, K), 1100–1200 m, *51614* (BM); LIWAGU/MESILAU RIVERS: 1200 m, *RSNB 3004* (K); LOHAN/MAMUT COPPER MINE: 900 m, *Beaman 10606* (K, L, MSC); NALUMAD: *Daim Andau 175* (K); SERINSIM: *Jibrin Sibil 101* (K).

59.3. PHRYNIUM Willd.

59.3.1. Phrynium capitatum Willd., *Sp. Pl.* 1: 17 (1797).

Rhizomatous perennial herb 1–2 m high; spherical inforescence from about half way up the petiole. Primary and secondary hill forest. Elevation: 500–900 m.

Material examined. BUNGOL: 900 m, *Clemens 51605* (BM); DALLAS: 900 m, *Clemens 27388* (BM); MELANGKAP TOMIS: *Lorence Lugas 729* (K); RANAU: 500 m, *Darnton 131* (BM).

59.3.2. Phrynium aff. **hirtum** Ridl., *J. Straits Branch Roy. Asiat. Soc.* 32: 181 (1899).

Rhizomatous perennial herb 1–2 m high; spherical inflorescence from about half way up petiole or higher. Hill forest.

Material examined. MELANGKAP TOMIS: *Lorence Lugas 2075* (K).

59.3.3. Phrynium placentarium (Lour.) Merr., *Philipp. J. Sci.* 115: 230 (1919).

Rhizomatous perennial herb 1–2 m high; spherical inflorescence from about half way up the petiole. Lowlands, hill forest. Elevation: 400–1300 m.

Material examined. BAMBANGAN RIVER: 800–900 m, *Tamura & Hotta 446* (L); EASTERN SHOULDER: 600–1000 m, *Kokawa & Hotta 4963* (L); LUGAS HILL: 1300 m, *Beaman 8461* (MSC), 1300 m, *10554* (MSC); MEKEDEU RIVER: 400–500 m, *Beaman 9656* (MSC); PORING: 600–900 m, *Kokawa & Hotta 4843* (L); PORING HOT SPRINGS: 600 m, *Beaman 7566* (K, MSC); SERINSIM: *Kinsun Bakia 233* (K).

59.3.4. Phrynium cf. **placentarium** (Lour.) Merr., *Philipp. J. Sci.* 115: 230 (1919).

Rhizomatous perennial herb 1–2 m high; spherical inflorescence from about half way up the petiole or higher. Lowland forest.

Material examined. SERINSIM: *Kinsun Bakia 73* (K).

59.4. STACHYPHRYNIUM K. Schum.

59.4.1. Stachyphrynium cylindricum (Ridl.) K. Schum., *Pflanzenr.* IV. 48 (Heft 11): 49 (1902).

Perennial herb with short rhizome; to 1 m tall with cylindrical, bracteate inflorescence on peduncle 10–20 cm long. Lowland forest.

Material examined. MENGGIS: *Matamin Rumutom 11* (K); PORING: *Good, J. B. & Minol SAN 122489* (K); SERINSIM: *Kinsun Bakia 74* (K); TEKUTAN: *Dius Tadong 224* (K).

60. MELANTHIACEAE

In collaboration with B. Mathew (K)

60.1. ALETRIS L.

Jessop, J. P. (1979). Liliaceae–1. *Fl. Males.* I, 9: 230–232.

60.1.1. Aletris foliolosa Stapf, *Trans. Linn. Soc. London, Bot.* 4: 240 (1894). Plate 17C. Type: KEMBURONGOH, 2300 m, *Haviland 1125b* (syntype K!); MARAI PARAI SPUR, *Haviland 1125a* (syntype K!).

Aletris rigida Stapf, *Trans. Linn. Soc. London, Bot.* 4: 241 (1894). Type: MOUNT KINABALU, 3500 m, *Haviland 1125c* (syntype K!), 2400–3400 m, *Low s.n.* (syntype K!).

Erect, stemless, rhizomatous herb. Lower montane forest, upper montane forest, in open areas, especially on ultramafics and granitics. Elevation: 1200–3800 m. Also in Sumatra and in the Philippines on Mt. Halcon.

Additional material examined. EAST MESILAU RIVER: *Collenette 21610* (L); EASTERN SHOULDER: 3800 m, *Collenette 634* (K); EASTERN SHOULDER, CAMP 4: 3200 m, *Chew & Corner RNSB 1133* (L); EASTERN SHOULDER, CAMP 3: 2700 m, *RSNB 1091* (K, L); EASTERN SHOULDER, CAMP 4: 3200 m, *Chew, Corner & Stainton RNSB 1133* (K); GURULAU SPUR: 2400–2700 m, *Clemens 50605* (K), 3200 m, *51413* (BM); KEMBURONGOH: *Clemens, M. S. 10535* (BM, K); KEMBURONGOH/PAKA-PAKA CAVE: *Clemens 30043* (K); KEMBURONGOH/SUMMIT AREA: 3000–3400 m, *Clemens 27090* (BM); MARAI PARAI: 2700 m, *Clemens 31679* (BM), 1500 m, *32226* (BM, K, L), 3400 m, *32363* (BM, K, L), 1500 m, *Collenette A 21* (BM); MARAI PARAI SPUR: 1200–1700 m, *Gibbs 4041* (BM); MEKEDEU VALLEY: 3700 m, *RSNB 5985* (K); MESILAU CAVE: 2000–2100 m, *Beaman 9134* (L, MSC); MESILAU CAVE TRAIL: 1700–1900 m, *Beaman 7989* (K, L, MSC); MESILAU RIVER: 2000 m, *Clemens 50605* (BM); MOUNT KINABALU: *Burbidge s.n.* (K), 2700–3700 m, *Gibbs 4234* (BM); *Haslam s.n.* (BM, K, L), 2600–3000 m, *Rickards 160* (K), 3600 m, *Sinclair et al. 9167* (L), 2400 m, *Whitehead s.n.* (BM); PAKA-PAKA CAVE: 3000 m, *Clemens s.n.* (BM); PANAR LABAN: 3500 m, *Beaman 8298* (MSC), 3200 m, *Hou 251* (K); SHEILA'S PLATEAU/ SHANGRI LA VALLEY: 3300 m, *Collenette 21520* (L); VICTORIA PEAK: *Clemens 51413* (BM).

60.2. PETROSAVIA Becc.

Jessop, J. P. (1979). Liliaceae–1. *Fl. Males.* I, 9: 198–200.

60.2.1. Petrosavia stellaris Becc., *Nuovo Giorn. Bot. Ital.* 3: 88 (1871). Plate 17D.

Saprophytic echlorophyllous herb to 15 cm. Lower montane forest, upper montane forest. Elevation: 1400–3000 m.

Material examined. BAMBANGAN/MESILAU RIVERS: 1600–1700 m, *Kokawa & Hotta 4217* (L); EAST MESILAU/MENTEKI RIVERS: 1700 m, *Beaman 8755* (MSC); EASTERN SHOULDER: 1400 m, *RSNB 601* (K); KEMBURONGOH: 2100 m, *Price 222* (K); KILEMBUN BASIN: 1700 m, *Clemens 32016* (L), 1700 m, *40242* (BM); LUMU-LUMU: *Clemens s.n.* (BM); MARAI PARAI: 1500 m, *Clemens s.n.* (BM), 1500 m, *32454* (BM, K, L); MESILAU: *Collenette 515* (K); PAKA-PAKA CAVE: 3000 m, *Molesworth-Allen s.n.* (K).

61. MUSACEAE

Schumann, K. (1900). Musaceae. *Pflanzenr.* IV. 45 (Heft 1): 1–45.

61.1. MUSA L.

Cheesman, E. E. (1948). Classification of the bananas. III. Critical notes on species. *Kew Bull.* 2: 145–153.
Hotta, M. (1967). Notes on the wild banana of Borneo. *J. Jap. Bot.* 42: 344–352, 1 pl. Simmonds,
N. W. (1957). Botanical results of the banana collecting expedition, 1954-5. *Kew Bull.* 11: 463–489.
Simmonds, N. W. (1962). *The Evolution of the Bananas.* Longmans, Green and Co. Ltd., London.

61.1.1. Musa acuminata Colla, *Mem. Gen. Musa*: 66 (1820).

Tree-like herb to 4 m from a thick rhizome. Cultivated widely on Mount Kinabalu, with many forms. The nomenclature and genetics of the cultivated bananas are
highly complex. Simmonds (1962) suggests that scientific names not be used for
cultivars.

Material examined. MELANGKAP TOMIS: *Lorence Lugas 462* (K).

61.1.1. Musa acuminata Colla

a. subsp. **microcarpa** (Becc.) N. W. Simmonds, *Kew Bull.* 11: 467 (1957).

Large perennial herb in tree form from a thick rhizome. Hill forest. Elevation:
900 m.

Material. MINITINDUK: 900 m, *Carr SFN 26737* (SING? n.v., cited by Simmonds).

61.1.2. Musa textilis Née, *Anales Ci. Nat.* 4: 123 (1801).

Large perennial herb in tree form from a thick rhizome. Apparently common at
lower elevations on Mount Kinabalu, but ranging up to 1700 m in valley bottoms.
Elevation: 1700 m.

Material examined. LIWAGU RIVER TRAIL: 1700 m, *Beaman 11364* (K).

61.1.3. Musa sp. Plate 18A.

Perennial herb to 2 m high from a thick rhizome. Lower montane forest.
Elevation: 1500 m. Hotta applied the manuscript name "*M. monticola*" to one or more
specimens of this taxon.

Material examined. BAMBANGAN CAMP: 1500 m, *RSNB 4576* (K).

62. ORCHIDACEAE

Wood, J. J., Beaman, R. S., and Beaman, J. H. (1993). *The Plants of Mount Kinabalu, 2: Orchids*. Royal Botanic Gardens, Kew.

The family is not included in this study, because it has already been treated in the series on *The Plants of Mount Kinabalu*. In that publication the Orchidaceae had the family number 55. Families added as a result of the *PEK* collections now require that the family be renumbered to 62.

63. PANDANACEAE

In collaboration with B. C. Stone†

Stone, B. C. (1975). Studies in Malesian Pandanaceae XV. Two new species of *Pandanus* from Mt. Kinabalu, Sabah (Borneo) and notes on the Pandanaceae of Mount Kinabalu. *Malaysian J. Sci.* 3A: 69–74. Warburg, O. (1900). Pandanaceae. *Pflanzenr.* IV. 9 (Heft 3): 1–97.

63.1. FREYCINETIA Gaudich.

Stone, B. C. (1970). Materials for a monograph of *Freycinetia* Gaud. (Pandanaceae). VI. Species of Borneo. *Gard. Bull. Singapore* 25: 209–233. Stone, B. C. (1975). Studies in Malesian Pandanaceae XV. Two new species of *Pandanus* from Mt. Kinabalu, Sabah (Borneo) and notes on the Pandanaceae of Mt. Kinabalu. *Malaysian J. Sci.* 3A: 69–74.

63.1.1. Freycinetia gitingiana Martelli in Elmer, *Leafl. Philipp. Bot.* 3: 1112 (1911).

Woody liana climbing to several m high. Hill forest. Elevation: 800 m.

Material examined. KOKOHITAN: 800 m, *Carr SFN 27342* (SING).

63.1.2. Freycinetia cf. **insignis** Blume, *Rumphia* 1: 158, t. 42 (1837).

Climber. Primary forest.

Material examined. NALUMAD: *Daim Andau 13* (K).

63.1.3. Freycinetia javanica Blume, *Rumphia* 1: 156 (1837). Plate 18B.

Woody liana; fruit-heads ternate, subsessile, red when ripe. Lower montane forest. Elevation: 1500 m.

Material examined. TENOMPOK: 1500 m, *Clemens 29985* (BM, K).

63.1.4. Freycinetia kalimantanica B. C. Stone, *Gard. Bull. Singapore* 25: 230 (1970).

Woody climber; leaves virtually unarmed; leaf auricles fragmenting into segments by lateral splits. Hill forest. Elevation: 900–1200 m.

Material examined. DALLAS: 900–1200 m, *Clemens 26242* (BM).

63.1.5. Freycinetia kinabaluana B. C. Stone, *Gard. Bull. Singapore* 25: 219 (1970).

Plate 18D. Type: MESILAU RIVER, 1500 m, *RSNB 7013* (holotype K!; isotype L!).

Woody liana; fruit heads 4-7 together, oblong-ellipsoid. Hill forest, lower montane forest. Elevation: 900–1600 m.

Additional material examined. BAMBANGAN RIVER: 1500 m, *RSNB 4405* (K, L); KIAU: *Stone & Littke 11443* (KLU); LUMU-LUMU: *Carr SFN 27104* (SING); MESILAU RIVER: 1500 m, *RSNB 4177* (K), *RSNB 4306* (K); MOUNT KINABALU: *Aban Gibot SAN 79595* (K); PARK HEADQUARTERS: 1600 m, *Stone 11430* (KLU, L); PENATARAN BASIN: 900 m, *Clemens 34217* (BM, L); TENOMPOK: 1500 m, *Clemens 28711* (BM, K, L).

63.1.6. Freycinetia lucida Martelli, *Webbia* 3: 168 (1910).

Climbing vine. Lowlands or hill forest. Elevation: 400 m.

Material examined. MELANGKAP TOMIS: *Lorence Lugas 2462* (K); TARAWAS: 400 m, *Shea & Aban SAN 76727* (K).

63.1.7. Freycinetia palawanensis Elmer, *Leafl. Philipp. Bot.* 1: 216 (1907).

a. var. palawanensis

Woody liana. Lower montane forest. Elevation: 1200–1500 m.

Material examined. PARK HEADQUARTERS: *Aban Gibot SAN 79600* (K, L), *Stone 12922* (K); PENIBUKAN: 1200 m, *Clemens 30842* (BM), 1200–1500 m, *31359* (BM).

63.1.8. Freycinetia rigidifolia Hemsl., *Kew Bull.* 1896: 166 (1896).

Woody liana; leaf auricles pectinate. Hill forest, lower montane forest. Elevation: 900–2700 m.

Material examined. BUNDU TUHAN: *Doinis Soibeh 719* (K); DALLAS: 900 m, *Clemens 26378* (BM, K); MARAI PARAI: 2700 m, *Clemens 32385* (BM, L); MESILAU RIVER: 1500 m, *RSNB 4269* (K); MESILAU TRAIL: 1400 m, *Sadau SAN 42888* (K); PARK HEADQUARTERS/POWER STATION: 2000 m, *Mikil SAN 38635* (K); PENIBUKAN: 1200 m, *Clemens 40369* (BM), 1100 m, *40443* (BM, K), 1100 m, *40452* (BM, K), 1200 m, *40510* (BM, K), 1200 m, *50086* (BM); POWER STATION: 1800 m, *Stone 11401* (KLU); TENOMPOK: *Clemens s.n.* (BM, BM), 1200 m, *26075* (BM, K), 1500 m, *28750* (BM, L), 1500 m, *29331* (BM, K), 1500 m, *30175* (K, L).

63.1.9. Freycinetia tenuis Solms, *Linnaea* 42: 87 (1879).

Slender woody liana, leaves small, 7–18 cm long, 2–3 cm wide; fruit heads ternate, small, subglobose, 1–2 cm. Hill forest, lower montane forest. Elevation: 900–1800 m.

Material examined. DALLAS: 900–1200 m, *Clemens 26210* (BM, K, L), 900 m, *26734* (BM); KILEMBUN BASIN: *Lorence Lugas 1861* (K); PENIBUKAN: 1200 m, *Clemens 31038* (BM, L), 1200 m, *32135* (BM); TINEKEK FALLS: 1800 m, *Clemens 50137* (BM), 1500 m, *50137a* (BM).

63.2. PANDANUS Parkinson

Stone, B. C. (1975). Studies in Malesian Pandanaceae XV. Two new species of *Pandanus* from Mt. Kinabalu, Sabah (Borneo) and notes on the Pandanaceae of Mt. Kinabalu. *Malaysian J. Sci.* 3A: 69–74. Stone, B. C. (1993). Studies in Malesian Pandanaceae 21. The genus *Pandanus* in Borneo. *Sandakania* 2: 35–84.

63.2.1. Pandanus beccatus B. C. Stone, *Sandakania* 2: 65 (1993). Type: MARAI PARAI, *Ansow, Kulis & Tan SNP 1054* (holotype SNP!).

Erect shrub to 4.5 m tall, trunks slender; apical ventral leaf pleats prickly; fruit head oblong-ellipsoid. Hill forest, lower montane forest, sometimes on ultramafic substrate. Elevation: 900 m.

Additional material examined. DALLAS: 900 m, *Clemens 26609 p.p.* (BO); LIWAGU RIVER: *Aban Gibot SAN 79596* (SAN); MARAI PARAI: *Clemens 32348* (G-DEL).

63.2.2. Pandanus gibbsianus Martelli in Gibbs, *J. Linn. Soc., Bot.* 42: 170 (1914).

Large, almost treelike shrub, usually over 3 m tall with stout trunk. Lowlands.

Material examined. PERANCANGAN: *Shea & Aban SAN 76376* (SAN).

63.2.3. Pandanus kinabaluensis H. St. John ex B. C. Stone, *Malaysian J. Sci. 3A*: 73 (1975). Type: POWER STATION, *Clemens 30176* (holotype G-DEL!); TENOMPOK, *Clemens 30176* (isotype? K!).

Low, erect rosette-forming shrub; apical ventral leaf pleats not prickly; fruit head oblong. Lower montane forest. Elevation: 1200–1600 m. The K and G-Del specimens of *Clemens 30176* show different localities, so it is uncertain if the K specimen is an isotype.

Additional material examined. MESILAU: 1500 m, *RSNB 8464* (K, L), 1500 m, *RSNB 8465* (K); PARK HEADQUARTERS: *Stone 11325* (KLU), *11326* (KLU, L), *11327* (KLU, L), *11429* (KLU), *11442* (KLU), 1600 m, *12923* (L); PENIBUKAN: 1200–1500 m, *Clemens 31376* (BM, K, L); TENOMPOK: 1500 m, *Clemens 29550* (BM), 1500 m, *29811* (BM), 1500 m, *30176* (K).

63.2.4. Pandanus leuconotus B. C. Stone, *J. Arnold Arbor.* 64: 319 (1983). Type: PORING HOT SPRINGS, 300 m, *Stone 12906* (holotype KLU!).

Short, erect, broad-leaved shrub; leaves white toward the base beneath; fruit-head short-oblong. Lowlands, hill forest. Elevation: 300–900 m. The elevation indicated for Poring is about 250 m too low.

Additional material examined. DALLAS: 900 m, *Clemens 26857* (BM, K), 900 m, *27751* (BM); PORING HOT SPRINGS: 300 m, *Stone 12905* (KLU, L).

63.2.5. Pandanus pectinatus Martelli, *Bol. Soc. Bot. Ital. n.s.* 11: 304 (1904).

Pandanus papilio B. C. Stone, *Malayan Scientist* 3: 24 (1967). Type: BAMBANGAN RIVER, 1500 m, *RSNB 4403* (holotype K!).

Epiphyte with creeping stem and short erect leafy tufts, stilt roots lacking, aerial feeding roots present. Hill forest, lower montane forest. Elevation: 900–2100 m.

Additional material examined. CASCADE TRAIL: *Peter William SNP 706* (SNP); DALLAS: 900–1200 m, *Clemens 26210* (K); EASTERN SHOULDER: 1500 m, *RSNB 1569a* (K); GURULAU SPUR: 1500 m, *Clemens 50546* (BM, K); KIAU VIEW TRAIL: 1600 m, *Stone & Littke 11444* (L); KILEMBUN RIVER: 1500–1800 m, *Clemens 34489* (BM); KULUNG HILL: *Meijer SAN 23451* (SAN); MESILAU RIVER: 1500 m, *RSNB 4158* (K), 1500 m, *RSNB 4206* (K); PARK HEADQUARTERS: *Cockburn SAN 71900* (SAN), *Stone et al. 11408* (KLU); PENATARAN RIVER: 2100 m, *Clemens 32549* (BM, K, L); PENIBUKAN: 1200–1500 m, *Clemens 30716* (BM, K, L), 1400 m, *40600* (K); PINOSUK PLATEAU: 1600 m, *RSNB 1780* (K); PORING: *Meijer SAN 34646* (SAN); SOSOPODON: 1200 m, *Pereira SAN 47214* (K); TAHUBANG RIVER: *Carr SFN 26384* (SING); TENOMPOK: 1500 m, *Clemens s.n.* (BM), 1500 m, *28726* (BM), *28801* (BO), 1500 m, *28832* (BM, L).

63.2.6. Pandanus pumilus H. St. John, *Pacif. Sci.* 19: 96, f. 206 (1965).

Shrub? Hill forest on ultramafic substrate.

Material examined. MELANGKAP TOMIS: *Lorence Lugas 1642* (K), *2528* (K).

63.2.7. Pandanus rusticus B. C. Stone, *Fed. Mus. J. n.s.* 14: 131 (1972).

Erect shrub with stems to 3 m high; leaves with prickly ventral pleats; fruit-head oblong, on a long slender peduncle. Hill forest, lower montane forest. Elevation: 900–1500 m.

Material examined. DALLAS: 900–1200 m, *Clemens 27053* (BM, K); MELANGKAP TOMIS: *Lorence Lugas 995* (K); PARK HEADQUARTERS: *Stone et al. 11425* (KLU); TENOMPOK: 1500 m, *Clemens 29550* (K).

63.2.8. Pandanus tectorius Parkinson in Z [J. P. du Roi], *Naturforscher (Halle)* 4: 250 (1774).

a. 'Laevis'

Branched tree to 6 m high. Cultivated. For explanation of authorship of the species name cf. Stone, B. C. (1981). *Notes Waimea Arbor. Bot. Gard.* 8(2): 4–10.

Material examined. MELANGKAP TOMIS: *Lorence Lugas 2472* (K).

63.2.9. Pandanus tunicatus B. C. Stone, *Malaysian J. Sci.* 3A: 70 (1975). Type: PARK HEADQUARTERS, 1500 m, *Stone et al. 11437* (isotypes K!, L!).

Short-trunked, rosette-forming shrub; leaves with prickly apical pleats; fruit-head ellipsoidal. Hill forest, lower montane forest, often on ultramafic substrate. Elevation: 900–1500 m.

Additional material examined. BAMBANGAN RIVER: 1500 m, *RSNB 4588* (K); DALLAS: 900–1200 m, *Clemens 26178* (BM, K, L); MARAI PARAI: 1500 m, *Clemens 32343* (BM); MARAI PARAI SPUR: 1500 m, *Clemens 32649* (BM, K, L); MELANGKAP TOMIS: *Lorence Lugas 2407* (K), *2545* (K); MESILAU RIVER: 1500 m, *RSNB 7020* (K); TEKUTAN: *Dius Tadong 237* (K); TENOMPOK: 1500 m, *Clemens 30176 p.p.* (L).

63.2.10. Pandanus indet.

Material examined. MELANGKAP TOMIS: *Lorence Lugas 2089* (K).

64. PHORMIACEAE

In collaboration with E. J. Cowley (K) and B. Mathew (K)

64.1. DIANELLA Lam.

Jessop, J. P. (1979). Liliaceae–1. *Fl. Males.* I, 9: 206–209. Schlittler, J. (1947–48). Unsere gegenwaertige Kenntnis ueber die Liliaceengattung *Dianella* in Malesien. *Blumea* 6: 200–228.

64.1.1. Dianella ensifolia (L.) DC. in Redouté, *Liliac.* 1, t. 1 (1802). Plate 18C.

Erect, rhizomatous perennial herb to 1 m. Hill forest, lower montane forest. Elevation: 800–3000 m. Many variants have been recognized at the rank of forma by Schlittler (1947–48).

Material examined. BAMBANGAN RIVER: 1600 m, *RSNB 4959* (K), 1600 m, *RSNB 4969* (K); DALLAS: 900 m, *Clemens 26741* (BM), 900 m, *26873* (BM); DALLAS/TENOMPOK: 900–1200 m, *Clemens 30046* (K, L); GURULAU SPUR: 1400–1500 m, *Kanis & Kuripin SAN 53968* (K); HEMPUEN HILL: 800–1000 m, *Beaman 7400* (L, MSC); KEMBURONGOH: 2100 m, *Price 196* (K); KIAU: *Jusimin Duaneh 508* (K); KULUNG HILL: 800 m, *Beaman 7806* (MSC); LUMU-LUMU: 2100–3000 m, *Clemens s.n.* (BM); MARAI PARAI: 1500 m, *Clemens 32234* (BM, L), 1500 m, *Collenette A 20* (BM); MELANGKAP TOMIS: *Lorence Lugas 65* (K), *565* (K), *915* (K), *1278* (K), *1895* (K); MENGGIS: *Matamin Rumutom 71* (K); MESILAU RIVER: 1500 m, *RSNB 4090* (K); MOUNT KINABALU: 800 m, *Puasa 1548* (L); NALUMAD: *Amin & Jarius SAN 114383* (K); PARK HEADQUARTERS: 1400 m, *Price 129* (K); PENIBUKAN: 1200 m, *Clemens 35030* (BM); PIG HILL: 2000–2300 m, *Beaman 9839* (K, L, MSC); POWER STATION/KEMBURONGOH: 2000 m, *Fuchs 21049* (K, L); SERINSIM: *Kinsun Bakia 435* (K); TEKUTAN: *Lomudin Tadong 487* (K).

64.1.2. Dianella javanica (Blume) Kunth, *Enum. Pl.* 5: 52 (1850). Plate 19A.

Caulescent, rhizomatous perennial herb to 1 m. Hill forest, lower montane forest, upper montane forest. Elevation: 1500–3000 m.

Material examined. BAMBANGAN RIVER: 1500 m, *RSNB 4401* (K), 1500 m, *Collenette 2377* (K); GURULAU SPUR: 2400 m, *Clemens 50929* (BM); KEMBURONGOH: 2100 m, *Price 203* (K), 2100 m, *208* (K); KILEMBUN RIVER: 2700–3000 m, *Clemens 33812* (BM, L); KILEMBUN RIVER HEAD: 2700–3000 m, *Clemens 33917* (BM); LAYANG-LAYANG: 2400–2600 m, *Hotta 3812* (L); LUBANG: 1700 m, *Gibbs 4157* (BM); MESILAU BASIN: 2700–3000 m, *Clemens s.n.* (BM); MESILAU CAVE: 2000–2100 m, *Beaman 9140* (L, MSC), 1900–2200 m, *9537* (K, L, MSC); MOUNT KINABALU: 2400 m, *Burbidge s.n.* (K).

65. POACEAE

In collaboration with S. Dransfield (K), bamboos, and J. F. Veldkamp (L), other grasses

Dransfield, S. (1992). *The Bamboos of Sabah*. Sabah Forest Records 14, xii + 94 pp. Gilliland, H. B. (1971). *A Revised Flora of Malaya. Vol. 3. Grasses of Malaya* (with contributions by R. E. Holttum & N. L. Bor, edited by H. M. Burkill). Botanic Gardens, Singapore. Holttum, R. E. (1958). The bamboos of the Malay Peninsula. *Gard. Bull. Singapore* 16: 1–135. Lazarides, M. (1980). The Tropical Grasses of Southeast Asia (Excluding Bamboos). *Phan. Monogr.* 12. J. Cramer, Vaduz.

65.1. AGROSTIS L.

Veldkamp, J. F. (1982). *Agrostis* (Gramineae) in Malesia and Taiwan. *Blumea* 28: 199–228.

65.1.1. Agrostis infirma Buse

a. var. **borneensis** (Stapf) Veldkamp, *Blumea* 41: 408 (1996).

Agrostis canina L. var. *borneensis* Stapf, *Trans. Linn. Soc. London, Bot.* 4: 246 (1894). Type: MOUNT KINABALU, 4000 m, *Haviland 1399* (holotype K!).
Agrostis rigidula Steud. var. *borneensis* (Stapf) J. M. Linden & Voskuil in Veldkamp, *Blumea* 28: 217 (1982).

Erect, tufted perennial grass 20–30 cm tall. Summit area on granitic rocks. Elevation: 4000–4100 m. Endemic to Mount Kinabalu.

Additional material examined. LOW'S PEAK: 4100 m, *Clemens 27072* (BM, K), 4000 m, *30273* (K), 4000 m, *Sinclair et al. 9162* (K, L); PAKA-PAKA CAVE/LOW'S PEAK: *Clemens, M. S. 10630* (K); SUMMIT AREA: 4000 m, *Clemens 29167* (BM, K, L); UPPER KINABALU: *Clemens s.n.* (L), *27072* (K), *29167a* (L), *29167b* (L).

b. var. **diffusissima** (Ohwi) Veldkamp, *Blumea* 41: 408 (1996).

Agrostis reinwardtii Buse var. *diffusissima* Ohwi, *Bull. Tokyo Sci. Mus.* 18: 8 (1947). Type: MARAI PARAI, 3200 m, *Clemens 33228* (holotype BO n.v.; isotypes BM!, K!, L!).
Agrostis rigidula Steud. var. *diffusissima* (Ohwi) Veldkamp, *Blumea* 28: 218 (1982).

Small tussock grass, tufts with 3 or 4 inflorescences, sometimes forming dense mats. Upper montane forest, summit area, on stream banks and in granitic depressions and crevices. Elevation: 2700–4000 m. Endemic to Mount Kinabalu.

Additional material examined. EASTERN SHOULDER: 3000 m, *RSNB 922* (K, L); KADAMAIAN RIVER: 3200 m, *Clemens 50914A* (K, L); KADAMAIAN RIVER HEAD: 3000–3400 m, *Clemens 50914* (BM, K, L); LAYANG-LAYANG: 2700 m, *Smith 458* (L); MEKEDEU RIVER: 3400–3800 m, *RSNB 5958* (K); MOUNT KINABALU: 3000 m, *Meijer SAN 22020* (K); PAKA-PAKA CAVE: 3100 m, *Cockburn & Aban SAN 82998* (K, L), 2700 m, *Forster F 42* (K), *Meijer SAN 54278* (L), 3000 m, *Sinclair et al. 9122* (K, L); PAKA-PAKA CAVE/LOW'S PEAK: *Clemens, M. S. 10631* (K); PANAR LABAN: 3400 m, *Stein Kipl 225* (L), 3400 m, *Stone 11329* (L); SAYAT-SAYAT: 3700 m, *Meijer SAN 24217* (K, L), 3800 m, *Smith 513* (L); SUMMIT AREA: 3700–4000 m, *Molesworth-Allen 3288* (K, L); UPPER KINABALU: *Clemens 27771* (K), *29175* (L).

c. var. **kinabaluensis** (Ohwi) Veldkamp, *Blumea* 41: 409 (1996).

Agrostis kinabaluensis Ohwi, *Bull. Tokyo Sci. Mus.* 18: 8 (1947). Type: LOW'S PEAK, 4000 m, *Clemens 30273* (holotype BO n.v.; isotypes K!, L!).
Agrostis rigidula Steud. var. *kinabaluensis* (Ohwi) Veldkamp, *Blumea* 28: 220 (1982).

Erect, densely tufted perennial grass, 10–40 cm tall, leaves subacicular. Summit area, in crevices on steep slopes and in shallow, moist places near the edge of scrub. Elevation: 3400–4100 m. Endemic to Mount Kinabalu.

Additional material examined. MEKEDEU VALLEY/KING GEORGE PEAK: 3500–4100 m, *RSNB 5968* (K); PAKA-PAKA CAVE/LOW'S PEAK: *Clemens, M. S. 10630* (K); SAYAT-SAYAT: 3800 m, *Smith 507* (L); SHANGRI LA VALLEY: 3400 m, *Collenette 21511A* (L); UPPER KINABALU: *Clemens 27072* (L), *27771* (L), *29167a* (K), *29167d* (K, L), *29175* (K), *30312* (K, L).

65.1.2. Agrostis stolonifera L.

a. var. **ramosa** (Gray) Veldkamp, *Blumea* 28: 223 (1982).

Perennial grass, branching at base, stoloniferous, rhizomes absent. Upper montane forest in disturbed, grassy areas. Elevation: 2700–3200 m.

Material examined. LAYANG-LAYANG: 2700 m, *Smith 461* (L); PAKA-PAKA CAVE: 3200 m, *Clemens 27987* (BM, K, L), 2700 m, *Forster F 43* (K).

65.2. ANISELYTRON Merr.

Korthof, A. M. and Veldkamp, J. F. (1985). A revision of *Aniselytron* with some new combinations in *Deyeuxia* in S. E. Asia (Gramineae). *Gard. Bull. Singapore* 37: 213–223.

65.2.1. Aniselytron treutleri (Kuntze) Sojàk, *Čas. Nár. Mus., Odd. Přr.* 148: 202 (1980).

Aulacolepis clemensiae Hitchc., *J. Wash. Acad. Sci.* 24: 290 (1934). Type: MESILAU RIVER, 2100 m, *Clemens 34448* (holotype US!; isotypes BM!, K!, L!).

Tufted perennial grass to 1 m high. Lower or upper montane forest among river boulders. Elevation: 2100 m.

Material examined. MESILAU BASIN: *Clemens 29692* (BM); MESILAU RIVER: 2100 m, *Clemens 34448A* (L).

65.3. ANTHOXANTHUM L.

Schouten, Y. & Veldkamp, J. F. (1985). A revision of *Anthoxanthum* including *Hierochloë* (Gramineae) in Malesia and Thailand. *Blumea* 30: 319–351.

65.3.1. Anthoxanthum horsfieldii (Kunth ex Benn.) Mez ex Reeder

a. var. borneense (Jansen) Y. Schouten in Y. Schouten & Veldkamp, *Blumea* 30: 335 (1985). Type: SUMMIT AREA, 3700–4000 m, *Clemens 29176* (holotype BO!; isotypes BM!, K!, L!).

Anthoxanthum angustum (Hitchc.) Ohwi var. *borneense* Ohwi ex Jansen, *Reinwardtia* 2: 227 (1953).

Erect, tufted perennial grass with simple scaberulous culms. Upper montane forest, summit area, in open, moist, sometimes disturbed areas. Elevation: 3000–4000 m. Endemic to Mount Kinabalu.

Additional material examined. MOUNT KINABALU: *Clemens s.n.* (K), 3000 m, *Meijer SAN 22080* (K); OYAYUBI IWU PEAK: 3900 m, *Smith 556* (L); PANAR LABAN: 3400 m, *Stein Kipl 227* (L); SAYAT-SAYAT: 3800 m, *Smith 514* (L); SUMMIT AREA: 3800–4000 m, *Molesworth-Allen s.n.* (K); VICTORIA PEAK: 3700 m, *Clemens 51402* (BM, K).

65.4. APOCOPIS Nees

65.4.1. Apocopis collinus Balansa, *J. Bot. (Morot)* 4: 84 (1890).

Tufted perennial grass. Hill forest on ultramafic substrate.

Material examined. MELANGKAP TOMIS: *Lorence Lugas 586* (K).

65.5. BAMBUSA Schreb.

65.5.1. Bambusa vulgaris J. C. Wendl., *Coll. Pl.* 2: 26, t. 47 (1808).

Open, tufted bamboo to 20 m tall; culms non-spiny at the base. Lowlands and hill forest in wet, disturbed situations. Planted or naturalized.

Material examined. KITUNTUL: *Tiong & George SAN 88042* (K), *SAN 88045* (K); PORING: *Meliden Giking 78* (K), *102* (K).

65.6. BROMUS L.

Veldkamp, J. F., Eriks, M. and Smit, S. S. (1991). *Bromus* (Gramineae) in Malesia. *Blumea* 35: 483–497.

65.6.1. Bromus formosanus Honda, *Bot. Mag. (Tokyo)* 42: 136 (1928).

Bromus insignis Buse var. *kinabaluensis* Jansen, *Reinwardtia* 2: 245 (1953). Type: VICTORIA PEAK, *Clemens 29174* (holotype L!; isotypes BM!, K!).
Bromus kinabaluensis (Jansen) Veldkamp in Veldkamp, Eriks & Smit, *Blumea* 35: 492 (1991).

Tufted perennial grass branching at the base; culms few, erect, often geniculate at base, simple, up to 1.3 m tall, unbranched. Summit area in low heath on rubbly scree. Elevation: 3700–4000 m.

Additional material examined. SUMMIT AREA: 3700–4000 m, *Clemens s.n.* (BM), 4000 m, *Smith 554* (L); UPPER KINABALU: *Clemens 30134* (K, L), *30134 bis* (K, L).

65.7. CENTOTHECA Desv.

Monod de Froideville, C. (1971). Notes on Malesian grasses IV: A synopsis of *Centotheca* and reduction of *Ramosia*. *Blumea* 19: 57–60.

65.7.1. Centotheca lappacea (L.) Desv., *Nouv. Bull. Sci. Soc. Philom. Paris* 2: 189 (1810).

Tufted perennial grass to 1 m high. Lowlands, hill forest, in secondary vegetation. Elevation: 1200–1500 m.

Material examined. HIMBAAN: *Doinis Soibeh 131* (K); MARAK-PARAK: *Kinsun Bakia 238* (K); MELANGKAP TOMIS: *Lorence Lugas 433* (K), *1124* (K), *1841* (K); NALUMAD: *Daim Andau 44* (K); SERINSIM: *Jibrin Sibil 80* (K); TAHUBANG FALLS: 1200–1500 m, *Clemens 40275* (BM), 1500 m, *40278* (BM); TEKUTAN: *Dius Tadong 800* (K), *Lomudin Tadong 167* (K).

65.8. COIX L.

65.8.1. Coix lacryma-jobi L., *Sp. Pl.*: 972 (1753).

Large annual or perennial grass with branched culm. Lowlands, hill forest in wet situations by rice paddys or along streams.

Material examined. BUNDU TUHAN: *Doinis Soibeh 703* (K); HIMBAAN: *Doinis Soibeh 152* (K); KIAU: *Jusimin Duaneh 248* (K), *544* (K); MARAK-PARAK: *Jibrin Sibil 142* (K); MELANGKAP TOMIS: *Lorence Lugas 46* (K), *824* (K), *1344* (K), *1666* (K), *2364* (K); NALUMAD: *Daim Andau 70* (K), *Sigin et al. 112266* (K); NAPUNG: *Sani Sambuling 497* (K); SAYAP: *Tungking Simbayan 29* (K), *Yalin Surunda 33* (K); SERINSIM: *Jibrin Sibil 235* (K), *Kinsun Bakia 192* (K), *210* (K), *323* (K); TEKUTAN: *Dius Tadong 168* (K), *Lomudin Tadong 307* (K).

65.9. CYMBOPOGON Spreng.

Soenarko, S. (1977). The genus *Cymbopogon* Sprengel (Gramineae). *Reinwardtia* 9: 225–375.

65.9.1. Cymbopogon winterianus Jowitt, *Ann. Roy. Bot. Gard. (Peradeniya)* 4: 188, 189 (1908).

Large, tufted, aromatic perennial grass 2–2.5 m high. Lowlands and hill forest in cultivation or in secondary situations. According to Soenarko (1977) the species is known only in cultivation; the source of citronella oil.

Material examined. MELANGKAP TOMIS: *Lorence Lugas 1026* (K); SERINSIM: *Kinsun Bakia 308* (K).

65.10. CYRTOCOCCUM Stapf in Prain

65.10.1. Cyrtococcum accrescens (Trin.) Stapf, *Hooker's Icon. Pl.* 31: sub t. 3096 (1922).

Decumbent perennial grass to 1 m, rooting from the lower nodes; inflorescence large and open. Hill forest, lower montane forest, in open areas. Elevation: 300–1200 m.

Material examined. DALLAS: 900 m, *Clemens 27595* (K, L), 900 m, *30281* (K, L); DALLAS/TENOMPOK: 1200 m, *Clemens 27595* (BM); KEBAYAU: 300 m, *Clemens 27687* (BM, K, L); NALUMAD: *Daim Andau 173* (K); PORING: *Meliden Giking 210* (K); SERINSIM: *Jibrin Sibil 73* (K); SINGGAREN: 500 m, *Forster F 37A* (K); TENOMPOK: 1200 m, *Clemens 51048* (BM, L).

65.10.2. Cyrtococcum patens (L.) A. Camus, *Bull. Mus. Hist. Nat. (Paris)* 27: 118 (1921).

Low, decumbent perennial grass; inflorescence relatively narrow and strict. Hill forest in open areas. Elevation: 800–900 m.

Material examined. Dallas: 900 m, *Clemens 28301* (BM, K, L), 900 m, *30280* (K, L); Kadamaian River: 800 m, *Clemens 51622* (BM, L); Serinsim: *Kinsun Bakia 413* (K).

65.11. DACTYLOCTENIUM Willd.

65.11.1. Dactyloctenium aegyptium (L.) Willd., *Enum. Pl.*: 1029 (1809).

Creeping stoloniferous annual or perennial grass, rooting at the nodes, with erect or suberect culms. Lowlands in secondary situations.

Material examined. Marak-Parak: *Kinsun Bakia 145* (K), *251* (K).

65.12. DANTHONIA DC. in Lam. & DC.

65.12.1. Danthonia oreoboloides (F. Muell.) Stapf, *Hooker's Icon. Pl.* 27: t. 2606 (1899).

Low mat-forming tussock grass. Upper montane forest, summit area, in open wet places, often in granitic crevices. Elevation: 2400–4100 m.

Material examined. Gurulau Spur: 3400 m, *Clemens 50905* (BM, K, L); Kemburongoh/ Layang-Layang: 2400 m, *Meijer SAN 21056* (L); King George Peak: 4100 m, *RSNB 5973* (K); Low's Peak: 4100 m, *Clemens 27075* (BM, K, L); Marai Parai: *Clemens 32343* (L), 3200 m, *32343A* (BM); Paka-paka Cave/Low's Peak: *Clemens, M. S. 10628* (BM, K); Paka-paka Cave/Panar Laban: 3200 m, *Sinclair et al. 9142* (K, L); Panar Laban: 3300 m, *Jermy & Rankin J 15103* (BM); Sayat-Sayat: 3800 m, *Meijer SAN 28567* (K), 3800 m, *Smith 517* (L); Summit Area: *Clemens s.n.* (BM), 3700–4000 m, *27778* (BM, K, L), 3800 m, *Hou 257* (K, L), 3800 m, *Sleumer 4721* (L); Upper Kinabalu: *Clemens 30268* (K, L), *30313* (L); Victoria Peak: 3800 m, *Clemens 51414* (BM, K, L), 3800 m, *51528* (BM).

65.13. DESCHAMPSIA Beauv.

65.13.1. Deschampsia flexuosa (L.) Trin.

a. var. **ligulata** Stapf, *Trans. Linn. Soc. London, Bot.* 4: 248 (1894). Type: Mount Kinabalu, 4000 m, *Haviland 1400* (holotype K!)

Erect, densely tufted perennial grass 10–50 cm tall. Upper montane forest, summit area where abundant in granitic crevices and shallow depressions. Elevation: 3000–4100 m. Also known from Mt. Pulog in the Philippines.

Material examined. King George Peak: 4000 m, *RSNB 5967* (K); Low's Peak: 4100 m, *Clemens 27073* (BM, K, L); Paka-paka Cave/Low's Peak: *Clemens, M. S. 10610* (K); Paka-paka Cave/Summit Area: 3000–3700 m, *Gibbs 4192* (BM, K); Panar Laban: 3300 m, *Smith 467* (L); Sayat-sayat: 3700 m, *Meijer SAN 24207* (K, L), 3800 m, *Smith 511* (L); Shangri La Valley: 3400 m, *Collenette 21511* (L); Summit Area: 3700–4000 m, *Clemens 27786* (BM, K, L), 4000–4100 m, *Fosberg 44105* (US), 3700–4000 m, *Gibbs 4185* (BM, K), 4000–4100 m, *Johnson, D. H. s.n.* (US), 3800–4000 m, *Molesworth-Allen 3289* (K, L), *Sinclair et al.* *9153* (K, L, US), 3600–3800 m, *Sleumer 4716* (K, L); SUMMIT TRAIL: 3700 m, *Fuchs 21089* (K, L, US), 3500 m, *Jacobs 5756* (K, L, US); UPPER KINABALU: *Clemens 30259* (K, L), *30260* (K, L); VICTORIA PEAK: 3700–3800 m, *Clemens 51400* (BM, L, US).

65.14. DIGITARIA Heist. ex Fabr.

Henrard, J. Th. (1950). *Monograph of the genus* Digitaria. Leiden. 999 pp. Veldkamp, J. F. (1973). A revision of *Digitaria* Haller (Gramineae) in Malesia. *Blumea* 21: 1–80 (1973).

65.14.1. Digitaria junghuhniana (Steud.) Henrard, *Meded. Rijks-Herb.* 61: 11 (1930).

Rhizomatous perennial grass. Open, disturbed areas in the lowlands and lower montane forest. Elevation: 1500 m.

Material examined. KAUNG: *Forster F 109* (K); TENOMPOK: 1500 m, *Clemens 30286* (K, L).

65.14.2. Digitaria setigera Roth ex Roem. & Schult., *Syst. Veg.* 2: 474 (1817).

a. var. setigera

Decumbent, stoloniferous perennial grass, sometimes rooting at the nodes, to 1 m high. Lowlands in secondary situations.

Material examined. SERINSIM: *Kinsun Bakia 237* (K).

65.14.3. Digitaria violascens Link, *Hort. Reg. Bot. Berol.* 1: 229 (1827).

Low, delicate stoloniferous perennial grass. Lowlands, hill forest, in waste places. Elevation: 300 m.

Material examined. KAUNG/KEBAYAU: 300 m, *Clemens 27673* (BM, K, L).

65.15. DIMERIA R. Br.

65.15.1. Dimeria ornithopoda Trin., *Fund. Agrost.*: 167, t. 14 (1820).

Delicate narrow-leaved annual grass. Lowlands, hill forest, lower montane forest, in disturbed situations such as fallow paddy soil. Elevation: 200–1500 m.

Material examined. KAUNG: 200 m, *Forster F 56* (K); KEBAYAU/KAUNG: 300 m, *Clemens 27661* (BM, K, L); TENOMPOK: 1500 m, *Clemens 30316* (K, L).

65.16. DINOCHLOA Buse

Dransfield, S. (1981). The genus *Dinochloa* in Sabah. *Kew Bull.* 36: 613–633.

65.16.1. Dinochloa sublaevigata S. Dransf., *Kew Bull.* 36: 626 (1981). Plate 19B.
Type: PORING, 600 m, *Dransfield, S. SD 720* (holotype K!; isotype L!).

Climbing bamboo with large leaf blades, a huge inflorescence and globose fruits. Lowlands, hill forest, in disturbed, secondary situations. Elevation: 600–1100 m.

Additional material examined. EASTERN SHOULDER, CAMP 1: 900 m, *RSNB 1179* (K, L, US); KULUNG HILL: *Sinclair et al. 9274* (K, US); LIWAGU RIVER: 1100 m, *RSNB 5724* (K, US); NALUMAD: *Daim Andau 4* (K); PERANCANGAN: *Dius Tadong 778* (K); PORING: 700 m, *Beaman 11030* (K), *Meliden Giking 39* (K); RANAU/PORING ROAD: 700 m, *Dransfield, J. et al. JD 5506* (K, L); TEKUTAN: *Dius Tadong 100* (K), *499* (K), *Lomudin Tadong 276* (K); TINEKEK RIVER: 900 m, *Haviland 1390* (K).

65.17. ECHINOCHLOA P. Beauv.

65.17.1. Echinochloa colona (L.) Link, *Hort. Berol.* 2: 209 (1833).

Weak stoloniferous grass, rooting at the nodes, to 0.7 m high. Lowlands in open, wet situations.

Material examined. KIAS: *Jibrin Sibil 297* (K), *Kinsun Bakia 212* (K); SERINSIM: *Kinsun Bakia 252* (K).

65.18. ELEUSINE Gaertn.

65.18.1. Eleusine indica (L.) Gaertn., *Fruct. Sem. Pl.* 1: 8 (1788).

Weedy stoloniferous grass to 0.7 m high. Lowlands and hill forest along road-sides, in cultivated areas and other open situations and thickets.

Material examined. HIMBAAN: *Doinis Soibeh 10* (K); MELANGKAP TOMIS: *Lorence Lugas 292* (K), *1249* (K), *2558* (K); MENGGIS: *Matamin Rumutom 236* (K); PORING: *Meliden Giking 202* (K); SAYAP: *Yalin Surunda 30* (K); SERINSIM: *Jibrin Sibil 10* (K), *115* (K), *185* (K), *205* (K), *283* (K), *Kinsun Bakia 206* (K); TEGIS: *Dius Tadong 131* (K); TEKUTAN: *Dius Tadong 273* (K), *303* (K), *Lomudin Tadong 38* (K).

65.19. ERAGROSTIS Wolf

65.19.1. Eragrostis atrovirens (Desf.) Trin. ex Steud., *Nomencl. Bot.,* ed. 2, 1: 562 (1840).

Tufted, erect perennial grass to 1.35 m high. Hill forest in open, rather barren secondary situations.

Material examined. MELANGKAP TOMIS: *Lorence Lugas 1857* (K).

65.19.2. Eragrostis unioloides (Retz.) Nees ex Steud., *Syn. Pl. Glumac.* 1: 264 (1854).

Tufted erect perennial grass to 0.6 m high. Lowlands in thickets.

Material examined. TEKUTAN: *Lomudin Tadong 312* (K).

65.20. GIGANTOCHLOA Munro

Widjaja, E. A. (1987). A revision of Malesian *Gigantochloa* (Poaceae-Bambusoideae). *Reinwardtia* 10: 291–380.

65.20.1. Gigantochloa levis (Blanco) Merr., *Amer. J. Bot.* 3: 61 (1916). Plate 19C.

Erect, medium to large bamboo up to 25 m tall, culms 10–15 cm in diameter. Lowlands, hill forest. Elevation: 900–1500 m. Planted or naturalized; the species with which the common name poring is generally associated.

Material examined. DALLAS: 900–1200 m, *Clemens 26324* (BM, K); KIAU: *Dransfield, S. SD 717* (K); MELANGKAP TOMIS: *Lorence Lugas 549* (K); PENATARAN BASIN: 900 m, *Clemens 34174* (BM, K); PORING: *Dransfield, S. SD 719* (K), *Meliden Giking 54* (K), *Sani Sambuling 425* (K); SERINSIM: *Jibrin Sibil 279* (K); TEKUTAN: *Dius Tadong 367* (K); TENOMPOK: 1500 m, *Clemens 29906* (BM).

65.21. ICHNANTHUS P. Beauv.

Stieber, M. T. (1987). Revision of *Ichnanthus* sect. *Foveolatus* (Gramineae: Panicoideae). *Syst. Bot.* 12: 187–216.

65.21.1. Ichnanthus pallens (Sw.) Munro ex Benth., *Fl. Hongk.*: 414 (1861).

Perennial grass, culms to 0.4 m long, much-branched. Hill forest, lower montane forest. Elevation: 900–2100 m. The varieties Stieber recognized are not separated here.

Material examined. DALLAS: 900 m, *Clemens 28272* (BM, K); DALLAS/TENOMPOK: 1400 m, *Clemens s.n.* (BM); MAMUT RIVER: 1200 m, *RSNB 1244* (K, L); MARAI PARAI: 1700 m, *Clemens 40245* (BM); MESILAU RIVER: 1500–2100 m, *Clemens 51055* (BM), 1500–2100 m, *51163* (BM); MT. NUNGKEK: 1200 m, *Clemens 32502* (BM, K, L); NALUMAD: *Daim Andau 170* (K); PENIBUKAN: *Nooteboom & Aban 1561* (L); PINOSUK PLATEAU: 1400 m, *Beaman 10700* (K, L, MSC); TAHUBANG FALLS: *Clemens 30694* (K), *30694a* (BM); TAHUBANG RIVER: 1500 m, *Clemens 40277* (BM, K, L); TENOMPOK: 1500 m, *Clemens 27472* (BM), 1500 m, *30276* (K, L); WASAI RIVER: 1100–1200 m, *Clemens s.n.* (K), 1100–1400 m, *34125* (BM), 900–1100 m, *34129* (L).

65.22. IMPERATA Cirillo

65.22.1. Imperata conferta (Presl) Ohwi, *Bot. Mag. (Tokyo)* 55: 549 (1941).

Rhizomatous perennial grass to 2-3 m tall with relatively broad inflorescence. Lowlands in open disturbed areas.

Material examined. MELANGKAP TOMIS: *Lorence Lugas 526* (K), *1223* (K); NALUMAD: *Daim Andau 2* (K); SERINSIM: *Jibrin Sibil 6* (K), *224* (K); TEKUTAN: *Dius Tadong 46* (K), *272* (K), *383* (K).

65.22.2. Imperata cylindrica (L.) P. Beauv., *Ess. Agrostogr.* 8: 165, 177, t. 5 (1812).

Erect perennial grass to 1 m tall with strong, fire-resistant rhizomes. Widely distributed on Kinabalu especially in disturbed areas in hill forest and lower montane forest. Elevation: 900–1200 m. This obnoxious weed is little collected on Mount Kinabalu, perhaps because no one collects such common pests.

Material examined. BUNDU TUHAN: *Doinis Soibeh 71* (K); KIAU: *Jusimin Duaneh 102* (K); PENATARAN RIVER: 900–1200 m, *Clemens 34188* (K).

65.23. ISACHNE R. Br.

Jansen, P. (1953). *Isachne* R. Br. *Reinwardtia* 2: 279–292.

65.23.1. Isachne albens Trin., *Sp. Gram.* 1: 85 (1828).

Erect, tufted perennial grass with sometimes ascending or scrambling culms. Hill forest, lower montane forest, upper montane forest, in wet situations on rocks or boulders, often on ultramafic substrate. Elevation: 900–2700 m.

Material examined. GURULAU SPUR: 2400 m, *Clemens 50935* (BM); KILEMBUN BASIN: 2700 m, *Clemens 33731* (BM, L), 2700 m, *33736* (BM), 1100 m, *34410* (BM, K, L); KILEMBUN RIVER: *Clemens 33841* (K, L); KILEMBUN RIVER HEAD: 2700 m, *Clemens 33844 bis* (BM); KINATEKI RIVER: 1800 m, *Clemens 32655* (BM); LUBANG: 1800 m, *Gibbs 4113* (BM, K); LUMU-LUMU: 1800 m, *Clemens 28389* (BM, K, L); MARAI PARAI: 2400 m, *Clemens 32326* (BM); MESILAU: 2700 m, *RSNB 5999* (K); MESILAU BASIN: 2100 m, *Clemens 29693* (BM); MESILAU CAVE: 1800 m, *RSNB 4724* (K, L), 1800 m, *Clemens 51240* (BM, K), 1800 m, *51291* (L); MESILAU RIVER: *Clemens 51632* (BM, K, L); PENATARAN BASIN: 1100 m, *Clemens 32612* (BM), 900 m, *34048* (BM, K, L); TAHUBANG FALLS: 1500 m, *Clemens 40274* (BM, K, L); TAHUBANG RIDGE: *Clemens 30696* (BM, L); TAHUBANG RIVER: 1500 m, *Clemens 40272* (BM), 900 m, *50074* (BM, K, L); TINEKEK FALLS: 2000 m, *Clemens 50035* (BM).

65.23.2. Isachne albomarginata Jansen, *Reinwardtia* 2: 279 (1953).

Perennial, subcaespitose grass. Hill forest. Elevation: 900–2100 m. Jansen's variety *hirsuta* is not recognized here.

Material examined. BAMBANGAN RIVER: 1500 m, *RSNB 1307* (K); DALLAS/BUNGOL: 900 m, *Clemens s.n.* (BM); DALLAS/TENOMPOK: 1200 m, *Clemens 28164* (BM, K, L); KIAU: 1100 m, *Clemens 51053* (BM), 1100 m, *51223* (BM); MINITINDUK GORGE: *Clemens 29576* (BM, K, L); PENATARAN BASIN: 1100 m, *Clemens 40175A* (BM, L); PENATARAN RIVER: 900–1200 m, *Clemens 34121B* (L); TENOMPOK: 1100 m, *Clemens 50053* (K), *Forster F 39* (K); TINEKEK FALLS: 2100 m, *Clemens 50036* (BM); WASAI RIVER (PENATARAN BASIN): 1200 m, *Clemens 34121A* (BM).

65.23.3. Isachne clementis Merr., *J. Straits Branch Roy. Asiat. Soc.* 76: 76 (1917).

Decumbent rhizomatous perennial grass, rooting at the nodes, 20–30 cm tall. Upper montane forest, generally on ultramafic substrate. Elevation: 1700–3400 m. Jansen's variety *vulcanica* is not recognized here.

Material examined. EASTERN SHOULDER: 2900 m, *RSNB 991* (K, L); GURULAU SPUR: 3400 m, *Clemens 50916* (BM, K); KEMBURONGOH: 2600 m, *Clemens 27077* (BM, K, L), 2400 m, *27986* (BM, K, L); KEMBURONGOH/PAKA-PAKA CAVE: 2400 m, *Sinclair et al. 9083* (K, L); KILEMBUN RIVER HEAD: 2700–3000 m, *Clemens 33900* (BM, L); LAYANG-LAYANG: 2700 m, *Smith 456* (L); MARAI PARAI: 2700 m, *Clemens 31688* (L), 1700 m, *33806* (L); MARAI PARAI SPUR: 1700 m, *Clemens 32849* (BM); MOUNT KINABALU: 2400 m, *Haviland 1408* (K); PAKA-PAKA CAVE: 2400 m, *Forster F 53* (K); SUMMIT TRAIL: 2500–2900 m, *Jacobs 5722* (K, L).

65.23.4. Isachne kinabaluensis Merr., *J. Straits Branch Roy. Asiat. Soc.* 76: 71 (1917). Type: LUBANG/PAKA-PAKA CAVE, *Clemens, M. S. 10704* (holotype PNH†; isotype K!).

Erect and rigid to dwarfed, densely tufted grass, variable in habit. Lower montane forest, upper montane forest, in open areas. Elevation: 1400–3200 m.

Additional material examined. EASTERN SHOULDER: 2900 m, *RSNB 994* (K, L); GURULAU SPUR: 2400 m, *Clemens 51038* (BM, L); KEMBURONGOH: 2600 m, *Clemens 27076* (BM, K), 2200 m, *Sinclair et al. 9066* (K, L); LAYANG-LAYANG: 2700 m, *Smith 460* (L); MAMUT COPPER MINE: 1400–1500 m, *Beaman 10361* (L, MSC); MARAI PARAI: 1500 m, *Collenette A 86* (BM); MARAI PARAI SPUR: 1700 m, *Gibbs 4069* (K); MESILAU HILL: 2100 m, *Poore H 199* (K); MOUNT KINABALU: 2400 m, *Forster F 45* (K), 2900–3200 m, *Smitinand 8182* (L), *Whitehead s.n.* (BM); MT. NUNGKEK: 1700 m, *Clemens 32014* (BM), 1700 m, *32237* (K, L); PENATARAN BASIN: 1800 m, *Clemens 40155* (BM, L); SUMMIT TRAIL: 2500–2900 m, *Jacobs 5723* (K); UPPER KINABALU: *Clemens 27076* (L), *27077* (K), *30269* (K, L).

65.23.5. Isachne miliacea Roth ex Roem. & Schult., *Syst. Veg.* 2: 476 (1817).

Low, delicate, decumbent perennial grass, rooting from the lower nodes. Lowlands, hill forest, lower montane forest? Elevation: 800–1500 m.

Material examined. TAMPASSUK RIVER: 800 m, *Clemens 51042* (BM, K, L); TENOMPOK: 1500 m, *Clemens 30283* (K).

65.23.6. Isachne indet.

Material examined. PIG HILL: 2000–2300 m, *Beaman 9855* (MSC).

65.24. ISCHAEMUM L.

65.24.1. Ischaemum barbatum Retz., *Observ. Bot.* 6: 35 (1791).

Erect or ascending perennial grass, culms often branched at the base. Lowlands in secondary forest or disturbed areas.

Material examined. PORING: *Sani Sambuling 707* (K).

65.24.2. Ischaemum polystachyum J. Presl, *Reliq. Haenk.* 1: 328 (1830).

Perennial grass to ca 1 m high; inflorescence branched. Lowlands in secondary forest or disturbed areas.

Material examined. KEBAYAU/MELANGKAP TOMIS: *Lorence Lugas 2648* (K).

65.25. KINABALUCHLOA K. M. Wong

65.25.1. Kinabaluchloa nebulosa K. M. Wong, *Kew Bull.* 48: 526 (1993). Type: MESILAU CAMP, 1900 m, *Mikil SAN 38475* (holotype K!).

Erect then scrambling bamboo, with thin-walled culms and long internodes (the longest 120 cm); culm-sheaths with black hairs. Lower montane forest. Elevation: 1700–2100 m.

Additional material examined. KIAU VIEW TRAIL: 1800 m, *Dransfield, S. SD 849* (K); LUMU-LUMU: 2100 m, *Clemens 27984* (BM, K, L), 1800 m, *29172* (BM, L); PARK HEADQUARTERS: 1800 m, *Dransfield, S. SD 758* (K), *SD 759* (K); TENOMPOK: 1700 m, *Clemens 28283* (BM, K).

65.26. LOPHATHERUM Brongn. in Duperrey

Yang, G., & Chen, S.-L. (1988–89). A synthetic taxonomic study on genus *Lophatherum* Brongn. [in Chinese with English abstract]. *Bull. Nanjing Bot. Gard.* 1988–1989: 14–20.

65.26.1. Lophatherum gracile Brongn. in Duperrey, *Voy. Coquille, Bot. Phan.*: 50 (1831).

Perennial grass from a short rhizome. Hill forest. Elevation: 600–1500 m.

Material examined. DALLAS: 900 m, *Clemens 27301* (BM, K, L); DALLAS/TENOMPOK: 1200 m, *Clemens 30275* (K, L); GURULAU SPUR: 1200 m, *Clemens 51564a* (BM, K, L), *Clemens, M. S. 10797* (BM, K), *10814* (BM, K), 1500 m, *Gibbs 4005* (BM, K), 1500 m, *4007* (K); MT. NUNGKEK: 600–800 m, *Clemens 32858* (BM, L); TAHUBANG RIVER: 1200 m, *Clemens 30700* (BM, K, L).

65.27. MELINIS P. Beauv.

65.27.1. Melinis repens (Willd.) Zizka, *Biblioth. Bot.* 138: 55 (1988).

Tufted erect perennial grass to 2 m high. Weed in hill forest in open, disturbed situations.

Material examined. KIAU: *Jusimin Duaneh 160* (K).

65.28. MICROSTEGIUM Nees

65.28.1. Microstegium geniculatum (Hayata) Honda, *J. Fac. Sci. Univ. Tokyo, Sect. 3, Bot.*: 410 (1930).

Erect or decumbent, rhizomatous, tufted perennial grass of moderate height, occurring in patches. Hill forest, lower montane forest, in open areas. Elevation: 900–2400 m.

Material examined. BIDAU-BIDAU FALLS: 1500 m, *Clemens 34046* (BM); DALLAS: 900 m, *Clemens 26061* (K, L), 900 m, *30282* (K); DALLAS/TENOMPOK: 1500–2400 m, *Clemens 30265* (K, L); EASTERN SHOULDER: 1200 m, *RSNB 268* (K, L); KEMBURONGOH: 2400 m, *Clemens 27798* (BM, K, L); LIWAGU/ MESILAU RIVERS: 1500 m, *RSNB 1921* (K), 1400 m, *RSNB 2768* (K, L); LUBANG: *Clemens, M. S. 10423* (K); MOUNT KINABALU: *RSNB s.n.* (K); PENATARAN RIVER: 900 m, *Clemens 32615* (K, L); PENIBUKAN: 1200 m, *Clemens 32181* (L); TENOMPOK: 1500 m, *Clemens s.n.* (BM, BM), 1500 m, *28083* (BM, K), 1500 m, *30186* (K, L).

65.28.2. Microstegium nudum (Trin.) A. Camus, *Ann. Soc. Linn. Lyon, n.s.* 68: 201 (1921).

Creeping, slender grass with branched culms, decumbent, rooting at the nodes; inflorescence subdigitate. Hill forest, possibly also in lower montane forest. Elevation: 900–1200 m.

Material examined. DALLAS: 900 m, *Clemens 28555* (BM); EASTERN SHOULDER: 1200 m, *RSNB 267* (K); KEMBURONGOH: 1200 m, *Forster F 40* (K); KUNDASANG: 1200 m, *Sinclair et al. 9237* (K).

65.29. MISCANTHUS Anderss.

Lee, Y. N. (1964). Taxonomic studies on the genus *Miscanthus* (5). Relationships among the section, subsection and species. Part 3. Enumeration of species and varieties. *J. Jap. Bot.* 39: 289–298.

65.29.1. Miscanthus floridulus (Labill.) Warb. ex K. Schum. & Lauterb.

a. var. malayanus Y. N. Lee, *J. Jap. Bot.* 39: 120 (1964). Plate 19D.

Large perennial grass with culms 2–3 m tall, hollow; blades up to 70 cm long; inflorescence a conspicuous terminal panicle. Hill forest, lower montane forest, infrequently upper montane forest, along roadsides and in open places. Elevation: 500–3000 m.

Material examined. DALLAS: 900 m, *Holttum SFN 25268* (K, L); DALLAS/TENOMPOK: 900 m, *Clemens 27483* (BM); KEMBURONGOH: 2000 m, *Meijer SAN 29114* (K); MAMUT RIVER: 1200 m, *RSNB 1223* (K); MARAI PARAI: 2000 m, *Clemens 32656* (L); MELANGKAP TOMIS: *Lorence Lugas 482* (K), *1847* (K), *2130* (K), *2595* (K); MESILAU RIVER: 1500 m, *RSNB 4149* (K, L); MINATUKAN SPUR: 3000 m, *Clemens 32656* (BM); MOUNT KINABALU: 1500 m, *Low s.n.* (K); PENATARAN RIVER: 500 m, *Beaman 9287* (L, MSC); PENIBUKAN: 1500 m, *Clemens 30978* (BM), 1200–1500 m, *31362* (BM); SOSOPODON: *Ampon & Saikeh SAN 71802* (L); TINEKEK FALLS: 2100 m, *Clemens 50130* (BM).

65.30. MNESITHEA Kunth

Veldkamp, J. F., de Koning, R. and Sosef, M. S. M. (1986). Generic delimitation of *Rottboellia* and related genera (Gramineae). *Blumea* 31: 281–307 (1986).

65.30.1. Mnesithea glandulosa (Trin.) de Koning & Sosef, *Blumea* 31: 290 (1986).

Erect perennial grass to 2 m high. Hill forest in open, secondary situations. Elevation: 900–1200 m. The species was formerly included in the genus *Coelorhachis*.

Material examined. DALLAS: 900 m, *Clemens 30277* (K, L); DALLAS/TENOMPOK: 900–1200 m, *Clemens 27474* (K, L); MELANGKAP TOMIS: *Lorence Lugas 1588* (K).

65.31. NEYRAUDIA Hook. f.

Conert, H. J. (1959). Beiträge zur Monographie der Gattungen *Cleistogenes* und *Neyraudia*. *Bot. Jahrb. Syst.* 78: 208–245.

65.31.1. Neyraudia curvipes Ohwi, *Bull. Tokyo Sci. Mus.* 18: 9 (1947). Type: PENATARAN River, *Clemens 34189* (holotype BO n.v.; isotypes BM!, L!).

Robust reed-like perennial grass with prominent feathery panicles. Lower montane forest on ultramafic substrate. Elevation: 1400–1800 m. The type specimen was incorrectly cited by Ohwi as *Clemens 31189*.

Additional material examined. KILEMBUN BASIN: 1500 m, *Clemens 34497* (K), 1500 m, *40090* (BM, L); PENATARAN BASIN: 1800 m, *Clemens 40147* (BM, L).

65.32. OPLISMENUS P. Beauv.

Davey, J. C., and Clayton, W. D. (1978). Some multiple discriminant function studies on *Oplismenus* (Gramineae). *Kew Bull.* 33: 147–157. Scholz, U. (1981). Monographie der Gattung *Oplismenus* (Gramineae). *Phan. Monogr.* 13: 1–213. J. Cramer, Vaduz. (Her varieties are not recognized here.)

65.32.1. Oplismenus compositus (L.) P. Beauv., *Ess. Agrostogr.*: 54, 168, 169 (1812).

Decumbent perennial creeping stoloniferous grass 0.2–1 m tall. Lowlands, hill forest, infrequently lower montane forest, in lightly shaded places, gregarious in disturbed areas. Elevation: 200–1800 m.

Material examined. DALLAS: 900 m, *Clemens 27594* (K, L); DALLAS/TENOMPOK: 1200 m, *Clemens s.n.* (BM), 1200 m, *27594* (BM); GURULAU SPUR: 1800 m, *Clemens 51051* (BM); KEBAYAU/KAUNG: 200–400 m, *Holttum SFN 25113* (K, L); KUNDASANG: 1200 m, *RSNB 1416* (K, L); KUNDASANG/MESILAU

CAVE: 1500 m, *Clemens 51621* (K, L); KUNDASANG/MESILAU RIVER: 1200 m, *Clemens 51619* (BM); LUBANG: *Clemens, M. S. 10424* (K), 1500 m, *Gibbs 4133* (K); MESILAU CAMP: *Poore & Ho s.n.* (K); MESILAU RIVER: *Clemens 51621* (BM); PENATARAN RIVER: 900 m, *Clemens 32613* (BM, L); PENIBUKAN: 1200 m, *Clemens 51567* (BM, K, L); PINOSUK PLATEAU: 1400 m, *Beaman 10748* (L, MSC); PORING HOT SPRINGS: 600 m, *Beaman 7572* (K, L, MSC); SOSOPODON: *Amin & Jarius SAN 116557* (L); TAHUBANG RIVER: 900 m, *Haviland 1409* (K); TEKUTAN: *Dius Tadong 664* (K); TENOMPOK: 1500 m, *Clemens 30288* (K, L), 1500 m, *30289* (K, L), 1400 m, *51045* (BM).

65.32.2. Oplismenus hirtellus (L.) P. Beauv., *Ess. Agrostogr.*: 54, 168, 170 (1812).

Decumbent perennial grass, rooting at the nodes, forming mats; flowering culms up to 70 cm tall. Hill forest, lower montane forest, at edge of streams, sometimes on ultramafic substrate. Elevation: 900–1500 m.

Material examined. KUNDASANG/MESILAU RIVER: 1200 m, *Clemens 51619a* (BM); LUBANG: *Clemens, M. S. 10422* (K); MESILAU RIVER: 1500 m, *RSNB 1338* (K, L); PENATARAN BASIN: 900–1100 m, *Clemens 34126A* (BM), 1100 m, *40175* (L); TAHUBANG FALLS: 1500 m, *Clemens 40276* (BM, K, L); TINEKEK FALLS: 1500 m, *Clemens 40813* (BM); WASAI FALLS: 1100–1400 m, *Clemens 34126* (BM, K, L).

65.33. ORYZA L.

Duistermaat, H. (1987). A revision of *Oryza* (Gramineae) in Malesia and Australia. *Blumea* 32: 157–193.

65.33.1. Oryza meyeriana (Zoll. & Moritzi) Baill., *Hist. Pl.* 12: 166 (1893).

a. var. meyeriana

Perennial, loosely tufted or stoloniferous grass, culms erect to ascending, sometimes branched at base, rooting at lower nodes. Hill forest. Elevation: 800 m.

Material examined. MENGGIS: *Matamin Rumutom 202* (K); SERINSIM: *Kinsun Bakia 169* (K); TEKUTAN: *Dius Tadong 450* (K), 800 m, *Shea & Aban SAN 77226* (K, L).

65.33.2. Oryza sativa L., *Sp. Pl.*: 333 (1753).

Tufted annual grass. Cultivated, apparently occasionally occurring spontaneously.

Material examined. MELANGKAP TOMIS: *Lorence Lugas 1188* (K), *2603* (K); SERINSIM: *Jibrin Sibil 294* (K).

65.34. OTTOCHLOA Dandy

65.34.1. Ottochloa nodosa (Kunth) Dandy, *J. Bot.* 69: 55 (1931).

Gregarious decumbent perennial grass, climbing over other plants. Hill forest in somewhat shaded, not too dry places. Elevation: 900–1200 m.

Material examined. DALLAS: 900 m, *Clemens 30274* (L); DALLAS/TENOMPOK: 1200 m, *Clemens 27591* (K); KIAU: 1100–1200 m, *Clemens 51222* (BM, K, L).

65.35. PANICUM L.

Veldkamp, J. F. (1996). Revision of *Panicum* and *Whiteochloa* in Malesia (Gramineae-Paniceae). *Blumea* 41: 181–216.

65.35.1. Panicum brevifolium L., *Sp. Pl.*: 59 (1753).

Perennial grass with freely branching culms, rooting at the nodes and scrambling over other vegetation. Hill forest in shady places. Elevation: 800–1200 m.

Material examined. BUNGOL: 800 m, *Clemens 51566* (K, L); DALLAS: 900 m, *Clemens 28274* (BM, K); DALLAS/TENOMPOK: 1200 m, *Clemens 27590* (BM, L); KADAMAIAN RIVER: 800 m, *Clemens 51566a* (BM).

65.35.2. Panicum humidorum Buch.-Ham. ex Hook. f., *Fl. Brit. India* 7: 53 (1896).

Erect perennial grass, culms 0.5–1 m high. Hill forest in old rice paddys. Elevation: 800–900 m.

Material examined. DALLAS: 900 m, *Clemens 28275* (BM, K, L), 900 m, *28275A* (K, L); KADAMAIAN RIVER: 800 m, *Clemens 51562* (BM, L).

65.35.3. Panicum notatum Retz., *Observ. Bot.* 4: 18 (1786).

Perennial grass with erect or scrambling rather woody culms and broad leaf blades, often rooting from the lower nodes. Hill forest in disturbed situations. Elevation: 400–1200 m.

Material examined. DALLAS: 900 m, *Clemens 27662* (L), 900 m, *28271* (BM, K); DALLAS/TENOMPOK: 1100 m, *Clemens 27589* (K, L), 1100 m, *27589A* (K, L); GURULAU SPUR: 1100 m, *Clemens 51224* (BM, K, L); KAUNG: 400 m, *Darnton 267* (L); KIAU: *Clemens, M. S. 10028* (K, L); LIWAGU/MESILAU RIVERS: 1200 m, *RSNB 2593* (K, L).

65.35.4. Panicum sarmentosum Roxb., *Fl. Ind.* 1: 311 (1820).

Robust perennial grass with strongly branching, creeping or scrambling culms, rooting and branching from the nodes. Lowlands, hill forest. Elevation: 300–1200 m.

Material examined. DALLAS: 900 m, *Clemens 28300* (L), 1100–1200 m, *51489* (BM, L); KEBAYAU/KAUNG: 300–600 m, *Clemens s.n.* (BM), 300–600 m, *28300a* (K, L); MELANGKAP TOMIS: *Lorence Lugas 160* (K), *594* (K), *2651* (K); NALUMAD: *Daim Andau 172* (K); SERINSIM: *Jibrin Sibil 97* (K); TEKUTAN: *Dius Tadong 182* (K), *500* (K), *697* (K), *711* (K), *718* (K), *Lomudin Tadong 13* (K).

65.36. PASPALUM L.

De Koning R., and Sosef, M. S. M. (1985). The Malesian species of *Paspalum* L. (Gramineae). *Blumea* 30: 279–318 (1985).

65.36.1. Paspalum conjugatum Bergius, *Acta Helv. Phys.-Math.* 7: 129, t. 8 (1762).

Vigorous, leafy, stoloniferous perennial grass with erect or ascending culms forming loose mats over extensive areas. Hill forest; common weed in cultivated or waste ground. Elevation: 900–1200 m.

Material examined. BUNDU TUHAN: *Doinis Soibeh 55* (K); KIAU: *Clemens, M. S. 10030* (BM); KUNDASANG: 1200 m, *RSNB 1418* (K, L); MAMUT RIVER: 1200 m, *RSNB 1220* (K, L); MELANGKAP TOMIS: *Lorence Lugas 1647* (K); TEKUTAN: *Lomudin Tadong 161* (K); TENOMPOK: *Clemens s.n.* (BM); WASAI FALLS: 1100–1200 m, *Clemens 34124* (L); WASAI RIVER: 900 m, *Clemens 32616* (BM).

65.36.2. Paspalum longifolium Roxb., *Fl. Ind.* 1: 283 (1820).

Tufted perennial grass to 1.5 m high. Hill forest.

Material examined. MELANGKAP TOMIS: *Lorence Lugas 543* (K).

65.36.3. Paspalum scrobiculatum L.

a. var. bispicatum Hack. in Kneuck., *Allg. Bot. Z. Syst.* 20: 146 (1914).

Loosely tufted or decumbent to creeping annual or perennial grass. Hill forest, lower montane forest. Elevation: 2400 m.

Material examined. KEMBURONGOH: 2400 m, *Clemens s.n.* (BM); MELANGKAP TOMIS: *Lorence Lugas 544* (K), *1856* (K); TENOMPOK: *Clemens s.n.* (BM); UPPER KINABALU: *Clemens 30264* (K, L).

65.36.4. Paspalum virgatum L., *Syst. Nat.* ed. 10: 855 (1759).

Tufted perennial grass to 1.5 m high. Lowlands. New record for Malesia; a cultivated specimen was collected in Java in the mid-19th century.

Material examined. MARAK-PARAK: *Kinsun Bakia 520* (K).

65.37. PENNISETUM Rich. in Pers.

65.37.1. Pennisetum clandestinum Hochst. ex Chiov., *Ann. 1st Bot. Roma* 8: 41, t. 5, f. 2 (1908).

Creeping, rhizomatous and stoloniferous perennial grass forming a coarse turf, branching and rooting freely. Lower montane forest. Elevation: 1600 m. Introduced, apparently for pasturage.

Material examined. DAIRY ANNEX: 1600 m, *Beaman 10794* (K, L, MSC).

65.37.2. Pennisetum polystachion (L.) Schult., *Mant.* 2: 146 (1824).

Tufted, erect perennial grass to 2 m high. Lowlands in disturbed areas or thickets.

Material examined. SERINSIM: *Kinsun Bakia 518* (K).

65.37.3. Pennisetum purpureum Schumach., *Beskr. Guin. Pl.*: 44 (1827).

Large tufted perennial grass 2–3 m high. Lowlands, only known from Mount Kinabalu from a house compound (cultivated?).

Material examined. SERINSIM: *Jibrin Sibil 72* (K).

65.38. POA L.

Veldkamp, J. F. (1994). *Poa* L. (Gramineae) in Malesia. *Blumea* 38: 409–457.

65.38.1. Poa annua L., *Sp. Pl.*: 68 (1753).

Tufted low grass. Upper montane forest, disturbed areas. Elevation: 3700–3800 m. Apparently a recent introduction.

Material examined. SAYAT-SAYAT: 3800 m, *Smith 510* (L); SUMMIT AREA: 3700 m, *Smith 500* (L).

65.38.2. Poa borneensis Jansen, *Reinwardtia* 2: 322 (1953). Type: VICTORIA PEAK, 3400–3800 m, *Clemens 51527* (holotype UC n.v.; isotypes BM!, K!, L!).

Densely caespitose grass with erect culms 20–30(40) cm tall. Upper montane forest, summit area, in open scrub and on wet rocks. Elevation: 3400–4100 m. Endemic to Mount Kinabalu.

Additional material examined. LOW'S PEAK: 4000 m, *Clemens 30273* (L); SUMMIT AREA: 3700–4000 m, *Clemens s.n.* (BM, BM), 4000 m, *Molesworth-Allen 3290* (K); UPPER KINABALU: *Clemens 27787* (L), *30314* (K, L), *30315* (L).

65.38.3. Poa epileuca (Stapf) Stapf, *Hooker's Icon. Pl.* 27: sub t. 2607 (1899).

Deyeuxia epileuca Stapf, *Trans. Linn. Soc. London, Bot.* 4: 247 (1894). Type: MOUNT KINABALU, 3200 m, *Haviland 1401* (holotype K!).

Low, tufted perennial grass. Summit area and upper edge of upper montane forest. Elevation: 3200–4100 m. Also in Sulawesi and New Guinea.

Additional material examined. Low's Peak: 4100 m, *Clemens 27074* (BM, L); Paka-paka Cave/Low's Peak: *Clemens, M. S. 10607* (BM); Sayat-sayat: 3800 m, *Smith 505* (L); Summit Area: *Clemens s.n.* (BM, BM), 4000 m, *27770* (BM), 4100 m, *Gibbs 4352* (BM); Victoria Peak: 3800 m, *Clemens 51198 p.p.* (BM).

65.38.4. Poa papuana Stapf, *Hooker's Icon. Pl.* 27: t. 2607 (1899).

Low, densely tufted perennial grass with erect panicles, culms 5–20 cm tall; blades setaceous. Upper montane forest, summit area, in marshy areas and in wet granitic crevices and depressions. Elevation: 3000–4100 m. Also in Sulawesi and New Guinea.

Material examined. Low's Peak: 4100 m, *Clemens 27074* (K, L), 4000 m, *27770* (K, L); Paka-paka Cave: 3200–3700 m, *Clemens 30266* (K, L), 3100 m, *Sleumer 4708* (L); Paka-paka Cave/Low's Peak: *Clemens, M. S. 10607* (K); Sayat-sayat: 3800 m, *Smith 506* (L); Summit Area: *Clemens 30267* (K, L), 4100 m, *Gibbs 4352* (K), 3800–4100 m, *Molesworth-Allen 3286* (K), 3800 m, *3287* (K), 4100 m, *Sinclair et al. 9160* (K, L); Summit Trail: 3000–3500 m, *Jacobs 5761* (K, L); Victoria Peak: 3800 m, *Clemens 51198 p.p.* (BM), *51408 p.p.* (BM), 3800 m, *51529* (K, L).

65.39. POGONATHERUM P. Beauv.

Chase, A. (1950). *Pogonatherum* Beauv. *J. Arnold Arbor.* 31: 130–132. Jansen, P. (1953). *Pogonatherum* P. B. *Reinwardtia* 2: 333–334.

65.39.1. Pogonatherum crinitum (Thunb.) Kunth, *Enum. Pl.* 1: 478 (1833).

Slender, tufted grass. Lowlands, hill forest, lower montane forest, sometimes on ultramafic substrate. Elevation: 400–1500 m. Material previously determined as *P. paniceum* (Lam.) Hack. has been included here.

Material examined. Bambangan River: 1500 m, *RSNB 1308* (K, L); Kaung: 400 m, *Darnton 341* (BM); Mamut River: 1200 m, *RSNB 1242* (K); Marak-Parak: *Aban Gibot SAN 100071* (L); Penataran Basin: 900 m, *Clemens 32614a* (BM); Penataran River: 500 m, *Beaman 9288* (K, L, MSC), 900 m, *Clemens 32614* (L); Tekutan: *Lomudin Tadong 313* (K).

65.40. RACEMOBAMBOS Holttum

Dransfield, S. (1983). The genus *Racemobambos* (Gramineae-Bambusoideae). *Kew Bull.* 37: 661–679. Holttum, R. E. (1956). *Racemobambos*, a new genus of bamboos. *Gard. Bull. Singapore* 15: 267–273.

65.40.1. Racemobambos gibbsiae (Stapf) Holttum, *Gard. Bull. Singapore* 15: 272 (1956). Plate 20A.

Bambusa (?) *gibbsiae* Stapf in Gibbs, *J. Linn. Soc., Bot.* 42: 189 (1914). Type: KEMBURONGOH, 2300–2700 m, *Gibbs 4232* (syntype K!; isosyntype BM!); MARAI PARAI SPUR, 1500–2400 m, *Gibbs 4091* (syntype K!; isosyntype BM!).

Scrambling bamboo, branch complements many with middle branch dominant, spikelets hairy. Upper montane mossy forest, rarely lower montane forest. Elevation: 1500–2900 m. Endemic to Mount Kinabalu.

Additional material examined. GURULAU SPUR: 2400–2700 m, *Clemens 51457* (BM, K); KEMBURONGOH: *Dransfield, S. SD 755* (K, L), 2200 m, *Price 195* (K), *Sinclair et al. 9204* (K, L); LAYANG-LAYANG: 2900 m, *Dransfield, S. SD 757* (K, L), *SD 759* (K); LUMU-LUMU: 1800–2100 m, *Molesworth-Allen 3261* (K); MESILAU CAVE: 2400 m, *Clemens 51161* (BM, K); MOUNT KINABALU: 2400 m, *Whitehead s.n.* (BM).

65.40.2. Racemobambos hepburnii S. Dransf., *Kew Bull.* 37: 670 (1983). Type: MESILAU RIVER, 1500 m, *RSNB 4111* (holotype K!).

Scrambling bamboo; branch complements many with middle branch dominant; leaves very narrow; spikelets glabrous. Lower montane forest. Elevation: 1500–2100 m.

Additional material examined. BAMBANGAN RIVER: 1500 m, *RSNB 4585* (K, L, US); EASTERN SHOULDER: 2000 m, *RSNB 167* (K, L, US); MESILAU CAVE TRAIL: 2100 m, *Meijer SAN 48117* (K, L); MESILAU RIVER: 1500 m, *RSNB 4108* (K, L, US), 1500 m, *RSNB 4124* (K, L, US); PARK HEADQUARTERS: *Wong FRI 35159* (K, L); PARK HEADQUARTERS/POWER STATION: *Dransfield, S. SD 718* (K), 1700–1800 m, *Kanis SAN 53807* (K); PARK HEADQUARTERS/POWER STATION ROAD: 1800 m, *Smith & Everard 147* (K); PINOSUK PLATEAU: *Poore H 124* (K), *H 258* (K, L), *2133* (K), *2600* (K), *3587* (K, L, US), *3726* (K); TENOMPOK/KEMBURONGOH: 1800–2000 m, *Fosberg 43960* (K), 1700 m, *Smythies S. 10609* (L).

65.40.3. Racemobambos hirsuta Holttum, *Gard. Bull. Singapore* 15: 272 (1956). Type: PENIBUKAN, 1200–1500 m, *Clemens s.n.* (holotype SING n.v.; isotype K!).

Scrambling bamboo; branch complements up to 4, middle branch not dominant; spikelet hairy. Hill forest, lower montane forest, on ultramafic substrate. Elevation: 1100–2000 m. This species occurs at lower elevations than *R. gibbsiae* and *R. rigidifolia*.

Additional material examined. HEMPUEN HILL: *Meijer SAN 20266* (L); KILEMBUN BASIN: 1100 m, *Clemens 40025* (BM, L); MARAI PARAI: *Dransfield, S. SD 715A* (K); PENATARAN BASIN: 2000 m, *Clemens 40149* (BM, L); PENIBUKAN: 1200 m, *Clemens 30751* (BM, K, L), 1200–1500 m, *31618* (K).

65.40.4. Racemobambos rigidifolia Holttum, *Gard. Bull. Singapore* 15: 273 (1956). Plate 20B, C. Type: PENIBUKAN, 1400 m, *Carr SFN 27438* (holotype SING n.v.; isotype BM!).

Scrambling bamboo; branch complements up to 4 with middle branch not dominant; spikelets glabrous. Lower montane forest on ultramafic substrate. Elevation: 1200–2100 m. Endemic to Mount Kinabalu; the species occurs at higher elevations than *R. hirsuta*.

Additional material examined. MARAI PARAI: 1500 m, *Clemens 32257* (BM, L), 1500 m, *32749* (BM, L), 1500 m, *33087* (BM, K, L), *Dransfield, S. SD 715* (K, L); MARAI PARAI SPUR: 1500–2100 m, *Clemens s.n.* (US), 1500–1800 m, *32511* (BM, L); PENIBUKAN: 1200–1500 m, *Clemens s.n.* (BM).

65.41. SACCIOLEPIS Nash in Britton

65.41.1. Sacciolepis indica (L.) Chase, *Proc. Biol. Soc. Wash.* 21: 8 (1908).

Erect or somewhat decumbent annual or short-lived perennial grass. Hill forest, lower montane forest, sometimes on ultramafics, especially poor wet soils in open areas. Elevation: 900–1500 m.

Material examined. DALLAS: 900 m, *Clemens 28273* (BM, K, L); MARAI PARAI: 1500 m, *Clemens 32238* (BM, L).

65.42. SCHIZACHYRIUM Nees

65.42.1. Schizachyrium brevifolium (Sw.) Nees ex Buse, *Pl. Jungh.* 3: 359 (1854).

Spreading, stoloniferous perennial grass with long runners. Lowlands, hill forest, in paddy fields. Elevation: 600–900 m. The infraspecific taxa of Henrard are not recognized here.

Material examined. DALLAS: 900 m, *Clemens 27540* (BM, K); KAUNG: *Forster F 57* (K); KITUNTUL: *Tiong SAN 88036* (L); PORING: 600 m, *Dransfield, S. SD 721* (L).

65.43. SCHIZOSTACHYUM Nees

65.43.1. Schizostachyum brachycladum Kurz, *J. Asiat. Soc. Bengal* 39: 89 (1870).

Erect bamboo to 7 m tall, culms 7 cm in diameter, thin-walled, green or yellow, blades of culm-sheaths triangular, erect. Planted, becoming naturalized. Elevation: 600 m.

Material examined. KIAU: *Clemens, M. S. 10285* (BM, K, US); KITUNTUL: *Tiong SAN 88036* (K); PORING: *Aban Gibot SAN 85733* (K), 600 m, *Dransfield, S. SD 721* (K), *Meliden Giking 55* (K); TEKUTAN: *Dius Tadong 135* (K), *474* (K).

65.43.2. Schizostachyum latifolium Gamble, *Ann. Roy. Bot. Gard. (Calcutta)* 7: 117 (1896).

Erect bamboo to 5 m tall; culm diameter 3 cm, green; blades of culm-sheaths ovate-lanceolate, deflexed or spreading. Lowlands, hill forest, in waste ground or disturbed situations. Elevation: 600 m.

Material examined. HIMBAAN: *Doinis Soibeh 100* (K); KIAU: *Topping 1579* (K, US); KITUNTUL: *Tiong SAN 88612* (K); PORING: 600 m, *Dransfield, S. SD 722* (K, L), *Meliden Giking 208* (K); TEKUTAN: *Dius Tadong 226* (K), *378* (K).

65.43.3. Schizostachyum lima (Blanco) Merr., *Sp. Blancoan.*: 77 (1918).

Erect bamboo to 5 m tall; culm 2.5 cm in diameter, green; blades of culm-sheaths long, lanceolate, deflexed. Hill forest, in disturbed areas. Elevation: 1200 m. Widespread in the Philippines, Borneo, to Maluku.

Material examined. PENIBUKAN: 1200 m, *Clemens 30529* (BM, K); TENOMPOK: *Kadir A 1675* (K).

65.43.4. Schizostachyum cf. lima (Blanco) Merr., *Sp. Blancoan.*: 77 (1918).

Erect bamboo to 5.5 m tall; culm 2.5 cm in diameter, green; blades of culm-sheaths long, lanceolate, deflexed. Hill forest in disturbed situations. Elevation: 800–1500 m.

Material examined. DALLAS/TENOMPOK: 1200 m, *Clemens s.n.* (BM); MINITINDUK: 800 m, *Carr SFN 26604* (BM); PENIBUKAN: 1200–1500 m, *Clemens s.n.* (BM).

65.43.5. Schizostachyum indet.

Material examined. DALLAS/TENOMPOK: 1200 m, *Clemens 27592* (BM); PORING: *Sani Sambuling 52* (K), *696* (K).

65.44. SCROTOCHLOA Judziewicz

65.44.1. Scrotochloa urceolata (Roxb.) Judziewicz, *Phytologia* 56: 299 (1984).

Bambusoid grass about 1 m tall; blades stalked. Lowlands, hill forest, possibly lower montane forest. Elevation: 500–1500 m.

Material examined. KIPUNGIT FALLS TRAIL: 700 m, *Beaman 7857* (K, L, MSC); MENGGIS: *Matamin Rumutom 135* (K), *269* (K); NALUMAD: *Daim Andau 211* (K); PINAWANTAI: 500 m, *Shea & Aban SAN 76749* (K, L); TEKUTAN: *Lomudin Tadong 422* (K); TENOMPOK: 1500 m, *Clemens 29988* (K).

65.45. SETARIA P. Beauv.

Veldkamp, J. F. (1994). Miscellaneous notes on Southeast Asian Gramineae. IX. *Setaria* and *Paspalidium*. *Blumea* 39: 373–384.

65.45.1. Setaria italica (L.) P. Beauv., *Ess. Agrostogr.*: 51 (1812).

Tufted erect annual grass to 2 m high. Hill forest, lower montane forest; cultivated, possibly escaped. Elevation: 900–1500 m.

Material examined. DALLAS: 900 m, *Clemens s.n.* (BM); KILEMBUN RIVER: *Lorence Lugas 1880* (K); TENOMPOK: 1500 m, *Clemens s.n.* (BM, BM).

65.45.2. Setaria palmifolia (J. König) Stapf, *J. Linn. Soc., Bot.* 42: 186 (1914).

Robust perennial grass up to 1 m tall, with lanceolate, plicate blades 20–50 cm long. Lowlands, hill forest, lower montane forest; in lightly shaded and moist ground, sometimes in disturbed areas. Elevation: 300–1500 m.

Material examined. BUNDU TUHAN: *Doinis Soibeh 56* (K); DALLAS: 900 m, *Clemens s.n.* (BM); KAUNG: 400 m, *Darnton 301* (BM); KAUNG/DALLAS: 300–600 m, *Clemens 26018* (BM); LIWAGU/ MESILAU RIVERS: 1200 m, *RSNB 2741* (K, L); MELANGKAP TOMIS: *Lorence Lugas 893* (K); PARK HEADQUARTERS: 1500 m, *Ogata 11097* (L), *Simpson & Casserly 89/196* (K); PENATARAN RIVER: 1000 m, *Clemens 34067* (K, L); PENIBUKAN: 1200 m, *Clemens 40389A* (BM, L); TENOMPOK: *Kadir A 1674* (K).

65.45.3. Setaria sphacelata (Schumach.) Stapf & C. E. Hubb., *Fl. Trop. Afr.* 9: 795 (1930).

Tufted annual grass to 1 m tall, usually much shorter. Lowlands and hill forest in open secondary or disturbed areas. Elevation: 500 m.

Material examined. KAUNG: *Amin & Francis SAN 121550* (K); KINASAPIAN: *Amin & Francis SAN 123421* (K); RANAU: 500 m, *Darnton 152* (BM).

65.46. SORGHUM Moench

Snowden, J. D. (1936). *The cultivated races of* Sorghum. Royal Botanic Gardens, Kew. Snowden, J. D. (1955). The wild fodder sorghums of the section *Eu-Sorghum. J. Linn. Soc., Bot.* 55: 191–260.

65.46.1. Sorghum bicolor (L.) Moench, *Methodus*: 207 (1794).

Tall, thick-culmed annual grass to 2–3 m. Lowlands and hill forest, cultivated and possibly escaped into open disturbed areas.

Material examined. MELANGKAP TOMIS: *Lorence Lugas 317* (K), *1837* (K); PORING: *Sani Sambuling 21* (K); TEKUTAN: *Dius Tadong 107* (K).

65.46.2. Sorghum nitidum (Vahl) Pers., *Syn. Pl.* 1: 101 (1805).

Tufted perennial grass 1–2 m tall. Hill forest in open, disturbed situations.

Material examined. RANAU/KUNDASANG: *Sinclair et al. 8959* (K).

65.46.3. Sorghum propinquum (Kunth) Hitchc., *Lingnan Sci. J.* 7: 249 (1931).

Perennial grass to 2 m tall. Lowlands, hill forest, probably in openings in either secondary or primary situations.

Material examined. HIMBAAN: *Doinis Soibeh 135* (K); MARAK-PARAK: *Kinsun Bakia 276* (K); MELANGKAP TOMIS: *Lorence Lugas 501* (K), *2352* (K).

65.47. SPHAEROBAMBOS S. Dransf.

Dransfield, S. (1989). *Sphaerobambos*, a new genus of bamboo (*Gramineae-Bambusoideae*) in Malesia. *Kew Bull.* 44: 428–434.

65.47.1. Sphaerobambos hirsuta S. Dransf., *Kew Bull.* 44: 428 (1989). Type: LOHAN RIVER, 700 m, *Dransfield, S. SD 844* (holotype K!).

Erect or scrambling bamboo 10 m high; culm sheath with stiff light brown hairs. Hill forest on ultramafic substrate. Elevation: 700 m. Endemic to Mount Kinabalu; known only from the Lohan area.

Additional material examined. LOHAN: 700 m, *Dransfield, S. SD 843* (K), *Tiong SAN 88660* (K, L); LOHAN RIVER: *Wong FRI 35154* (K).

65.48. THEMEDA Forssk.

65.48.1. Themeda villosa (Poir.) A. Camus in Lecomte, *Fl. Indo-Chine* 7: 364 (1922).

Large clumped perennial grass 2–4 m tall. Hill forest, in disturbed secondary or primary situations and on river banks.

Material examined. HIMBAAN: *Doinis Soibeh 384* (K); KIAU: *Jusimin Duaneh 177* (K); MELANGKAP TOMIS: *Lorence Lugas 161* (K), *1660* (K), *2479* (K).

65.49. THYSANOLAENA Nees

65.49.1. Thysanolaena latifolia (Roxb.) Honda, *J. Fac. Sci. Univ. Tokyo, Sect. 3, Bot.* 3: 312 (1930).

Robust perennial grass with solitary or clustered stems; leaves broad; the large panicle with numerous small spikelets. Hill forest. Elevation: 500–1400 m.

Material examined. DALLAS: 900 m, *Clemens 27482* (BM), 900 m, *30262* (K), 900 m, *30263* (K); MAMUT: *Amin & Suin SAN 121476* (K); MELANGKAP TOMIS: *Lorence Lugas 1424* (K), *1838* (K);

PENDIRUAN RIVER: 500 m, *Shea & Aban SAN 76898* (K); SAYAP: *Tungking Simbayan 30* (K); SOSOPODON: *Ampon & Saikeh SAN 71802* (K); TAHUBANG FALLS: 1400 m, *Clemens 40271* (BM, K); TEKUTAN: *Dius Tadong 205* (K).

65.50. TRISETUM Pers.

Chrtek, J. (1970). *Trisetum spicatum* (L.) Richt. in northern Borneo. *Folia Geobot. Phytotax.* 5: 447–448. Hultén, E. (1959). The *Trisetum spicatum* complex. *Svensk Bot. Tidskr.* 53: 203–228. Veldkamp, J. F., and van der Have, J. C. (1983). The genus *Trisetum* (Gramineae) in Malesia and Taiwan. *Gard. Bull. Singapore* 36: 125–135.

65.50.1. Trisetum spicatum (L.) K. Richt.

a. subsp. **kinabaluense** Chrtek, *Folia Geobot. Phytotax.* 5: 447 (1970). Plate 21B. Type: SUMMIT AREA, 4000 m, *Clemens 51668* (holotype BM!; isotypes K!, L!).

Tussocky perennial grass, up to 40 cm high. Upper montane forest to summit area, in seepage areas and disturbed places, sometimes weedy. Elevation: 2700–4000 m. Endemic to Mount Kinabalu.

Additional material examined. LAYANG-LAYANG: 2700 m, *Smith 459* (L); PAKA-PAKA CAVE: 3100 m, *Cockburn & Aban SAN 82997* (K, L), 3000 m, *Sinclair et al. 9185* (K, L); PANAR LABAN: 3400 m, *Stone 11342* (L); SAYAT-SAYAT: 3700 m, *Meijer SAN 24208* (K), 3800 m, *Smith 512* (L), *Stein Ki S 50* (L); SUMMIT AREA: 3700 m, *Fuchs 21090* (L); UPPER KINABALU: *Clemens 30261* (K, L).

65.51. YUSHANIA Keng f.

65.51.1. Yushania tessellata (Holttum) S. Dransf., *Kew Bull.* 37: 678 (1983).

Racemobambos tessellata Holttum, *Gard. Bull. Singapore* 26: 211 (1973). Type: MOUNT KINABALU, 2000 m, *Holttum 46* (holotype K!).

Erect bamboo about 7 m tall. Lower montane forest, upper montane forest. Elevation: 1800–2500 m.

Additional material examined. PARK HEADQUARTERS/POWER STATION: 1800 m, *Dransfield, S. SD 756* (K, L); SUMMIT TRAIL: 2500 m, *Dransfield, S. SD 754* (K, L); UPPER KINABALU: *Clemens 29173* (K).

65.52. ZEA L.

65.52.1. Zea mays L., *Sp. Pl.*: 971 (1753).

Robust annual grass with solid culm and broad leaves; male inflorescence terminal, female inflorescence lateral. Cultivated, possibly occurring spontaneously.

Material examined. MELANGKAP TOMIS: *Lorence Lugas 327* (K), *1254* (K); SERINSIM: *Jibrin Sibil 59* (K), *191* (K).

66. SMILACACEAE

66.1. SMILAX L. Plate 21A.

Koyama, T. (1960). Materials toward a monograph of the genus *Smilax*. *Quart. J. Taiwan Mus.* 13: 1–61, 4 pl.

66.1.1. Smilax borneensis A. DC., *Monogr. Phan.* 1: 202 (1878).

Woody climber to 10 m, stems spiny; leaves with dense tomentum below. Lowlands, hill forest. Elevation: 500–900 m.

Material examined. MELANGKAP TOMIS: 500–900 m, *Beaman 8706* (MSC, NY), *Lorence Lugas 1117* (K), *1853* (K); PORING: *Amin et al. SAN 123178* (K); PORING/NALUMAD: 500–600 m, *Kokawa & Hotta 5253* (L); TEKUTAN: *Dius Tadong 99* (K), *798* (K), *Lomudin Tadong 209* (K).

66.1.2. Smilax corbularia Kunth, *Enum. Pl.* 5: 262 (1850).

Unarmed climber to 5 m, infrequently with tendrils; leaves glaucous beneath. Hill forest, lower montane forest. Elevation: 800–2100 m.

Material examined. BAMBANGAN RIVER: 1500 m, *RSNB 1313* (K, L); DALLAS/TENOMPOK: 1200 m, *Clemens s.n.* (BM); EAST MESILAU/MENTEKI RIVERS: 1700–2000 m, *Beaman 9596* (K, MSC, NY); EASTERN SHOULDER: 1200 m, *RSNB 74* (K, L); GURULAU SPUR: 1500 m, *Clemens 50584* (BM); HEMPUEN HILL: *Amin et al. SAN 117207* (K), 800–1200 m, *Beaman 7701* (MSC, NY); KIAU VIEW TRAIL: *Aban Gibot SAN 79601* (K, L); KINATEKI RIVER HEAD: 1800–2100 m, *Clemens 31755* (BM, K, L); MAMUT COPPER MINE: 1400–1500 m, *Beaman 10333* (K, MSC, NY); MAMUT MINE: *Amin et al. SAN 123309* (K); MELANGKAP TOMIS: *Lorence Lugas 386* (K), *973* (K), *1258* (K); MINITINDUK GORGE: 900–1200 m, *Clemens 29663* (BM, K, L); MOUNT KINABALU: 900 m, *Puasa 1542* (K); MT. TEMBUYUKEN: 1800 m, *Aban Gibot SAN 68573* (K, L); NALUMAD: *Daim Andau 205* (K); PARK HEADQUARTERS: 1700 m, *Kanis & Sinanggul SAN 50101* (K), *Lassan SAN 76697* (K, L); PENIBUKAN: 1200–1500 m, *Clemens 31605* (BM), 1200 m, *35050* (BM, L), 1200 m, *40537* (BM), 1200 m, *40645* (BM, K); SOSOPODON: 1500 m, *Battah SAN 33918* (K, L), 1300 m, *Meijer SAN 42470* (K, L); TENOMPOK: 1500 m, *Clemens 28443* (BM, K), 1500 m, *28622* (BM, K, L); TENOMPOK RIDGE: 1400–1500 m, *Beaman 8195* (MSC, NY).

66.1.3. Smilax lanceifolia Roxb., *Fl. Ind.* Ed. 1832, 3: 792 (1832).

Woody climber; stems unarmed or sparsely armed; leaves narrowly to broadly lanceolate, rarely ovate, stipules small, tendrils infrequent. Hill forest, lower montane forest, upper montane forest. Elevation: 700–2700 m.

Material examined. BAMBANGAN RIDGE: 1800–2000 m, *Beaman 11391* (K, MSC); BAMBANGAN RIVER: 1500 m, *RSNB 4474* (K), 1500 m, *RSNB 1330* (K); DALLAS: 900 m, *Clemens 26522* (K, L), 900 m, *26844* (BM, K, L), 900 m, *26916* (BM, K, L), 900 m, *26916A* (BM), 900 m, *26919* (BM, K); EAST MESILAU/MENTEKI RIVERS: 1700–2000 m, *Beaman 9583* (MSC, NY); EASTERN SHOULDER, CAMP 3: 2400 m, *RSNB 1086* (K, L), 2400 m, *RSNB 1114* (K, L); GOLF COURSE SITE: 1800 m, *Beaman 7485* (MSC); HEMPUEN HILL: *Madani SAN 89500* (K, L); HIMBAAN: *Doinis Soibeh 253* (K); KEMBURONGOH: 2400 m, *Clemens 27762* (BM, K, L); KIAU: *Jusimin Duaneh 300* (K); KILEMBUN BASIN: 2300 m, *Clemens 33816* (BM); LIWAGU RIVER TRAIL: 1700 m, *Beaman 11361* (K, MSC); LIWAGU RIVER TRAIL HEAD: 1800–1900 m, *Beaman 11285* (K, MSC); LUMU-LUMU: 2100 m, *Clemens s.n.* (BM), *27848* (BM, K);

MAMUT/BAMBANGAN RIVER: 1600–1800 m, *Kokawa & Hotta 5854* (L); MARAI PARAI: 1500 m, *Clemens s.n.* (BM), 1500 m, *35033* (BM); MARAI PARAI SPUR: 1500 m, *Clemens 32515* (BM); MELANGKAP KAPA: 700–1000 m, *Beaman 8806* (MSC, NY); MELANGKAP TOMIS: *Lorence Lugas 24* (K), *229* (K), *581* (K), *1232* (K), *1319* (K), *2189* (K); MESILAU BASIN: 2100–2400 m, *Clemens 29704* (BM); MESILAU CAVE: 2000–2100 m, *Beaman 8136* (MSC), 2000 m, *11158* (K, MSC), 1800 m, *RSNB 4686* (K), 2000 m, *Collenette 902* (K); MESILAU CAVE TRAIL: 1700–1900 m, *Beaman 7975* (K, MSC, NY), 1700–1900 m, *9087* (K, MSC, NY); MOUNT KINABALU: 2400 m, *Haviland 1145* (K), 1400 m, *Stevens et al. 625* (L), 1500 m, *651* (L); MT. NUNGKEK: 1200 m, *Darnton 465* (BM); PARK HEADQUARTERS: 1600 m, *Amin & Zaini SAN 117271* (K), 1700 m, *Kanis & Sinanggul SAN 51490* (K, L); PENATARAN BASIN: 2700 m, *Clemens 33662* (BM); PENIBUKAN: 1200–1500 m, *Clemens s.n.* (BM, BM), 1200 m, *32152* (BM), 1500 m, *50320* (BM); PINOSUK PLATEAU: 1700 m, *RSNB 1848* (K, L), 1600 m, *De Vogel 8012* (L); PORING: *Amin et al. SAN 123172* (K), *Sani Sambuling 77* (K), *474* (K), *641* (K); POWER STATION/KEMBURONGOH: 1900 m, *Fuchs 21045* (L); TEKUTAN: *Dius Tadong 194* (K), *Lomudin Tadong 12* (K); TENOMPOK: 1500 m, *Clemens s.n.* (BM, BM, K), 1500 m, *27708* (BM, L), 1500 m, *28120* (L), 1700 m, *28126* (BM), 1500 m, *28177* (BM), 1500 m, *30344* (K, L).

66.1.4. Smilax leucophylla Blume, *Enum. Pl. Javae* 1: 18 (1827).

Large climber with woody, armed or unarmed stems; leaves usually ovate, coriaceous, with large stipules;. Hill forest, lower montane forest, upper montane forest, often on ultramafic substrate. Elevation: 600–2900 m. tendrils conspicuous. A highly variable, widely distributed species.

Material examined. BAMBANGAN CAMP: 1500 m, *RSNB 4573* (K); BAMBANGAN RIVER: 1500 m, *RSNB 1283* (K); DALLAS: 900–1200 m, *Clemens s.n.* (BM), 900 m, *26016* (BM, K, L); EASTERN SHOULDER: 2900 m, *RSNB 945* (K, L); GURULAU SPUR: 2400 m, *Clemens 50937* (BM); HEMPUEN HILL: 800–1000 m, *Beaman 7385* (K, L, MSC); HIMBAAN: *Doinis Soibeh 379* (K); KAGAPON HILL: 600 m, *Brand & Anak SAN 25301* (L); KIAU: *Jusimin Duaneh 410* (K); KILEMBUN BASIN: 2400 m, *Clemens 33759* (BM, L), 2100 m, *33779* (BM), 1100 m, *34421* (BM); KINATEKI RIVER: 1500 m, *Clemens 31067* (BM); KULUNG HILL: *Lajangah SAN 44614* (K, L); KUNDASANG: *Tikau SAN 28920* (K); LOHAN: *Lajangah SAN 28772* (L); LOHAN RIVER: 800–1000 m, *Beaman 9069* (MSC, NY); MAMUT RIVER: 1200–1400 m, *Kokawa & Hotta 5804* (L); MARAI PARAI: 1700 m, *Clemens 32450* (BM, L), 1500 m, *32872* (BM, L); MELANGKAP TOMIS: *Lorence Lugas 68* (K), *300* (K), *793* (K); MENGGIS: *Matamin Rumutom 317* (K); MENTEKI RIVER: 1600 m, *Beaman 10778* (MSC, NY); MESILAU CAVE: 1900–2200 m, *Beaman 9561a* (K, MSC, NY), 1800 m, *RSNB 4703* (K); MESILAU CAVE TRAIL: 1700–1900 m, *Beaman 8001* (MSC); MESILAU RIVER: 1500 m, *RSNB 4062* (K), 1400 m, *RSNB 1345* (K); MESILAU/BAMBANGAN RIVERS: *Kokawa & Hotta 4244* (L); MOUNT KINABALU: 1700 m, *Haviland 1226* (K); MT. NUNGKEK: 1200–1500 m, *Clemens 32/99* (BM); NALUMAD: *Amin et al. SAN 115912* (K); PARK HEADQUARTERS: *Aban Gibot SAN 60564* (K); PENATARAN BASIN: 1700 m, *Clemens 34309* (L); PENIBUKAN: *Clemens s.n.* (BM), 1200 m, *30489* (BM, L), 1200 m, *30646 p.p.* (BM), 1200 m, *32014* (BM, L), 1200 m, *32117* (BM, L), 1500–1800 m, *35032* (BM), 1200 m, *40505* (BM), 1200 m, *51680 bis* (BM); PINOSUK: *Amin et al. SAN 123155* (K); PINOSUK PLATEAU: 1400–1500 m, *Beaman 9196* (MSC, NY); SOSOPODON: 1400 m, *Meijer SAN 42771* (K), 1400 m, *Mikil SAN 28149* (K); SUMMIT TRAIL: 1800 m, *Ogata 11146* (L); TENOMPOK: 1500 m, *Clemens s.n.* (BM, BM, BM), 1500 m, *28440* (BM, K, L), 1500 m, *28442* (BM, K), 1500 m, *29364* (BM, K), 1500 m, *29575* (BM, K, L), 1600 m, *29603* (BM, K, L), 1500 m, *30037* (K, L); TENOMPOK RIDGE: 1400–1500 m, *Beaman 8182* (MSC).

66.1.5. Smilax megacarpa A. DC., *Monogr. Phan.* 1: 186 (1878).

Large woody climber, stems smooth or sparsely armed; mature fruits up to 3 cm in diameter. Hill forest. Elevation: 900 m.

Material examined. MEMPAIT: *Amin & Francis SFN 116245* (K); MINITINDUK: 900 m, *Carr SFN 26657* (SING n.v.).

66.1.6. Smilax odoratissima Blume, *Enum. Pl. Javae* 1: 19 (1827).

Climber with stems verruculose and sparsely to abundantly armed; leaves mostly broadly elliptical; pedicels short. Lowlands, hill forest. Elevation: 500–1500 m.

Material examined. DALLAS: 1100–1200 m, *Clemens s.n.* (BM); HIMBAAN: *Doinis Soibeh 199* (K); MARAI PARAI: 1500 m, *Clemens 35034* (BM); MELANGKAP KAPA: 700–1000 m, *Beaman 8790* (MSC, NY); NALUMAD: *Daim Andau 168* (K); TAHUBANG RIVER: 900 m, *Haviland 1293* (K), 1000 m, *Nooteboom & Aban 1534* (K, L); TEKUTAN/NALUMAD: 500 m, *Shea & Aban SAN 77260* (L).

66.1.7. Smilax sp. 1

Climber with conspicuously winged stems, sparsely armed; leaves ovate, 3(–5)-nerved, thin, wrinkled when dry. Hill forest, lower montane forest. Elevation: 1200–1600 m.

Material examined. LUBANG: *Clemens, M. S. 10417* (K), *Topping 1789* (K); PARK HEADQUARTERS: *Amin & Galuis et al. SAN 117330* (K); PENIBUKAN: 1200 m, *Clemens 30646 p.p.* (BM), 1200 m, *50103* (BM); SOSOPODON: 1400 m, *Battah SAN 33912* (K, L), 1400–1600 m, *Kokawa & Hotta 5182* (L); TENOMPOK: 1500 m, *Clemens s.n.* (BM), 1500 m, *30032* (K), 1500 m, *30038* (K).

66.1.8. Smilax sp. 2

Climber with conspicuously winged, zig-zaged stems, sparsely armed; leaves ovate, 3(–5)-nerved, thick, coriaceous. Upper montane forest. Elevation: 2100–3400 m.

Material examined. KINATEKI RIVER HEAD: 2700 m, *Clemens 31967* (BM); LAYANG-LAYANG: 2600 m, *Beaman 11420* (K, MSC); LUMU-LUMU: 2100 m, *Clemens s.n.* (BM); LUMU-LUMU & KEMBURONGOH: *Clemens s.n.* (BM); MARAI PARAI/KEMBURONGOH: 2100 m, *Clemens 33152* (BM, L); PAKA-PAKA CAVE: 3400 m, *Clemens s.n.* (BM).

67. TACCACEAE

In collaboration with B. Mathew (K)

Drenth, E. (1972). A revision of the family Taccaceae. *Blumea* 20: 375–406. Drenth, E. (1976). Taccaceae. *Fl. Males.* I, 7: 806–819.

67.1. TACCA J. R. Forst. & G. Forst.

67.1.1. Tacca integrifolia Ker Gawl., *Bot. Mag.* 35: t. 1488 (1812). Plate 21C.

Terrestrial perennial scapose herb from a tuberous rhizome; leaves entire. Lower montane forest, probably on ultramafic substrate. Elevation: 1500 m.

Material examined. GURULAU SPUR: 1500 m, *Clemens 50501* (BM).

67.1.2. Tacca palmata Blume, *Enum. Pl. Javae* 1: 83 (1827).

Terrestrial perennial scapose herb from a tuberous rhizome; leaves deeply palmately lobed. Hill forest, generally on ultramafic substrate. Elevation: 500–1200 m.

Material examined. DALLAS: 900–1200 m, *Clemens 26370* (BM, L); LOHAN RIVER: 800–1000 m, *Beaman 9979* (K, MSC); MELANGKAP TOMIS: *Lorence Lugas 272* (K); MENGGIS: *Matamin Rumutom 54* (K); MENGGIS?: *Matamin Rumutom 417* (K); NALUMAD: *Daim Andau 568* (K); PENATARAN RIVER: 500 m, *Beaman 8871* (L, MSC); TEKUTAN: *Lomudin Tadong 516* (K).

68. TRIURIDACEAE

In collaboration with M. Cheek (K)

Beccari, O. (1890). Le Triuridaceae della Malesia. *Malesia* 3: 318–344, pl. 39–42. Giesen, H. (1938). Triuridaceae. *Pflanzenr.* IV. 18 (Heft 104): 1–84. Meerendonk, J. P. M. van de. (1984). Triuridaceae. *Fl. Males.* I, 10: 109–121.

68.1. SCIAPHILA Blume

68.1.1. Sciaphila arfakiana Becc., *Malesia* 3: 337, t. 41: 6-14 (1890).

Andruris clemensae (Hemsl.) Giesen var. *borneensis* Giesen, *Pflanzenr.* IV. 18 (Heft 104): 25 (1938). Type: PENIBUKAN, 900–1500 m, *Clemens 51263A* (holotype B n.v.; isotype BM!).

Tiny echlorophyllous saprophytic herb. Hill forest, lower montane forest, on ultramafic substrate. Elevation: 900–1500 m.

Additional material examined. MARAI PARAI SPUR: *Clemens, M. S. 11029A* (L); PENIBUKAN: 1200 m, *Clemens 32025* (BM), 900–1500 m, *51263* (BM, K).

68.1.2. Sciaphila secundiflora Thwaites ex Benth., *Hooker's J. Bot. Kew Gard. Misc.* 7: 10 (1855). Plate 22A, B, C.

Tiny echlorophyllous saprophytic herb. Hill forest on ultramafic substrate. Elevation: 1200 m. This record based on *Carr SFN 26537* (SING), cited by Giesen as *S. major* Becc.

Material examined. PENIBUKAN: 1200 m, *Carr SFN 26537* (SING n.v.).

68.1.3. Sciaphila tenella Blume, *Bijdr.* 10: 515 (1826).

Tiny echlorophyllous saprophytic herb. Hill forest on ultramafic substrate. Elevation: 1200 m.

Material examined. PENIBUKAN: 1200 m, *Clemens 51264* (BM, L).

68.1.4. Sciaphila winkleri Schltr., *Bot. Jahrb. Syst.* 48: 88 (1912).

Tiny red echlorophyllous saprophytic herb. Hill forest, mostly on ultramafic substrate. Elevation: 700–1500 m.

Material examined. MARAI PARAI: *Clemens, M. S. 11030a* (BO); MELANGKAP KAPA: 700–1000 m, *Beaman 8778* (L); PENIBUKAN: 1300 m, *Clemens 40730 p.p.* (B), 900–1500 m, *51263a bis* (B).

68.1.5. Sciaphila indet.

Material examined. GURULAU SPUR: 1500 m, *Clemens 50539* (BM).

69. ZINGIBERACEAE

In collaboration with E. J. Cowley (K) and R. M. Smith (E)

Burtt, B. L., and Smith, R. M. (1972). Key species in the taxonomic history of Zingiberaceae. *Notes Roy. Bot. Gard. Edinburgh* 31: 177–228. Burtt, B. L., and Smith, R. M. (1972). Tentative keys to the subfamilies, tribes and genera of Zingiberaceae. *Notes Roy. Bot. Gard. Edinburgh* 31: 171–176. Holttum, R. E. (1950). The Zingiberaceae of the Malay Peninsula. *Gard. Bull. Singapore* 13: 1–250. Schumann, K. (1904). Zingiberaceae. *Pflanzenr.* IV. 46: 1–458.

69.1. ALPINIA Roxb.

Smith, R. M. (1985). A review of Bornean Zingiberaceae: 1 (Alpineae p.p.). *Notes Roy. Bot. Gard. Edinburgh* 42: 261–314 [*Alpinia* on pp. 273–288]. Smith, R. M. (1990). *Alpinia* (Zingiberaceae): A proposed new infrageneric classification. *Edinburgh J. Bot.* 47: 1–75.

69.1.1. Alpinia aquatica (Retz.) Roscoe, *Trans. Linn. Soc. London* 8: 346 (1807).

Rhizomatous perennial herb; inflorescence terminal. Hill forest.

Material examined. KINOPULIDAN: *Amin et al. SAN 116022* (K); MARAK-PARAK: *Amin & Mansus SAN 118997* (K); MELANGKAP TOMIS: *Lorence Lugas 103* (K), *270* (K), *1750* (K), *2216* (K), *2544* (K); MENGGIS: *Matamin Rumutom 335* (K); MT. TEMBUYUKEN: *Collector unknown SNP 144* (SNP); NALUMAD: *Daim Andau 296* (K); PERANCANGAN: *Amin et al. SAN 115942* (K); PORING: *Sani Sambuling 70* (K), *379* (K); SERINSIM: *Jibrin Sibil 84* (K), *319* (K), *Kinsun Bakia 152* (K); TEKUTAN: *Dius Tadong 64* (K), *327* (K), *362* (K), *408* (K), *426* (K), *801* (K).

69.1.2. Alpinia beamanii R. M. Sm., *Notes Roy. Bot. Gard. Edinburgh* 45: 341 (1989 "1988").

Rhizomatous perennial herb; inflorescence terminal. Hill forest. Elevation: 1200 m. The name honours Reed Beaman, who collected the type in the Crocker Range.

Material examined. PENIBUKAN: 1200 m, *Clemens s.n.* (BM).

69.1.3. Alpinia cf. **brevilabris** Presl, *Reliq. Haenk.* 1: 110, t. 17 (1827).

Rhizomatous perennial herb to 4 m high; inflorescence terminal. Lowland secondary forest. Elevation: 500 m.

Material examined. PINAWANTAI: 500 m, *Shea & Aban SAN 76969* (K).

69.1.4. Alpinia capitellata Jack, *Malayan Misc.* 2, n. 7: 4 (1822).

Rhizomatous perennial herb; flowers terminal on the leafy shoot. Hill forest, sometimes on ultramafic substrate. Elevation: 500 m.

Material examined. MELANGKAP TOMIS: *Lorence Lugas 2090* (K); RANAU: 500 m, *Darnton 181* (BM).

69.1.5. Alpinia galanga (L.) Willd., *Sp. Pl.*: 12 (1797).

Perennial herb; flowers borne terminally on the leafy shoot. Lowlands, cultivated and possibly occurring spontaneously in secondary situations.

Material examined. MELANGKAP TOMIS: *Lorence Lugas 180* (K), *289* (K), *1239* (K), *1427* (K); PORING: *Sani Sambuling 368* (K).

69.1.6. Alpinia glabra Ridl., *J. Straits Branch Roy. Asiat. Soc.* 32: 116 (1899).

Rhizomatous perennial herb to 1.5 m; inflorescence terminal on leafy stem. Hill forest, possibly also in lower montane forest. Elevation: 1200–1500 m. A variable species in need of more critical study.

Material examined. DALLAS/TENOMPOK: 1200 m, *Clemens s.n.* (BM); PENIBUKAN: 1200–1500 m, *Clemens s.n.* (BM), 1200–1500 m, *31325* (BM), 1200 m, *32115* (BM, L); PENIBUKAN TRAIL: *Collector unknown SNP 1045* (KNP), *SNP 1048* (KNP), *SNP 1146* (KNP); TAHUBANG RIVER: 1500 m, *Clemens 31022* (BM).

69.1.7. Alpinia havilandii K. Schum., *Pflanzenr.* IV. 46 (Heft 20): 329 (1904). Plate 23A. Type: MOUNT KINABALU, 1500 m, *Haviland 1233* (holotype K!).

Rhizomatous perennial herb to 2 m; inflorescence pushed out laterally below uppermost leaves but nonetheless terminal. Lower montane forest. Elevation: 1400–2000 m. Occurring also in the Crocker Range and on Mt. Trus Madi.

Additional material examined. DALLAS/TENOMPOK: 1400 m, *Clemens s.n.* (BM); EAST MESILAU RIVER: 2000 m, *Collenette 894* (K, L); EAST MESILAU/MENTEKI RIVERS: 1700–2000 m, *Beaman 9603* (E, MSC); EASTERN SHOULDER: 1500 m, *RSNB 1566* (K); EASTERN SHOULDER, CAMP 2: 2000 m, *RSNB 1066* (K); GOLF COURSE SITE: 1700–1800 m, *Beaman 7202* (E, K, MSC); LIWAGU/MESILAU RIVERS: 1500 m, *RSNB 1370* (K, L), 1500 m, *RSNB 2880* (K); MAMUT RIDGE: 1500–1800 m, *Kokawa & Hotta 5667* (L); MELANGKAP TOMIS: *Lorence Lugas 799* (K), *2111* (K); MENTEKI RIVER: 1600 m, *Beaman 10756* (E, MSC); PARK HEADQUARTERS: *Lajangah SAN 33113* (K), 1600 m, *Stone 11412* (L); PARK HEADQUARTERS/POWER STATION: 1600–1800 m, *Kokawa 6176* (L); TENOMPOK: 1500 m, *Clemens 28481* (K, L).

69.1.8. Alpinia latilabris Ridl., *J. Straits Branch Roy. Asiat. Soc.* 32: 268 (1899). Plate 22D.

Rhizomatous perennial herb to 2 m with inflorescence up to 20 cm long. Hill forest. Elevation: 900 m.

Material examined. DALLAS: 900 m, *Clemens s.n.* (BM, BM); KIAU: *Clemens, M. S. 10157* (BM, K); MELANGKAP TOMIS: *Lorence Lugas 307* (K), *1646* (K); PORING: *Lamb s.n.* (E).

69.1.9. Alpinia ligulata K. Schum., *Bot. Jahrb. Syst.* 27: 275 (1899).

Rhizomatous perennial herb with terminal inflorescence on leafy stem. Lowlands in primary forest and thickets, rarely in secondary situations. Elevation: 500 m.

Material examined. LIWAGU RIVER: *Amin & Suin SAN 123287* (K); MELANGKAP TOMIS: *Lorence Lugas 105* (K), *822* (K); MENGGIS: *Matamin Rumutom 25* (K); NALUMAD: *Daim Andau 891* (K); PINAWANTAI: 500 m, *Shea & Aban SAN 76759* (K); PORING: *Sani Sambuling 22* (K); SERINSIM: *Kinsun Bakia 76* (K); TEKUTAN: *Dius Tadong 435* (K).

69.1.10. Alpinia nieuwenhuizii Valeton, *Bull. Inst. Bot. Buitenzorg* 20: 86 (1904).

Alpinia flava Ridl. in Gibbs, *J. Linn. Soc., Bot.* 42: 163 (1914). Type: GURULAU SPUR, 1500 m: *Gibbs 4011* (holotype K!).

Rhizomatous perennial herb to 1.5 m; inflorescence terminal on leafy stem. Infrequently in the lowlands, mostly in hill forest, occasionally in lower montane forest. Elevation: 400–1700 m.

Material examined. BUNDU TUHAN: *Doinis Soibeh 85* (K); DALLAS: 900 m, *Clemens 26898* (BM, K, L); EASTERN SHOULDER, CAMP 1: 1200 m, *RSNB 1203* (K); GOLF COURSE: *Smith 28/86* (E); KAUNG: 400 m, *Darnton 371* (BM); KIAU: *Jusimin Duaneh 271* (K); KIBUNDU: *Amin & Jarius SAN 121173* (K); LOHAN/MAMUT COPPER MINE: 900 m, *Beaman 10634* (MSC), 1000 m, *10644* (MSC); LUGAS HILL: 1300 m, *Beaman 8440* (E, MSC); MELANGKAP TOMIS: *Lorence Lugas 1198* (K), *1423* (K), *2521* (K); MESILAU

CAMP/KUNDASANG: 1400–1600 m, *Kokawa & Hotta 4482* (L); MT. NUNGKEK: 1100 m, *Darnton 448* (BM); PENIBUKAN: 1200–1500 m, *Clemens s.n.* (BM, L); PENIBUKAN/TAHUBANG RIVER: 1200 m, *Clemens 31120* (BM, L), 1100 m, *Nooteboom & Aban 1585* (L); PINOSUK: 1200 m, *Meijer SAN 20298* (K); SOSOPODON/KUNDASANG: 1500 m, *Mikil SAN 29245* (K); TENOMPOK: 1500 m, *Clemens 26898 bis* (K, L).

69.2. AMOMUM L.

Smith, R. M. (1985). A review of Bornean Zingiberaceae: 1 (Alpineae p.p.). *Notes Roy. Bot. Gard. Edinburgh* 42: 261–314 [*Amomum* on pp. 295–314].

69.2.1. Amomum anomalum R. M. Sm., *Bot. J. Linn. Soc.* 85: 61 (1982).

Rhizomatous perennial herb to 5 m; inflorescence from the rhizome. Hill forest, possibly also lower montane forest. Elevation: 1100–1500 m.

Material examined. EASTERN SHOULDER: 1100 m, *RSNB 672* (K); HAYE-HAYE RIVER: *Smith et al. 50/86* (E); MAMUT RIVER: 1200 m, *RSNB 1652* (K, L); TENOMPOK: 1500 m, *Clemens 27537* (K).

69.2.2. Amomum aff. anomalum R. M. Sm., *J. Linn. Soc., Bot.* 85: 61 (1982).

Rhizomatous perennial herb to 3 m high; inflorescence from the base. Hill forest.

Material examined. DALLAS: *Clemens 27537* (BM).

69.2.3. Amomum compactum Sol. ex Maton, *Trans. Linn. Soc. London* 10: 251 (1811).

Rhizomatous perennial herb with inflorescence borne separately from the leafy shoot at or just above ground level. Lower montane forest.

Material examined. LIWAGU RIVER TRAIL: *Smith & Phillipps 25* (E); PARK HEADQUARTERS: *Smith 12/86* (E).

69.2.4. Amomum flavoalbum R. M. Sm., *Notes Roy. Bot. Gard. Edinburgh* 42: 310 (1985).

Rhizomatous perennial herb with the inflorescence borne separately from the leafy shoot at or just above ground level. Lower montane forest. Elevation: 1500 m.

Material examined. MESILAU RIVER: 1500 m, *RSNB 4042* (K), 1500 m, *RSNB 4094* (K).

69.2.5. Amomum kinabaluense R. M. Sm., *Notes Roy. Bot. Gard. Edinburgh* 44: 233 (1987). Type: PARK HEADQUARTERS, 1500 m, *Smith 16/86* (holotype E!).

Rhizomatous perennial herb to 1 m; inflorescence basal. Lower montane forest. Elevation: 1400–1800 m. Apparently endemic to Mount Kinabalu, but possibly occurring also in the adjacent Crocker Range.

Additional material examined. LIWAGU RIVER TRAIL: *Smith & Phillipps 27* (E); MARAI PARAI: 1500 m, *Clemens 33100* (BM); MESILAU CAMP: 1400 m, *Meijer SAN 38062* (K); PARK HEADQUARTERS: *Smith 15/86* (E); PENATARAN RIVER: 1800 m, *Clemens 33099* (BM); TENOMPOK: 1500 m, *Carr SFN 27050* (BM), 1500 m, *Clemens s.n.* (BM); TENOMPOK/TOMIS: 1500–1800 m, *Clemens s.n.* (BM), 1600 m, *29264* (BM, K); TINEKEK FALLS: 1700 m, *Clemens 40894* (BM).

69.2.6. Amomum longipedunculatum R. M. Sm., *Notes Roy. Bot. Gard. Edinburgh* 42: 309 (1985). Type: MESILAU RIVER, *Argent 1360* (holotype E!).

Rhizomatous perennial herb to 1.5 m with stilt roots; inflorescence basal. Lower montane forest, upper montane forest. Elevation: 2100–2700 m. Endemic to Mount Kinabalu.

Additional material examined. KEMBURONGOH: 2100 m, *Clemens 29798* (BM, K); KINATEKI RIVER: *Clemens s.n.* (BM); KINATEKI RIVER HEAD: 2700 m, *Clemens 31969* (BM); PARK HEADQUARTERS: *Smith 40/86* (E).

69.2.7. Amomum oliganthum K. Schum., *Bot. Jahrb. Syst.* 27: 321 (1899). Plate 23B.

Rhizomatous perennial herb to 1 m; inflorescence borne separately from the leafy shoot at or just above ground level. Hill forest.

Material examined. HAYE-HAYE RIVER: *Smith et al. 51/86* (E); PORING HOT SPRINGS/LANGANAN FALLS: *Smith et al. 44/86* (E).

69.2.8. Amomum sceletescens R. M. Sm., *Edinburgh J. Bot.* 47: 367 (1990). Plate 23C. Type: LIWAGU/MESILAU RIVERS, 1200 m, *RSNB 2649* (holotype K!).

Rhizomatous perennial herb to 3 m, inflorescence borne separately from the leafy shoot at or just above ground level. Hill forest, lower montane forest on scree slopes. Elevation: 900–1500 m.

Additional material examined. BAMBANGAN CAMP: 1500 m, *RSNB 4580* (K); DALLAS: 900 m, *Clemens 26838* (BM, K), 900 m, *27624* (BM); EASTERN SHOULDER: 1100 m, *RSNB 672* (K); MAMUT RIVER: 1200 m, *RSNB 1714* (K, L); PENIBUKAN: 1200–1500 m, *Clemens 31547* (BM); TENOMPOK: 1500 m, *Clemens 28430* (BM).

69.2.9. Amomum uliginosum J. König in Retz., *Observ. Bot.* 3: 56 (1783).

Rhizomatous perennial herb; inflorescence from the base. Lowland primary forest.

Material examined. SERINSIM: *Kinsun Bakia 451* (K).

69.2.10. **Amomum** indet.

Material examined. SAYAP: *Yalin Surunda 38* (K).

69.3. BOESENBERGIA Kuntze

Smith, R. M. (1987). A review of Bornean Zingiberaceae: III (Hedychieae). *Notes Roy. Bot. Gard. Edinburgh* 44: 203–232 [*Boesenbergia* on pp. 215–232].

69.3.1. **Boesenbergia** aff. **grandis** R. M. Sm., *J. Linn. Soc., Bot.* 85: 49 (1982).

Clump-forming perennial herb with several-leaved fertile shoots; inflorescence bracts 2-ranked. Lowland primary forest.

Material examined. SERINSIM: *Kinsun Bakia 450* (K).

69.3.2. **Boesenbergia** cf. **parva** (Ridl.) Merr., *Bibl. Enum. Born. Pl.*: 122 (1921).

Clump-forming perennial herb; leaves very small, to 10 cm long; inflorescence bracts 2-ranked. Primary lowland and hill forest.

Material examined. BUNGOL: *Clemens s.n.* (BM); SERINSIM: *Kinsun Bakia 444* (K).

69.3.3. **Boesenbergia pulchella** (Ridl.) Merr., *Bibl. Enum. Born. Pl.*: 122 (1921).

Clump-forming erect perennial herb; leaf shoots with 2-several leaves; inflorescence bracts 2-ranked. Lowlands on semi-shaded river banks. Elevation: 500 m.

Material examined. KOTA BELUD/KAUNG: 500 m, *Darnton 501* (BM).

69.4. BURBIDGEA Hook. f.

Smith, R. M. (1985). A review of Bornean Zingiberaceae: 1 (Alpineae p.p.). *Notes Roy. Bot. Gard. Edinburgh* 42: 261–314 [*Burbidgea* on pp. 262–265].

69.4.1. **Burbidgea schizocheila** Hackett, *Gard. Chron.* 36: 301 (1904).

Burbidgea pubescens Ridl., *J. Bot.* 75: 203 (1937). Type: KIAU, *Clemens, M. S. 9939* (holotype K!).

Rhizomatous perennial herb to 2.5 m, epiphytic or terrestrial; inflorescence terminal on the leafy stem. Hill forest, lower montane forest. Elevation: 800–1600 m.

Additional material examined. BAMBANGAN RIVER: 1500 m, *RSNB 4467* (K); MAMUT MINE: 1400 m, *Collenette 1003* (K, L); MENTEKI RIVER: 1600 m, *Beaman 10757* (E, MSC); PENIBUKAN: 1200 m, *Clemens 31858* (BM); SINGH'S PLATEAU: 900 m, *RSNB 1016* (K, L); TAHUBANG FALLS: 1200 m, *Clemens 30677* (BM); TAHUBANG RIVER: 1200–1500 m, *Clemens 31654* (BM), 900–1100 m, *Kanis SAN 51474* (K); TINEKEK RIVER: 800 m, *Collenette A 1* (BM).

69.4.2. Burbidgea indet.

Material examined. TAHUBANG RIVER: *Clemens 31654* (L).

69.5. CURCUMA L.

69.5.1. Curcuma zedoaria (Christm.) Roscoe, *Trans. Linn. Soc.* 8: 354 (1807).

Rhizomatous perennial herb with terminal inflorescence. Lowlands and hill forest. Elevation: 400–900 m. Introduced and naturalized but apparently very rare; origin unknown, cf. Burtt (1977).

Material examined. KAUNG: 400 m, *Carr SFN 27319* (BM); KIAU: 900 m, *Gibbs 3956* (BM).

69.6. ETLINGERA Giseke

Smith, R. M. (1986). A review of Bornean Zingiberaceae: II (Alpineae, concluded). *Notes Roy. Bot. Gard. Edinburgh* 43: 439–466 [*Etlingera* on pp. 440–457].

69.6.1. Etlingera cf. brachychila (Ridl.) R. M. Sm., *Notes Roy. Bot. Gard. Edinburgh* 43: 243 (1986).

Large perennial herb from rhizome; inflorescence borne separately from the leafy shoot. Lower montane forest. Elevation: 1500 m. The collection differs in flower colour from other specimens of this species.

Material examined. TENOMPOK: 1500 m, *Clemens 28306* (K, L).

69.6.2. Etlingera brevilabris (Valeton) R. M. Sm., *Notes Roy. Bot. Gard. Edinburgh* 43: 243 (1986).

Rhizomatous perennial herb to 1 m with few-flowered basal inflorescence from the rhizome. Hill forest. Elevation: 900 m.

Material examined. DALLAS: 900 m, *Clemens s.n.* (BM); KAUNG/KIAU: *Clemens, M. S. 9912* (UC).

69.6.3. Etlingera elatior (Jack) R. M. Sm., *Notes Roy. Bot. Gard. Edinburgh* 43: 244 (1986).

Long-leaved rhizomatous perennial herb; petioles long; inflorescence scapose from the base. Lowlands and hill forest, in wet places in primary and secondary situations. Elevation: 900 m. The torch ginger.

Material examined. DALLAS: 900 m, *Clemens s.n.* (BM); PORING: *Meliden Giking 77* (K); TEKUTAN: *Dius Tadong 379* (K).

69.6.4. Etlingera fimbriobracteata (K. Schum.) R. M. Sm., *Notes Roy. Bot. Gard. Edinburgh* 43: 245 (1986). Plate 24A.

Rhizomatous perennial herb to 3 m; inflorescence borne separately from the leafy shoot. Lowlands, hill forest. Elevation: 1200–1400 m.

Material examined. KINUNUT RIVER HEAD: 1400 m, *Carr SFN 27172* (BM); LIWAGU/MESILAU RIVERS: 1300 m, *RSNB 1976* (K, L); MAMUT RIVER: 1200 m, *RSNB 1744* (K); MELANGKAP TOMIS: *Lorence Lugas 609* (K), *610* (K), *1148* (K); PORING HOT SPRINGS: *Smith 35/86* (E).

69.6.5. Etlingera littoralis (J. König) Giseke, *Prael. Ord. Nat. Pl.* 209, 229, 251 (1792).

Rhizomatous perennial herb; inflorescence borne at the base, separate from the leafy shoot. Lowlands, hill forest.

Material examined. MT. TEMBUYUKEN: *Argent & Walpole 1455* (E); PORING HOT SPRINGS: *Smith 31/86* (E).

69.6.6. Etlingera cf. metriocheilos (Griff.) R. M. Sm., *Notes Roy. Bot. Gard. Edinburgh* 43: 247 (1986). Plate 23D.

Rhizomatous perennial herb; inflorescence borne separately from the leafy shoot. Hill forest. Elevation: 1300 m.

Material examined. LUGAS HILL: 1300 m, *Beaman 9519* (E, MSC).

69.6.7. Etlingera muluensis R. M. Sm., *Notes Roy. Bot. Gard. Edinburgh* 43: 455 (1986).

Erect rhizomatous perennial herb to 4 m high; inflorescence borne separately from the leafy shoot. Hill forest, lower montane forest.

Material examined. PARK HEADQUARTERS: *Smith 11/86* (E); RANAU: *Stevens et al. 641* (E).

69.6.8. Etlingera aff. muluensis R. M. Sm., *Notes Roy. Bot. Gard. Edinburgh* 43: 455 (1986).

Erect rhizomatous perennial herb to 4 m high; inflorescence borne separately from the leafy shoot. Lower montane forest. Elevation: 1200–1500 m.

Material examined. MESILAU RIVER: 1500 m, *RSNB 4139* (K); PENATARAN RIVER: 1200 m, *Clemens 34157* (BM, K, L); TENOMPOK: 1500 m, *Clemens 28053* (BM).

69.6.9. Etlingera punicea (Roxb.) R. M. Sm. sens. lat., *Notes Roy. Bot. Gard. Edinburgh* 43: 249 (1986). Plate 24B.

Rhizomatous perennial herb to 4 m high; inflorescence borne separately from the leafy shoot. Hill forest, infrequently lower montane forest, in primary or secondary situations. Elevation: 600–1000 m.

Material examined. MELANGKAP KAPA: 700–1000 m, *Beaman 8783* (E, MSC); MESILAU CAMP: *Poore H 261* (K); PORING HOT SPRINGS: 600 m, *Price 179* (K); SINGGAREN: 600 m, *Darnton 511* (BM); SINGH'S PLATEAU: 900 m, *RSNB 1018* (K).

69.6.10. Etlingera pyramidosphaera (K. Schum.) R. M. Sm., *Notes Roy. Bot. Gard. Edinburgh* 43: 249 (1986).

Large rhizomatous perennial herb; petioles short; inflorescence borne separately from the leafy shoot. Primary hill forest.

Material examined. MELANGKAP TOMIS: *Lorence Lugas 880* (K).

69.6.11. Etlingera velutina (Ridl.) R. M. Sm., *Notes Roy. Bot. Gard. Edinburgh* 43: 250 (1986).

Large rhizomatous perennial herb with runners; inflorescence from the base. Lower montane forest. Elevation: 1500 m.

Material examined. BAMBANGAN RIVER: 1500 m, *RSNB 4390* (K, L).

69.7. GLOBBA L.

Smith, R. M. (1988). A review of Bornean Zingiberaceae: IV (Globbeae). *Notes Roy. Bot. Gard. Edinburgh* 45: 1–19.

69.7.1. Globba atrosanguinea Teijsm. & Binn., *Natuurk. Tijdschr. Ned.-Indië* 27: 22 (1864).

Perennial herb to 1 m high; inflorescence terminal on the leafy shoot. Hill forest. Elevation: 600–900 m. Produces many bulbils.

Material examined. DALLAS/KIAU: 900 m, *Clemens 26200* (BM, K, L); KINATEKI RIVER: 600 m, *Collenette A 129* (BM, L); LOHAN/MAMUT COPPER MINE: 900 m, *Beaman 10620* (E, MSC); NALUMAD: *Daim Andau 597* (K); SINGH'S PLATEAU: 900 m, *RSNB 1030* (K, L).

69.7.2. Globba franciscii Ridl., *J. Linn. Soc., Bot.* 42: 162 (1914).

Perennial herb to 1 m high; inflorescence terminal on the leafy shoot. Hill forest, lower montane forest. Elevation: 600–1500 m.

Material examined. KIAU: *Jusimin Duaneh 231* (K); KULUNG HILL: 600 m, *Darnton 166* (BM, L); LOHAN RIVER: 700–900 m, *Beaman 9209* (E, MSC); MENGGIS: *Matamin Rumutom 3* (K); MOUNT KINABALU: *Clemens, M. S. 10177* (K, UC); NALUMAD: *Daim Andau 15* (K); NAPUNG: *Sani Sambuling 500* (K); PORING: *Sani Sambuling 50* (K), *336* (K), *Smith 33/86* (E); RANAU/PORING: 600 m, *Darnton 166* (BM); SERINSIM: *Kinsun Bakia 70* (K); SINGH'S PLATEAU: 900 m, *RSNB 1031* (K); TEKUTAN: *Dius Tadong 677* (K), *Lomudin Tadong 136* (K); TENOMPOK: 1500 m, *Clemens 30278* (K, L).

69.7.3. Globba pendula Roxb., *Asiat. Res.* 11: 359 (1810).

Perennial herb to 1 m high; inflorescence terminal on the leafy shoot. Hill forest. Elevation: 900–1500 m. Bulbils sometimes produced.

Material examined. DALLAS: 900 m, *Clemens 26073* (BM, K, L); KIAU: *Smith 48/86* (E); LUGAS HILL: 1300 m, *Beaman 10544* (E, K, MSC), *Collector unknown SNP 1017* (SNP); MELANGKAP TOMIS: *Lorence Lugas 1897* (K); PENIBUKAN: 1200–1500 m, *Clemens s.n.* (BM); PORING: *Sani Sambuling 623* (K); TINEKEK RIVER: 1200 m, *Carr SFN 27928* (BM).

69.7.4. Globba propinqua Ridl., *J. Straits Branch Roy. Asiat. Soc.* 46: 230 (1906).

Small perennial herb with arching terminal inflorescence on the leafy shoot. Lowlands, hill forest, lower montane forest. Elevation: 400–1500 m. Bulbils frequent.

Material examined. DALLAS/TENOMPOK: 1100 m, *Clemens s.n.* (BM), *26801* (BM); KAUNG: 400 m, *Carr SFN 26275* (BM); KINATEKI RIVER: 800 m, *Collenette A 121* (BM); LIWAGU RIVER TRAIL: *Smith & Phillipps 24a* (E); MENGGIS: *Matamin Rumutom 12* (K); PENATARAN BASIN: 1400 m, *Clemens 40198* (BM); PENIBUKAN: 1200–1500 m, *Clemens s.n.* (BM), *1200* m, *32094* (BM), 1200 m, *35206* (BM).

69.7.5. Globba tricolor Ridl.

a. var. gibbsiae (Ridl. in Gibbs) R. M. Sm., *Notes Roy. Bot. Gard. Edinburgh* 45: 11 (1988). Plate 23E.

Perennial herb; inflorescence terminal on the leafy shoot. Hill forest. Elevation: 600–1500 m.

Material examined. DALLAS: 900 m, *Clemens 26822* (BM, K, L); KIAU: *Jusimin Duaneh 533* (K); KIPUNGIT HILL: 700 m, *Beaman 7665* (MSC), 700 m, *7826* (E, MSC); MELANGKAP KAPA: 600–700 m, *Beaman 8565* (MSC); MELANGKAP TOMIS: *Lorence Lugas 1193* (K); PENIBUKAN: 1500 m, *Clemens s.n.* (BM), 1200 m, *50125* (BM); SAYAP: 800–1000 m, *Beaman 9805* (E, K, MSC).

69.8. HEDYCHIUM J. König

Smith, R. M. (1987). A review of Bornean Zingiberaceae: III (Hedychieae). *Notes Roy. Bot. Gard. Edinburgh* 44: 203–232 [*Hedychium* on pp. 204–205].

69.8.1. Hedychium cylindricum Ridl., *J. Malayan Branch Roy. Asiat. Soc.* 1: 98 (1923). Plate 25A.

Tall rhizomatous perennial herb to 1.5 m; inflorescence terminal on the leafy stem. Hill forest, lower montane forest, sometimes on ultramafic substrate. Elevation: 900–1800 m.

Material examined. BAMBANGAN RIVER: 1500 m, *RSNB 4540* (K, L); DALLAS: 900 m, *Clemens s.n.* (BM); GOLF COURSE SITE: 1700–1800 m, *Beaman 7203* (E, MSC), 1700 m, *8547* (E, MSC), 1700–1800 m, *10677* (E, MSC); MAMUT MINE: 1400 m, *Collenette 1012* (K); MARAI PARAI: 1500 m, *Clemens 32315* (BM); MESILAU: *Amin et al. SAN 123335* (K), *SAN 123510* (K), *Mason 2681* (K); MESILAU RIVER: 1500 m, *RSNB 1365* (K); PARK HEADQUARTERS: *Price 118* (K), 1400 m, *250* (K); TENOMPOK: 1500–1600 m, *Beaman 9456* (E, MSC), 1500 m, *Clemens s.n.* (BM), 1500 m, *28170* (BM, K, L), 1500 m, *28451* (K); WASAI FALLS: 1200 m, *Clemens 34147* (BM).

69.8.2. Hedychium muluense R. M. Sm., *Bot. J. Linn. Soc.* 85: 57 (1982).

Rhizomatous perennial herb; inflorescence terminal on the leafy stem. Hill forest, lower montane forest. Elevation: 600–1500 m.

Material examined. DALLAS: 900 m, *Clemens 26045* (BM), 900 m, *26804A* (BM, L); DALLAS/TENOMPOK: 1200 m, *Clemens 26804* (BM); KIAU: *Clemens, M. S. 10206* (K); MAMUT RIVER: 1400 m, *RSNB 1673* (K); PARK HEADQUARTERS: *Carson SAN 28025* (K); PORING: 600 m, *Darnton 153* (BM); TENOMPOK: 1200 m, *Clemens 26804* (K), 1500 m, *28451* (BM, K, L).

69.9. HORNSTEDTIA Retz.

Smith, R. M. (1985). A review of Bornean Zingiberaceae: I (Alpineae p.p.). *Notes Roy. Bot. Gard. Edinburgh* 42: 261–314 [*Hornstedtia* on pp. 289–295].

69.9.1. Hornstedtia gracilis R. M. Sm., *Notes Roy. Bot. Gard. Edinburgh* 44: 233 (1987). Plate 25B. Type: LANGANAN FALLS, *Smith et al. 36/86* (holotype E!).

Rhizomatous perennial herb to 3 m; inflorescence basal. Hill forest, lower montane forest. Elevation: 1200–1800 m.

Additional material examined. BAMBANGAN CAMP: 1500 m, *RSNB 4577* (K, L); LIWAGU RIVER TRAIL: 1800 m, *Smith & Phillipps 24* (E); LIWAGU/MESILAU RIVERS: 1200 m, *RSNB 2536* (K); MAMUT RIVER: 1200 m, *RSNB 1648* (K, L, SING); PARK HEADQUARTERS: 1600 m, *Stevens et al. 661* (E, L).

69.9.2. Hornstedtia havilandii (K. Schum.) K. Schum., *Pflanzenr.* IV. 46 (Heft 20): 193 (1904).

Rhizomatous perennial herb to 5 m; inflorescence basal, long pedunculate. Lowlands, hill forest. Elevation: 500 m.

Material examined. MELANGKAP TOMIS: *Lorence Lugas 1425* (K); PORING: 500 m, *Darnton 162* (BM), 500 m, *Meijer et al. SAN 122520* (K); PORING HOT SPRINGS: *Smith 34/86* (E); SAYAP: *Tungking Simbayan 27* (K); TEKUTAN: *Dius Tadong 77* (K), *347* (K).

69.9.3. Hornstedtia incana R. M. Sm., *Notes Roy. Bot. Gard. Edinburgh* 42: 290 (1985). Page v & Plate 25C.

Rhizomatous perennial herb to 3 m, rhizome on stilt roots; inflorescence basal. Lower montane forest. Elevation: 2000 m.

Material examined. EASTERN SHOULDER: 2000 m, *RSNB 171* (K, L); MESILAU RIVER: *Argent & Lamb 1366* (E); TIBABAR FALLS: *Smith 21/86* (E).

69.9.4. Hornstedtia reticulata (K. Schum.) K. Schum., *Pflanzenr.* IV. 46 (Heft 20): 193 (1904).

Rhizomatous perennial herb 2–3 m high, with stilt roots; inflorescence from the base. Primary hill forest. Elevation: 900 m.

Material examined. DALLAS: 900 m, *Clemens 27253* (BM, K); MELANGKAP TOMIS: *Lorence Lugas 2087* (K).

69.10. PLAGIOSTACHYS Ridl.

Smith, R. M. (1985). A review of Bornean Zingiberaceae: 1 (Alpineae p.p.). *Notes Roy. Bot. Gard. Edinburgh* 42: 261–314 [*Plagiostachys* on pp. 265–272].

69.10.1. Plagiostachys albiflora Ridl., *J. Straits Branch Roy. Asiat. Soc.* 50: 150 (1908).

Rhizomatous perennial herb to 2 m high; inflorescence terminal but appearing lateral from somewhat above the base. Hill forest. Elevation: 800 m.

Material examined. EASTERN SHOULDER: 800 m, *RSNB 999* (K); MELANGKAP TOMIS: *Lorence Lugas 678* (K), *2087* (K); PORING: *Sani Sambuling 617* (K).

69.10.2. Plagiostachys crocydocalyx (K. Schum.) B. L. Burtt & R. M. Sm., *Notes Roy. Bot. Gard. Edinburgh* 31: 315 (1972).

Rhizomatous perennial herb; inflorescence terminal on the leafy shoot. Hill forest along streams.

Material examined. MELANGKAP TOMIS: *Lorence Lugas 730* (K).

69.10.3. Plagiostachys "poringensis" Cowley ined.

Rhizomatous perennial herb; inflorescence terminal on the leafy shoot but appearing pushed out from the side. Hill forest. Elevation: 700–800 m.

Material examined. KIPUNGIT HILL: 700 m, *Beaman 7664* (E), 700 m, *7830* (MSC); PORING: 700–800 m, *Beaman 10932* (K).

69.10.4. Plagiostachys strobilifera (Baker) Ridl., *J. Straits Branch Roy. Asiat. Soc.* 32: 151 (1899).

Perennial herb to 2 m; inflorescence basal, scarlet. Hill forest. Elevation: 900 m.

Material examined. EASTERN SHOULDER: 900 m, *RSNB 1018A* (K); MELANGKAP TOMIS: *Lorence Lugas 1209* (K).

69.10.5. Plagiostachys indet.

Material examined. DALLAS/TENOMPOK: 1200 m, *Clemens 27560* (BM, K); PARK HEADQUARTERS: 1700 m, *Vermeulen & Chan 392* (L).

69.11. ZINGIBER Boehm.

Smith, R. M. (1989). A review of Bornean Zingiberaceae: V (*Zingiber*). *Notes Roy. Bot. Gard. Edinburgh* 45: 409–423.

69.11.1. Zingiber coloratum N. E. Br., *Gard. Chron.* 2: 166 (1879). Plate 25D.

Rhizomatous perennial herb; inflorescence from the base. Hill forest. Elevation: 700–1200 m.

Material examined. KIPUNGIT HILL: 700 m, *Beaman 7838* (E, MSC); LIWAGU/MESILAU RIVERS: 1200 m, *RSNB 2519* (K).

69.11.2. Zingiber pseudopungens R. M. Sm., *Notes Roy. Bot. Gard. Edinburgh* 45: 413 (1988 [1989]). Type: MAMUT RIVER, 1200 m, *RSNB 1710* (holotype K!).

Rhizomatous perennial herb to 2 m high; inflorescence from the base. Hill forest on ultramafic substrate. Elevation: 400–1200 m.

Additional material examined. HEMPUEN HILL: 800–1000 m, *Beaman 7382* (MSC), 800–1200 m, *7702* (MSC); KINATEKI RIVER: 800 m, *Collenette A 118* (BM); MELANGKAP TOMIS: *Lorence Lugas 845* (K); PINAWANTAI: 400 m, *Shea & Aban SAN 76770* (K).

69.11.3. Zingiber indet.

Material examined. DALLAS, 900–1200 m, *Clemens 26632* (BM); KIAU: *Clemens, M. S. 10083* (K); MENGGIS: *Matamin Rumutom 26* (K); MESILAU CAVE: 2000 m, *Collenette 21659* (L); NALUMAD: *Daim Andau 605* (K); PORING: *Sani Sambuling 345* (K).

INDEX TO NUMBERED MONOCOTYLEDON COLLECTIONS

Prefixes for collectors' numbers are not included.
Projek Etnobotani Kinabalu collector.

Badak 32388 (40.1.2).

Bailes & Cribb 503 (39.6.1); 846 (40.16.2).

Battah 33912 (66.1.7); 33918 (66.1.2).

Beaman 7201 (39.14.1); 7202 (69.1.7); 7203 (69.8.1); 7205 (44.1.1); 7206 (40.16.7); 7382 (69.11.2); 7385 (66.1.4); 7400 (64.1.1); 7428 (45.1.1); 7445 (40.16.7); 7450 (39.6.1); 7451 (39.2.4); 7452 (44.1.1); 7453 (39.2.4); 7462 (41.1.2); 7473 (39.10.3); 7485 (66.1.3); 7490 (47.1.4); 7492 (39.14.1); 7506 (53.1.2); 7550 (46.1.4); 7566 (59.3.3); 7572 (65.32.1); 7664 (69.10.3); 7665 (69.7.5a); 7693 (39.5.1); 7701 (66.1.2); 7702 (69.11.2); 7786 (39.5.1); 7806 (64.1.1); 7807 (39.14.4); 7826 (69.7.5a); 7830 (69.10.3); 7838 (69.11.1); 7856 (59.1.1); 7857 (65.44.1); 7861 (52.1.1); 7975 (66.1.3); 7989 (60.1.1); 8001 (66.1.4); 8007 (40.16.7); 8136 (66.1.3); 8182 (66.1.4); 8195 (66.1.2); 8221 (39.4.1); 8298 (60.1.1); 8299 (47.20.1); 8300 (47.19.1); 8361 (39.11.3); 8397 (49.1.1); 8401a (39.6.3); 8440 (69.1.10); 8451 (44.1.4); 8461 (59.3.3); 8463 (39.11.3); 8547 (69.8.1); 8550 (47.20.2); 8551 (47.1.9b); 8552 (39.6.1); 8565 (69.7.5a); 8580 (44.6.2); 8587 (39.6.3); 8593 (45.1.1); 8594 (39.10.1); 8633 (47.1.7); 8637 (44.1.1); 8638 (39.6.1); 8653 (47.21.2); 8694 (40.16.7); 8706 (66.1.1); 8747 (39.14.1); 8754 (41.1.2); 8755 (60.2.1); 8777 (41.1.3); 8778 (68.1.4); 8781 (47.8.1); 8783 (69.6.9); 8790 (66.1.6); 8792 (52.1.1); 8797 (45.1.1); 8806 (66.1.3); 8810 (44.6.1); 8811 (39.10.1); 8856 (62.1.1); 8871 (67.1.2); 8975 (47.8.1); 8976 (39.5.1); 9016 (40.16.7); 9036 (41.1.2); 9039 (40.16.7); 9069 (66.1.4); 9072 (48.1.9); 9087 (66.1.3); 9107 (39.14.1); 9115 (40.16.7); 9134 (60.1.1); 9140 (64.1.2); 9186 (40.16.7); 9196 (66.1.4); 9200 (39.4.1); 9209 (69.7.2); 9224 (47.1.3a); 9227 (44.8.1); 9245 (39.6.3); 9287 (65.29.1a); 9288 (65.39.1); 9293 (39.8.5); 9299 (49.1.1); 9456 (69.8.1); 9461 (44.6.1); 9504 (45.1.1); 9513 (44.1.1); 9519 (69.6.6); 9537 (64.1.2); 9542 (47.19.4); 9561a (66.1.4); 9582 (39.14.1); 9583 (66.1.3); 9596 (66.1.2); 9603 (69.1.7); 9656 (59.3.3); 9669 (47.14.7); 9754 (44.1.4); 9759 (44.1.1); 9768 (39.8.8); 9769 (39.13.1); 9799 (39.14.4); 9802 (39.14.2); 9805 (69.7.5a); 9839 (64.1.1); 9855 (65.23.6); 9925 (39.14.1); 9945 (41.1.2); 9979 (67.1.2); 9984 (39.14.1); 10333 (66.1.2); 10361 (65.23.4); 10544 (69.7.3); 10547 (49.1.2); 10554 (59.3.3); 10555 (40.7.8); 10567 (40.16.5); 10570 (52.1.1); 10588 (39.6.2); 10604 (44.6.2); 10605 (39.3.3); 10606 (59.2.1); 10617 (47.3.6); 10620 (69.7.1); 10627 (44.8.1); 10632 (39.13.1); 10634 (69.1.10); 10638 (39.8.8); 10644 (69.1.10); 10660 (40.16.7); 10677 (69.8.1); 10700 (65.21.1); 10705 (52.1.1); 10721 (47.14.5); 10738 (40.16.7); 10743 (40.16.4); 10748 (65.32.1); 10756 (69.1.7); 10757 (69.4.1); 10759 (40.16.7); 10778 (66.1.4); 10785 (44.1.1); 10794 (65.37.1); 10923 (44.1.4); 10932 (69.10.3); 10933 (39.14.3); 10948 (40.4.2); 10950 (40.4.2); 10963 (44.6.2); 10964 (39.8.2a); 11019 (40.7.5); 11030 (65.16.1); 11158 (66.1.3); 11285 (66.1.3); 11361 (66.1.3); 11364 (61.1.2); 11391 (66.1.3); 11420 (66.1.8).

*Benedict Busin 2a (40.10.1); 2b (40.3.11); 3 (40.16.7); 4 (40.4.2).

Brand & Anak 25301 (66.1.4); 25302 (40.3.18).

Carr 26275 (69.7.4); 26384 (63.2.5); 26537 (68.1.2); 26604 (65.43.4); 26657 (66.1.5); 26737 (61.1.1a); 27050 (69.2.5); 27104 (63.1.5); 27172 (69.6.4); 27319 (69.5.1); 27342 (63.1.1); 27438 (65.40.4); 27565 (54.1.1); 27928 (69.7.3).

Carson 26786 (40.7.6); 28025 (69.8.2); 28035 (40.1.2).

Chai & Ilias 6010 (39.6.1); 6036 (40.16.5).

Chew & Corner 4000 (40.16.7); 4001 (44.1.1); 4035 (40.16.7); 4042 (69.2.4); 4056 (40.3.20); 4062 (66.1.4); 4081 (39.6.1); 4083 (39.14.6); 4090 (64.1.1); 4094 (69.2.4); 4108 (65.40.2); 4111 (65.40.2); 4124 (65.40.2); 4125 p.p. (40.3.14, 40.7.7); 4136 (47.1.5); 4139 (69.6.8); 4149 (65.29.1a); 4158 (63.2.5); 4159 (40.3.8); 4177 (63.1.5); 4194 (40.17.1a); 4206 (63.2.5); 4233 (52.1.1); 4269 (63.1.8); 4280 (40.3.11); 4306 (63.1.5); 4325 (40.7.6); 4345 (40.3.10); 4353 (39.14.6); 4356 (40.3.13); 4373 (47.1.9b); 4374 (47.19.4); 4385 (40.7.7); 4390 (69.6.11); 4391 (40.7.7); 4401 (64.1.2); 4403 (63.2.5); 4405 (63.1.5); 4407 (47.1.3a); 4435 (40.7.6); 4438 (40.19.2); 4440 (40.3.11); 4461 (40.3.8); 4467 (69.4.1); 4474 (66.1.3); 4487 (40.16.4); 4488 (40.16.4); 4502 (47.7.2); 4520 (40.3.8); 4540 (69.8.1); 4562 (39.14.6); 4573 (66.1.4); 4576 (61.1.3); 4577 (69.9.1); 4580 (69.2.8); 4581 (40.3.18); 4585 (65.40.2); 4588 (63.2.9); 4589 (40.16.4); 4654 (40.16.4); 4655

(40.16.7); 4686 (66.1.3); 4691 (47.1.7); 4696 (47.1.5); 4703 (66.1.4); 4722 (40.7.7); 4723 (40.3.10); 4724 (65.23.1); 4801 (39.14.6); 4835 (40.3.10); 4863 (41.1.2); 4864 (39.14.1); 4899 (39.13.1); 4920 (40.3.10); 4921 (39.6.1); 4926 (45.1.1); 4938 (40.3.10); 4939 (39.14.7); 4959 (64.1.1); 4966 (47.2.1); 4969 (64.1.1); 4977 (39.11.1); 4995 (40.3.10); 5724 (65.16.1); 5958 (65.1.1b); 5964 (47.13.3); 5967 (65.13.1a); 5968 (65.1.1c); 5971 (47.1.11); 5973 (65.12.1); 5975 (47.19.1); 5977 (47.20.1); 5985 (60.1.1); 5998 (47.1.11); 5999 (65.23.1); 7013 (63.1.5); 7018 (40.1.2); 7020 (63.2.9); 7025 (40.3.10); 7051 (40.1.2); 7052 (39.6.1); 7069 (40.3.14); 8444 (47.14.5); 8464 (63.2.3); 8465 (63.2.3).

Chew, Corner & Stainton 21 (40.3.13); 24 (40.7.12); 66 (49.1.2); 74 (66.1.2); 167 (65.40.2); 168 (40.3.10); 169 (40.16.7); 171 (69.9.3); 192 (47.1.5); 199 (40.3.8); 202 (40.16.4); 218 (40.7.5); 225 (49.1.2); 235 (40.7.7); 256 (40.7.6); 267 (65.28.2); 268 (65.28.1); 287 (40.16.3); 289 (47.21.5); 299 (52.1.1); 565 (40.16.10a); 574 (39.2.5); 591 (40.16.1); 601 (60.2.1); 635 (46.1.2); 639 (45.1.1); 642 (39.13.7); 662 (46.1.1); 672 (69.2.1, 69.2.8); 708 (39.10.2); 716 (47.1.11); 717 (47.9.1); 718 (47.1.2a); 720 (47.1.9a); 732 (47.1.5); 734 (47.1.2a); 736 (47.1.11); 788 (40.3.8); 852 (47.1.9b); 856 (47.20.1); 856a (47.19.1); 867 (47.22.1); 912 (47.1.5); 922 (65.1.1b); 945 (66.1.4); 966 (39.8.5); 991 (65.23.3); 993 (47.13.3); 994 (65.23.4); 996 (54.1.1); 999 (69.10.1); 1009 (47.1.9a); 1016 (69.4.1); 1018 (69.6.9); 1018A (69.10.4); 1021 (47.14.5); 1026 (39.13.1); 1030 (69.7.1); 1031 (69.7.2); 1032 (47.21.5); 1035 (47.21.2); 1049 (53.1.1); 1066 (69.1.7); 1086 (66.1.3); 1091 (60.1.1); 1092 (50.1.1a); 1114 (66.1.3); 1123 (47.1.5, 47.1.11); 1123b (47.1.5); 1130 (47.19.1); 1133 (60.1.1); 1179 (65.16.1); 1197 (44.1.1); 1203 (69.1.10); 1209 (45.1.1); 1220 (65.36.1); 1223 (65.29.1a); 1242 (65.39.1); 1244 (65.21.1); 1283 (66.1.4); 1292 (47.13.2); 1307 (65.23.2); 1308 (65.39.1); 1313 (66.1.2); 1316 (47.1.9b); 1330 (66.1.3); 1338 (65.32.2); 1345 (66.1.4); 1365 (69.8.1); 1370 (69.1.7); 1383 (39.14.6); 1416 (65.32.1); 1418 (65.36.1); 1430 (39.4.1); 1434 (39.1.2); 1471 (44.3.2); 1543 (40.7.7); 1566 (69.1.7); 1569a (63.2.5); 1630 (47.14.5); 1648 (69.9.1); 1652 (69.2.1); 1653 (40.3.20); 1669 (40.7.6); 1673 (69.8.2); 1706 (47.1.3a); 1707 (47.20.2); 1708 (47.1.6); 1710 (69.11.2); 1714 (69.2.8); 1715 (52.1.1); 1725 (40.16.4); 1735 (40.3.11); 1744 (69.6.4); 1745 (47.14.5); 1780 (63.2.5); 1809 (55.1.1a); 1817 (40.3.14); 1848 (66.1.3); 1857 (39.6.1); 1892 (40.3.20); 1909 (41.3.1); 1921 (65.28.1); 1954 (39.2.4); 1957 (40.3.14); 1976 (69.6.4); 2501 (39.10.1); 2517 (40.7.6); 2519 (69.11.1); 2536 (69.9.1); 2593 (47.1.3a, 65.35.3); 2599 (40.16.4); 2613a (44.5.2); 2649 (69.2.8); 2651 (40.16.7); 2741 (65.45.2); 2768 (65.28.1); 2770 (39.14.2); 2814 (45.2.1a); 2824 (44.6.1); 2877a (45.1.1); 2880 (69.1.7); 2882 (52.1.1); 2940 (39.14.2); 2941 (39.14.4); 2964 (49.1.1); 3004 (59.2.1); 4387 (55.1.1a).

Chow & Leopold 76433 (44.1.1).

Clemens 26015 (55.1.1a); 26018 (65.45.2); 26024 (39.13.1); 26025 (39.13.1); 26034 (45.1.1); 26035 (52.1.1); 26039 (44.1.2); 26045 (69.8.2); 26061 (65.28.1); 26073 (69.7.3); 26075 (63.1.8); 26086 (39.8.5); 26121 (39.14.1); 26141 (44.1.3); 26142 (39.1.1), 26143 (48.1.9); 26171 (39.14 4); 26178 (63.2.9), 26199 (39.8.2a); 26200 (69.7.1); 26210 (63.1.9, 63.2.5); 26213 (39.2.4); 26217 (39.5.1); 26220 (49.1.3); 26224 (40.4.1); 26226 (39.8.1); 26242 (63.1.4); 26243 (39.14.1); 26268 (39.2.4); 26272 (59.2.1); 26324 (65.20.1); 26333 (48.1.7); 26360 (40.3.13); 26369 (59.2.1); 26370 (67.1.2); 26378 (63.1.8); 26379 (46.1.4); 26380 (40.19.3); 26389 (48.1.2); 26396 (44.1.1); 26433 (48.1.2); 26442 (48.1.9); 26443 (59.2.1); 26453 (39.12.2); 26455 (48.1.1); 26476 (44.1.2); 26495 (39.12.1); 26496 (40.3.16); 26522 (66.1.3); 26542 (39.11.2); 26609 p.p. (63.2.1); 26628 (39.6.3); 26632 (69.11.3); 26643 (40.16.4); 26659 (40.19.2); 26682 (40.16.4); 26683 (40.3.19); 26684 (40.16.4); 26686 (40.16.4); 26698 (48.1.9); 26699 (48.1.9); 26703 (39.1.1); 26713 (39.14.1); 26720 (39.12.3); 26728 (48.1.7); 26729 (48.1.9); 26730 (48.1.7, 39.8.1); 26730A (39.8.1); 26731 (39.12.2); 26732 (39.8.5); 26733 (39.4.1); 26734 (63.1.9); 26739 (62.1.1); 26741 (39.5.1, 64.1.1, 48.1.9); 26755 (40.3.16); 26793 (40.7.6); 26801 (69.7.4); 26804 (69.8.2); 26804A (69.8.2); 26807 (40.7.6); 26808 (39.11.6); 26820 (39.1.2); 26822 (69.7.5a); 26830 (47.8.1); 26838 (69.2.8); 26840 (39.6.1); 26844 (66.1.3); 26857 (63.2.4); 26873 (64.1.1); 26875 (39.14.1); 26876 (39.12.2); 26879 (40.2.1); 26886 (47.14.4); 26888 (40.7.12); 26898 (69.1.10); 26898 bis (69.1.10); 26900 (39.12.2); 26916 (66.1.3); 26916A (66.1.3); 26919 (66.1.3); 26921 (39.12.3); 26937 (48.1.3); 26972 (40.7.4); 26986 (39.8.5); 27009 (40.3.16); 27010 (40.3.16); 27012 (40.7.12); 27028 (47.14.2); 27030 (39.12.3); 27041 (48.1.9); 27053 (63.2.7); 27072 (65.1.1a65.1.1c); 27073 (65.13.1a); 27074 (65.38.3, 65.38.4); 27075 (65.12.1); 27076 (65.23.4); 27077 (65.23.3, 65.23.4); 27078 (47.20.1); 27079 (47.19.1); 27080 (47.1.11); 27089 (50.1.2); 27090 (60.1.1); 27156 (40.7.6); 27156A (40.7.6); 27157A (40.7.12); 27161 (48.1.13); 27188 (40.7.6); 27204 (59.2.1); 27216 (40.7.12); 27219 (40.4.1); 27226 (40.4.1);

27230 (39.8.2a); 27253 (69.9.4); 27259 (40.16.2); 27268 (40.7.6); 27269 (40.7.6); 27276 (40.16.4); 27301 (65.26.1); 27303 (47.21.2); 27306 (49.1.1); 27312 (39.12.3); 27320 (40.3.1, 40.3.10); 27339 (40.3.20); 27376 (39.5.1); 27472 (65.21.1); 27474 (65.30.1); 27482 (65.49.1); 27483 (65.29.1a); 27537 (69.2.2, 69.2.1); 27540 (65.42.1); 27560 (69.10.5); 27566 (41.3.1); 27573 (47.14.4); 27575 (52.1.1); 27589 (65.35.3); 27589A (65.35.3); 27590 (65.35.1); 27591 (65.34.1); 27592 (65.43.5); 27594 (65.32.1); 27595 (65.10.1); 27624 (69.2.8); 27661 (65.15.1); 27662 (65.35.3); 27673 (65.14.3); 27674 (47.3.1); 27679 (47.17.2); 27681 (47.5.6); 27687 (65.10.1); 27691 (51.1.1); 27708 (66.1.3); 27724 (48.1.9); 27751 (63.2.4); 27762 (66.1.3); 27770 (65.38.3, 65.38.4); 27771 (65.1.1b65.1.1c); 27776 (43.1.1); 27777 (50.1.2); 27778 (65.12.1); 27786 (65.13.1a); 27787 (65.38.2); 27798 (65.28.1); 27813 (50.1.1a); 27816 (47.20.1); 27835 (40.3.10); 27848 (66.1.3); 27896 (40.16.4); 27899 (40.3.20); 27901 (40.16.4); 27984 (65.25.1); 27986 (65.23.3); 27987 (65.1.2a); 28021 (47.1.2a); 28022 (47.1.11); 28023 (47.1.1a); 28024 (47.13.3); 28028 (47.1.1a); 28053 (69.6.8); 28056 (48.1.6, 48.1.9); 28073 (47.3.8); 28074 (47.12.1); 28083 (65.28.1); 28120 (66.1.3); 28126 (66.1.3); 28142 (39.14.1); 28163 (44.2.1); 28164 (65.23.2); 28170 (69.8.1); 28177 (66.1.3); 28209 (47.14.5); 28239 (47.18.2); 28269 (47.21.3a); 28271 (65.35.3); 28272 (65.21.1); 28273 (65.41.1); 28274 (65.35.1); 28275 (65.35.2); 28275A (65.35.2); 28283 (65.25.1); 28300 (65.35.4); 28300a (65.35.4); 28301 (65.10.2); 28306 (69.6.1); 28333 (40.2.1); 28351 (40.7.7); 28375 (40.3.13); 28389 (65.23.1); 28399 (40.16.4); 28407 (40.16.4); 28408 (40.3.20); 28430 (69.2.8); 28440 (66.1.4); 28442 (66.1.4); 28443 (66.1.2); 28451 (69.8.1, 69.8.2); 28481 (69.1.7); 28487 (39.12.3); 28514 (39.11.4); 28552 (39.13.7); 28555 (65.28.2); 28560 (40.7.7); 28564 (40.3.10); 28564a (40.3.10); 28565 (40.3.13); 28566 (40.3.13, 40.7.7); 28566 bis (40.3.13); 28622 (66.1.2); 28647 (39.13.7); 28650 (40.3.13); 28670 (40.16.5); 28711 (63.1.5); 28726 (63.2.5); 28750 (63.1.8); 28761 (40.1.2); 28761a (40.1.2); 28769 (40.19.2); 28800 (40.3.8); 28801 (63.2.5); 28813 (39.12.1); 28819 (40.19.2); 28832 (63.2.5); 28844 (40.3.13); 28923 (47.1.9b); 28925 (47.1.11); 28940 (43.1.1); 29001 (47.7.2); 29002 (47.20.1); 29004 (47.22.1); 29005 (47.1.1b); 29006 (47.15.1); 29007 (47.1.5); 29090 (47.20.1); 29133 (39.1.2); 29134 (39.8.4); 29135 (39.9.1); 29136 (39.10.1); 29137 (39.8.3); 29138 (39.8.1); 29139 (39.8.2a); 29140 (39.8.1); 29141 (39.7.1); 29142 (39.8.1); 29143 (39.13.7); 29144 (39.4.1); 29145 (39.8.1); 29146 (39.12.1); 29146a (39.12.1); 29148 (39.2.4); 29151 (39.8.6); 29152 (39.8.5); 29153 (39.13.6); 29154 (39.13.6); 29155 (39.11.4); 29156 (39.6.1); 29157 (39.13.7); 29167 (65.1.1a); 29167a (65.1.1a65.1.1c); 29167b (65.1.1a); 29167d (65.1.1c); 29172 (65.25.1); 29173 (65.51.1); 29174 (65.6.1); 29175 (65.1.1b65.1.1c); 29176 (65.3.1a); 29181 (39.14.1); 29182 (39.14.2); 29183 (39.13.1); 29184 (39.13.1); 29185 (39.13.1); 29186 (39.13.6); 29187 (39.13.6); 29188 (39.13.1); 29189 (40.7.4); 29190 (40.19.2); 29191 (40.7.7); 29191a (40.7.7); 29192 (40.7.7); 29193 (40.7.6); 29194 (40.7.6); 29195 (40.3.11); 29196 (40.3.8); 29196a (40.3.8); 29197 (40.3.8); 29197b (40.3.8); 29198 (40.3.8); 29198a (40.3.8); 29199 (40.3.8); 29200 (40.3.8); 29201 (40.3.8); 29201a (40.3.8); 29202 (40.1.2); 29203 (40.3.20); 29204 (40.16.7); 29204a (40.16.7); 29205 (40.16.7); 29205a (40.16.7); 29227 (39.12.3); 29264 (69.2.5); 29272 (39.13.5); 29331 (63.1.8); 29335 (39.14.5); 29355 (40.16.4); 29364 (66.1.4); 29493 (39.13.4); 29547 (40.16.4); 29550 (63.2.3, 63.2.7); 29575 (66.1.4); 29576 (65.23.2); 29603 (66.1.4); 29617 (39.14.2); 29620 (46.1.1); 29663 (66.1.2); 29669 (49.1.1); 29682 (44.7.1); 29692 (65.2.1); 29693 (65.23.1); 29704 (66.1.3); 29713 (41.1.2); 29726 (44.6.1); 29798 (69.2.6); 29811 (63.2.3); 29815 (52.1.1); 29893 (39.14.2); 29905 (41.3.1); 29906 (65.20.1); 29985 (63.1.3); 29988 (65.44.1); 30032 (66.1.7); 30037 (66.1.4); 30038 (66.1.7); 30043 (60.1.1); 30046 (64.1.1); 30059 (50.1.2); 30061 (54.1.1); 30062 (47.20.1); 30063 (47.13.3); 30064 (47.14.2); 30065 (47.14.3a, 47.14.4); 30067 (47.19.1); 30068 (47.1.3a, 47.1.9b); 30091 (48.1.6); 30092 (48.1.9); 30134 (65.6.1); 30134 bis (65.6.1); 30175 (63.1.8); 30176 (63.2.3); 30176 p.p. (63.2.9); 30186 (65.28.1); 30259 (65.13.1a); 30260 (65.13.1a); 30261 (65.50.1a); 30262 (65.49.1); 30263 (65.49.1); 30264 (65.36.3a); 30265 (65.28.1); 30266 (65.38.4); 30267 (65.38.4); 30268 (65.12.1); 30269 (65.23.4); 30270 (65.23.2); 30273 (65.1.1a65.1.1c, 65.38.2); 30274 (65.34.1); 30275 (65.26.1); 30276 (65.21.1); 30277 (65.30.1); 30278 (69.7.2); 30280 (65.10.2); 30281 (65.10.1); 30282 (65.28.1); 30283 (65.23.5); 30286 (65.14.1); 30288 (65.32.1); 30289 (65.32.1); 30312 (65.1.1c); 30313 (65.12.1); 30314 (65.38.2); 30315 (65.38.2); 30316 (65.15.1); 30344 (66.1.3); 30454 (39.3.3); 30489 (66.1.4); 30508 (39.6.1); 30524 (40.3.8); 30529 (65.43.3); 30646 p.p. (66.1.4, 66.1.7); 30654 (47.14.4); 30672 (40.7.7); 30677 (69.4.1); 30681 (44.1.1); 30687 (52.1.1); 30691 (47.1.3a); 30694 (65.21.1); 30694a (65.21.1); 30696 (65.23.1); 30700 (65.26.1); 30716 (63.2.5); 30751 (65.40.3); 30752 (47.13.2); 30805 (47.19.4); 30806 (47.2.1); 30830 (40.7.7); 30830 bis (40.3.10); 30842 (63.1.7a); 30887 (40.7.6); 30889 (44.1.4); 30900 (47.13.4); 30925 (48.1.6); 30973 (48.1.6); 30978 (65.29.1a); 30986 (47.1.9b); 31017 (39.5.1); 31022 (69.1.6); 31023 (39.8.1); 31028 (40.3.20); 31038 (63.1.9); 31039 (40.17.1a); 31067 (66.1.4); 31094 (39.14.6); 31120

(69.1.10); 31126 (39.11.4); 31149 (39.1.2); 31150 (39.12.1); 31161 (39.2.4); 31199 (49.1.2); 31237 (40.2.1); 31250 (39.14.2); 31280 (40.7.6); 31280a (40.7.6); 31284 (40.7.7); 31295 (44.1.6); 31296 (52.1.1); 31307 (47.14.4); 31315 (39.14.5); 31325 (69.1.6); 31343 (46.1.2); 31345 (39.13.7); 31359 (63.1.7a); 31362 (65.29.1a); 31376 (63.2.3); 31394 (40.16.7); 31400 (40.16.7); 31412 (47.1.9b); 31413 (47.1.9b); 31433 (40.3.10); 31438 (40.3.8); 31497 (40.3.13); 31498 (40.3.13); 31518 (44.6.2); 31547 (69.2.8); 31548 (39.2.1); 31581 (40.7.6); 31587 (40.3.3); 31604 (40.7.7); 31605 (66.1.2); 31618 (65.40.3); 31637 (39.14.5); 31654 (69.4.2, 69.4.1); 31674 (47.13.3); 31679 (60.1.1); 31688 (65.23.3); 31722 (40.3.8); 31749 (47.1.5); 31755 (66.1.2); 31758 (40.7.6); 31778 (47.1.9b); 31785 (40.3.8); 31858 (69.4.1); 31876 (39.10.1); 31910 (40.3.8); 31967 (66.1.8); 31969 (69.2.6); 32007 (40.17.1a); 32008 (40.3.13); 32014 (65.23.4, 66.1.4); 32016 (60.2.1); 32025 (68.1.1); 32084 (39.8.5); 32085 (44.1.6); 32094 (69.7.4); 32096 (47.14.6); 32115 (69.1.6); 32117 (66.1.4); 32135 (63.1.9); 32152 (66.1.3); 32163 (48.1.12); 32166 (40.7.6); 32179 (39.5.1); 32181 (65.28.1); 32192 (41.1.3); 32205 (39.13.7); 32226 (60.1.1); 32234 (64.1.1); 32237 (65.23.4); 32238 (65.41.1); 32257 (65.40.4); 32269 (40.3.10); 32290 (39.10.1); 32315 (69.8.1); 32326 (65.23.1); 32327 (40.3.13); 32329 (47.20.1); 32333 (47.14.5, 47.22.1); 32336 (50.1.2); 32341 (47.22.1); 32342 (47.1.11); 32342A (47.1.5); 32343 (65.12.1, 63.2.9); 32343A (65.12.1); 32344 (47.1.1b); 32348 (63.2.1); 32352 (47.1.9b); 32359 (47.19.3); 32363 (60.1.3); 32382 (47.19.2); 32385 (63.1.8); 32389 (41.1.2); 32396 (40.3.22); 32396A (40.3.10); 32404 (47.13.1); 32415 (43.1.1); 32450 (66.1.4); 32451 (40.16.7); 32454 (60.2.1); 32466 (40.3.10); 32477 (39.6.1); 32500 bis (40.7.7); 32502 (65.21.1); 32508 (39.8.2a); 32511 (40.3.10, 65.40.4); 32515 (66.1.3); 32520 (40.3.10); 32532 (40.16.7); 32549 (63.2.5); 32573 (47.1.5); 32579 (47.21.5); 32580 (47.1.5); 32581 (47.1.3a); 32588 (40.16.4); 32599 (47.21.5); 32600 (40.3.8); 32612 (65.23.1); 32613 (65.32.1); 32614 (40.16.4, 65.39.1); 32614a (65.39.1); 32615 (65.28.1); 32616 (65.36.1); 32627 (47.2.1); 32628 (47.13.1); 32629 (50.1.1a); 32649 (63.2.9); 32655 (65.23.1); 32656 (65.29.1a); 32672 (40.3.8); 32700 (47.1.9b); 32727 (40.16.7); 32736 (47.13.4); 32749 (65.40.4); 32750 (40.16.4); 32757 (41.2.1); 32763 (47.1.3a); 32795 (47.21.5); 32799 (66.1.4); 32806 (47.13.3); 32842 (40.16.7); 32844 (47.1.5); 32849 (65.23.3); 32858 (65.26.1); 32865 (40.7.6); 32872 (66.1.4); 32928 (47.1.9b); 32929 (47.1.5); 32931 (54.1.1); 32961 (47.1.5); 33056 (45.2.1a); 33059 (44.5.2); 33061 (48.1.3); 33087 (65.40.4); 33099 (69.2.5); 33100 (69.2.5); 33152 (66.1.8); 33157 (47.1.3b); 33228 (65.1.1b); 33231 (47.1.11); 33232 (47.19.1); 33651 (47.1.9b); 33652 (47.1.9a); 33662 (66.1.3); 33663 (40.3.10); 33730 (47.20.1); 33731 (65.23.1); 33736 (65.23.1); 33759 (66.1.4); 33779 (66.1.4); 33795 (40.16.4); 33806 (65.23.3); 33808 (47.1.11); 33812 (64.1.2); 33816 (66.1.3); 33837 (40.16.4); 33841 (65.23.1); 33842 (47.20.1); 33844 bis (65.23.1); 33853 (47.13.3); 33884 (39.7.1); 33900 (65.23.3); 33917 (64.1.2); 33995 (52.1.1); 34045 (47.1.3a47.1.3b); 34046 (65.28.1); 34048 (65.23.1); 34067 (65.45.2); 34090 (50.1.1a); 34099 (39.14.5); 34121A (65.23.2); 34121B (65.23.2); 34122 (47.13.2); 34123 (47.1.3a); 34124 (65.36.1); 34125 (65.21.1); 34126 (65.32.2); 34126A (65.32.2); 34129 (65.21.1); 34147 (69.8.1); 34157 (69.6.8); 34163 (40.3.20); 34164 (40.3.14); 34169 (47.20.1); 34174 (65.20.1); 34188 (65.22.2); 34189 (65.31.1); 34200 (44.1.1); 34213 (47.7.2); 34214 (47.2.1); 34216 (44.6.1); 34217 (63.1.5); 34272 (44.1.1); 34273 (53.1.1); 34274 (47.13.4); 34297 (47.1.6); 34298 (39.12.1); 34309 (66.1.4); 34351 (40.3.8); 34410 (65.23.1); 34421 (66.1.4); 34431 (47.1.9b); 34448 (65.2.1); 34448A (65.2.1); 34473 (45.2.1a); 34489 (63.2.5); 34497 (65.31.1); 35000 (39.6.1); 35001 (39.3.3); 35003 (39.11.4); 35004 (39.14.1); 35005 (39.14.6); 35006 (39.8.1); 35007 (39.13.7); 35009 (53.1.2); 35016 (47.14.4); 35030 (64.1.1); 35031 (49.1.3); 35032 (66.1.4); 35033 (66.1.3); 35034 (40.7.6, 66.1.6); 35035 (40.3.8); 35050 (66.1.2); 35063 (47.1.9b); 35064 (47.1.11); 35128 (39.14.1); 35153 (39.6.1); 35206 (69.7.4); 35643 (47.1.9b); 40025 (65.40.3); 40033 (47.13.4); 40041 (47.14.5); 40044 (47.1.9b); 40090 (65.31.1); 40104 (39.5.1); 40104 bis (39.8.1); 40147 (65.31.1); 40149 (65.40.3); 40155 (65.23.4); 40175 (65.32.2); 40175A (65.23.2); 40198 (69.7.4); 40206 (40.3.20); 40232 (47.2.1); 40233 (47.11.1); 40242 (60.2.1); 40245 (65.21.1); 40271 (65.49.1); 40272 (65.23.1); 40274 (65.23.1); 40275 (65.7.1); 40276 (65.32.2); 40277 (65.21.1); 40278 (65.7.1); 40281 (47.1.10a); 40298 (45.2.1a); 40303 (44.6.3); 40369 (63.1.8); 40380 (40.7.7); 40389A (65.45.2); 40421 (40.7.6); 40426 (47.1.9b); 40443 (63.1.8); 40452 (63.1.8); 40500 (48.1.6); 40505 (66.1.4); 40510 (63.1.8); 40520 (40.3.3); 40537 (66.1.2); 40588 (39.2.3); 40589 (40.3.17); 40600 (63.2.5); 40601 (39.14.5); 40645 (66.1.2); 40708 (55.1.1a); 40730 p.p. (68.1.1, 68.1.4); 40754 (40.7.6); 40813 (65.32.2); 40834 (40.19.2), 40881 (40.16.7); 40884 (40.19.2); 40894 (69.2.5); 40932 (40.1.2); 40939 (40.7.7); 40985 (39.1.2); 40992 (39.13.7); 50026 (40.16.7); 50035 (65.23.1); 50036 (65.23.2); 50037 (47.1.5); 50038 (47.21.5); 50053 (65.23.2); 50074 (65.23.1); 50079 (39.13.7); 50086 (63.1.8); 50101 (41.2.1); 50103 (66.1.7); 50125 (69.7.5a); 50130 (65.29.1a); 50137 (63.1.9); 50137a (63.1.9); 50140 (40.7.7); 50213 (40.7.6); 50244 (40.7.7); 50279 (39.8.5); 50309 (47.13.4); 50320 (66.1.3); 50343 (40.7.6); 50397

(40.3.20); 50405a (44.1.6); 50449 (45.1.1); 50462 (40.7.7); 50501 (67.1.1); 50504 (47.14.4); 50539 (68.1.5); 50546 (63.2.5); 50547 (40.7.6); 50570 (48.1.10); 50584 (66.1.2); 50604 (47.13.3); 50605 (60.1.1); 50636 (47.13.3); 50642 (54.1.1); 50643 (50.1.1a); 50644 (47.1.5); 50710 (39.11.4); 50743 (40.7.7); 50756 (40.3.8); 50757 (40.3.13); 50794 (40.3.13); 50805 (40.3.8); 50836 (40.16.4); 50881 (47.1.9b); 50883 (43.1.1); 50905 (65.12.1); 50914 (65.1.1b); 50914A (65.1.1b); 50916 (65.23.3); 50929 (64.1.2); 50935 (65.23.1); 50937 (66.1.4); 50940 (47.1.11); 50983 (47.19.1); 50985 (40.3.13); 51019 (39.8.3); 51034 (54.1.1); 51038 (65.23.4); 51041 (47.12.1); 51042 (65.23.5); 51045 (65.32.1); 51048 (65.10.1); 51051 (65.32.1); 51053 (65.23.2); 51055 (65.21.1); 51059 (47.11.1); 51060 (47.19.1); 51061 (47.20.1); 51062 (47.2.1); 51118 (40.16.4); 51120 (50.1.2); 51128 (39.14.1, 39.14.6); 51161 (65.40.1); 51163 (65.21.1); 51180 (47.1.1a47.1.1b); 51181 (47.11.1); 51182 (47.13.3); 51183 (47.1.5); 51198 p.p. (65.38.3, 65.38.4); 51222 (65.34.1); 51223 (65.23.2); 51224 (65.35.3); 51228 (47.19.1); 51229 (47.20.1); 51229A (47.20.1); 51233 (47.10.2); 51240 (65.23.1); 51263 (68.1.1); 51263A (68.1.1); 51263a bis (68.1.4); 51264 (68.1.3); 51279 (47.1.9b); 51291 (65.23.1); 51341 (47.1.5); 51342 (47.1.11); 51347 (47.1.11); 51348 (47.15.1); 51349 (47.1.1a); 51371 (56.2.1); 51381 (40.16.7); 51400 (65.13.1a); 51402 (65.3.1a); 51404 (43.1.1); 51408 p.p. (65.38.4); 51413 (60.1.1); 51414 (65.12.1); 51416 (40.3.8); 51420 (39.4.1); 51455 (47.1.5); 51457 (65.40.1); 51489 (65.35.4); 51527 (65.38.2); 51528 (65.12.1); 51529 (65.38.4); 51558 (47.5.1); 51559 (47.8.1); 51562 (65.35.2); 51564a (65.26.1); 51566 (65.35.1); 51566a (65.35.1); 51567 (65.32.1); 51614 (59.2.1); 51617 (47.19.1); 51618 (47.20.1); 51619 (65.32.1); 51619a (65.32.2); 51621 (65.32.1); 51622 (65.10.2); 51632 (65.23.1); 51657 (47.20.1); 51668 (65.50.1a); 51674 (56.1.2); 51677 (40.16.7); 51680 (41.1.3); 51680 bis (66.1.4); 51694 (39.13.7).

Clemens, M. S. 9912 (69.6.2); 9939 (69.4.1); 10003 (39.1.2); 10025 (40.3.10); 10026 (48.1.7); 10028 (65.35.3); 10030 (65.36.1); 10052 (40.11.8); 10083 (69.11.3); 10142 (39.12.2); 10157 (69.1.8); 10177 (69.7.2); 10206 (69.8.2); 10285 (65.43.1); 10417 (66.1.7); 10422 (65.32.2); 10423 (65.28.1); 10424 (65.32.1); 10442 (39.1.2); 10503 (65.23.3); 10535 (60.1.1); 10578 (47.1.11); 10607 (65.38.3, 65.38.4); 10608 (47.13.3); 10609 (47.20.1); 10610 (65.13.1a); 10611 (50.1.2); 10615 (47.1.11); 10625 (43.1.1); 10628 (65.12.1); 10630 (65.1.1a65.1.1c); 10631 (65.1.1b); 10704 (65.23.4); 10709 (47.7.2); 10772 (39.12.2); 10773 (39.4.1); 10797 (65.26.1); 10814 (65.26.1); 10874 (50.1.1a); 10885 (47.1.3a, 50.1.1a); 10897 (47.2.1); 11029 (41.2.1); 11029A (68.1.1); 11030a (68.1.4); 11087 (47.7.1); 11088 (47.13.1); 11092 (47.14.4); 11108 (52.1.1).

Cockburn 70108 (39.12.3); 71900 (63.2.5); 76812 (40.3.8).

Cockburn & Aban 82997 (65.50.1a); 82998 (65.1.1b).

Collector unknown 144 (69.1.1); 1017 (69.7.3); 1042 (50.1.1a); 1045 (69.1.6); 1048 (69.1.6); 1146 (69.1.6).

Collenette 1 (69.4.1); 7 (39.6.3); 8 (39.6.3); 20 (64.1.1); 21 (60.1.1); 36 (54.1.1); 83 (47.11.1); 86 (65.23.4); 98 (47.2.1); 112 (41.1.2); 118 (69.11.2); 121 (69.7.4); 125 (46.1.1); 129 (69.7.1); 131 (44.7.1); 132 (39.6.3); 515 (60.2.1); 588 (45.2.1a); 634 (60.1.1); 638 (41.3.1); 639 (39.6.1); 894 (69.1.7); 902 (66.1.3); 1003 (69.4.1); 1012 (69.8.1); 2377 (64.1.2); 21511 (65.13.1a); 21511A (65.1.1c); 21512 (47.1.5); 21520 (60.1.1); 21529 (47.1.11); 21585 (40.16.7); 21600 (47.20.1); 21610 (60.1.1); 21619 (47.1.3b); 21622 (47.2.1); 21634 (39.9.1); 21659 (69.11.3).

Cox 951 (44.1.6).

*Daim Andau 2 (65.22.1); 4 (65.16.1); 13 (63.1.2); 14 (59.1.1); 15 (69.7.2); 16 (47.21.4); 30 (40.7.6); 33 (55.1.1a); 44 (65.7.1); 60 (47.5.2); 63 (46.1.4); 70 (65.8.1); 81 (40.7.7); 82 (40.3.21); 83 (40.7.12); 84 (40.7.11); 85 (40.2.4); 108 (40.3.4); 148 (40.1.3); 167 (44.6.2); 168 (66.1.6); 170 (65.21.1); 171 (47.5.3); 172 (65.35.4); 173 (65.10.1); 175 (59.2.1); 176 (47.5.5); 194 (40.4.1); 205 (66.1.2); 211 (65.44.1); 229 (40.3.3); 296 (69.1.1); 495 (39.13.7); 505 (40.3.1); 566 (44.1.5); 567 (44.1.4); 568 (67.1.2); 579 (39.11.3); 597 (69.7.1); 605 (69.11.3); 607 (52.1.2); 632 (39.12.1); 680 (40.16.7); 715 (39.13.7); 769 (44.6.2); 770 (45.1.1); 791 (39.4.1); 856 (39.5.1); 891 (69.1.9); 906 (49.1.1); 934 (45.1.1); 935 (48.1.11); 953 (44.1.1); 980 (49.1.2).

*Danson Kandaong 1 (40.3.13).

Darnton 131 (59.3.1); 147 (44.3.1); 152 (65.45.3); 153 (69.8.2); 158 (53.1.2); 162 (69.9.2); 166 (69.7.2); 177 (40.3.10); 181 (69.1.4); 187 (39.10.1); 267 (65.35.3); 269 (40.7.12); 301 (65.45.2); 341 (65.39.1); 371 (69.1.10); 377 (49.1.1); 424 (39.8.5); 448 (69.1.10); 461 (39.6.1); 465 (66.1.3); 481 (39.1.2); 501 (69.3.3); 503 (41.1.1); 511 (69.6.9); 535 (44.5.2); 560 (40.16.7); 579 (47.14.1); 601 (47.14.4).

de Vogel 8009 (53.1.2); 8010 (53.1.2); 8012 (66.1.3).

*Dius Tadong 1 (40.3.17); 4 (40.7.6); 5 (40.12.3); 7 (40.3.12a); 8 (40.7.10a); 9 (40.3.21); 10 (40.1.3); 11 (40.3.2); 12 (40.7.3); 19 (40.11.4); 29 (40.7.5); 46 (65.22.1); 50 (46.1.4); 59 (40.4.1); 60 (40.19.4); 61 (40.7.1); 62 (40.16.4); 63 (40.3.12a); 64 (69.1.1); 77 (69.9.2); 92 (40.7.11); 93 (40.7.12); 99 (66.1.1); 100 (65.16.1); 107 (65.46.1); 111 (40.11.7); 131 (65.18.1); 135 (65.43.1); 149 (49.1.2); 162 (39.14.3); 166 (40.3.19); 168 (65.8.1); 182 (65.35.4); 183 (47.1.3a); 189 (40.7.6); 190 (40.3.21); 194 (66.1.3); 198 (40.2.4); 205 (65.49.1); 206 (40.7.5); 207 (40.11.7); 211 (59.1.1); 212 (40.19.2); 224 (59.4.1); 225 (40.3.18); 226 (65.43.2); 236 (40.3.4); 237 (63.2.9); 254 (51.1.1); 272 (65.22.1); 273 (65.18.1); 282 (48.1.4); 291 p.p. (40.3.4); 292 (40.7.11); 294 (40.16.10a); 303 (65.18.1); 318 (40.11.4); 320 (40.3.2); 321 (40.3.12a); 322 (40.12.4); 323 (40.11.3, 40.11.8); 327 (69.1.1); 343 (47.5.7); 346 (59.1.1); 347 (69.9.2); 349 (47.6.1); 355 (40.4.1); 358 (49.1.1); 362 (69.1.1); 367 (65.20.1); 378 (65.43.2); 379 (69.6.3); 383 (65.22.1); 384 (47.21.3a); 408 (69.1.1); 417 (40.2.2); 418 (40.19.5); 426 (69.1.1); 428 (49.1.1), 433 (40.2.2); 435 (69.1.9); 450 (65.33.1a); 471 (47.21.3a); 474 (65.43.1); 476 (40.3.4); 477 (40.3.15); 478 (40.7.11); 479 (40.17.3); 480 (40.12.3); 482 (59.1.1); 499 (65.16.1); 500 (65.35.4); 502 (40.5.1); 504 (40.7.12); 506 (40.3.18); 511 (40.11.7); 520 (40.3.22); 521 (40.7.5); 522 (40.3.2); 540 (59.1.1); 551 (47.21.3a); 573 (47.21.4); 574 (47.17.1); 583 (40.7.3); 584 (40.3.18); 585 (40.11.7); 586 (40.2.4); 587 (40.7.5); 588 (59.1.1); 600 (40.11.7); 635 (47.17.1); 642 (47.21.4); 646 (47.21.3a); 660 (47.10.3); 664 (65.32.1); 677 (69.7.2); 697 (65.35.4); 701 (47.21.4); 711 (65.35.4); 716 (40.3.2); 717 (40.3.17); 718 (65.35.4); 746 (47.5.7); 778 (65.16.1); 797 (39.1.1); 798 (66.1.1); 800 (65.7.1); 801 (69.1.1); 803 (47.21.3a); 822 (59.1.1).

*Doinis Soibeh 1 (40.3.12a); 10 (65.18.1); 24 (44.1.1); 29 (44.2.1); 55 (65.36.1); 56 (65.45.2); 70 (47.21.5); 71 (65.22.2); 85 (69.1.10); 93 (36.1.1); 100 (65.43.2); 111 (47.3.6); 131 (65.7.1); 135 (65.46.3); 136 (40.3.18); 137 (40.3.10); 138 (47.18.1); 141 (47.10.4); 152 (65.8.1); 189 (45.1.1); 199 (66.1.6); 206 (40.7.12); 207 (40.16.4); 208 (40.12.3); 235 (40.4.1); 240 (49.1.2); 253 (66.1.3); 256 (40.7.6); 257 (40.1.2); 258 (40.16.7); 259 (40.16.5); 320 (59.1.1); 325 (40.7.8); 327 (40.7.8); 328 (40.1.2); 329 (40.1.2); 330 (40.1.2); 341 (40.3.18); 342 (40.7.8); 343 (40.3.18); 344 (40.11.8); 345 (40.7.6); 346 (40.7.12); 347 (40.7.8); 348 (40.7.8); 352 (40.19.2); 353 (40.3.20); 355 (40.2.1); 378 (48.1.2); 379 (66.1.4); 384 (65.48.1); 417 (48.1.9); 424 (40.3.9); 611 (40.3.1); 612 (40.3.18); 613 (40.19.1); 614 (40.3.8); 615 (40.7.6); 616 (40.7.6); 617 (40.3.6); 681 (40.16.4); 682 (40.16.5); 683 (40.11.1); 685 (40.11.8); 686 (40.7.11); 688 (40.7.1); 689 (39.2.2); 703 (65.8.1); 719 (63.1.8); 722 (39.12.3); 724 (39.14.5); 727 (40.16.7); 728 (40.3.9); 729 (40.1.2); 730 (40.12.4); 731 (40.1.2); 732 (40.3.17); 767 (53.1.2); 777 (40.8.1); 778 (40.1.1); 785 (40.3.10); 786 (40.3.18); 787 (40.4.1); 827 (40.7.3); 828 (40.11.9); 855 (48.1.2).

Dransfield, J. et al. 5505 (40.11.7); 5506 (65.16.1); 5507 (40.7.12); 5549 (40.16.4); 5551 (40.7.6); 5552 (40.7.6); 5553 (40.16.7); 5554 (40.3.20); 5555 (40.3.20); 5556 (40.3.14); 5557 (40.3.14); 5559 (40.7.6); 5560 (40.3.18); 5561 (40.3.1); 5562 (40.3.3); 5563 (40.11.9); 5564 (40.7.1); 5565 (40.7.5); 5566 (40.11.8); 5567 (40.7.10a); 5568 (40.7.10a); 5569 (40.7.3); 5570 (40.3.17); 5571 (40.7.3); 5573 (40.18.1); 5574 (40.11.7); 5575 (40.7.11); 5576 (40.3.21); 5577 (40.3.2); 5578 (40.3.18); 5579 (40.3.12b); 5580 (40.3.9); 5581 (40.16.6); 5582 (40.3.13); 5583 (40.18.2); 5668 (40.3.8); 5669 (40.3.8); 5670 (40.3.8); 5672 (40.3.10); 5673 (40.3.10); 5674 (40.3.8); 5675 (40.16.7); 5676 (40.16.4); 5677 (40.3.8); 5681 (40.16.4); 5682 (40.3.8); 5683 (40.3.8); 5690 (40.7.7); 5691 (40.3.11); 5692 (40.16.7); 5693 (40.7.7); 5694 (40.7.7); 5695 (40.7.7); 5696 (40.3.14); 5697 (40.7.6); 5698 (40.3.14); 5699 (40.3.10); 5700 (40.7.7); 5701 (40.3.10); 5702 (40.16.4); 5704 (40.19.2); 5705 (40.16.4); 5706 (40.3.20); 5707 (40.3.20); 5708 (40.1.2).

Dransfield, S. 715 (65.40.4); 715A (65.40.3); 717 (65.20.1); 718 (65.40.2); 719 (65.20.1); 720 (65.16.1); 721 (65.42.1, 65.43.1); 722 (65.43.2); 754 (65.51.1); 755 (65.40.1); 756 (65.51.1); 757 (65.40.1); 758 (65.25.1); 759 (65.25.1, 65.40.1); 843 (65.47.1); 844 (65.47.1); 849 (65.25.1).

Edwards 2162 (39.9.1).

Forster 22 (47.20.1); 37A (65.10.1); 39 (65.23.2); 40 (65.28.2); 42 (65.1.1b); 43 (65.1.2a); 45 (65.23.4); 53 (65.23.3); 56 (65.15.1); 57 (65.42.1); 109 (65.14.1); 519 (47.19.1); 523 (47.1.9b).

Fosberg 43960 (65.40.2); 44105 (65.13.1a).

Fuchs 21045 (66.1.3); 21049 (64.1.1); 21071 (47.7.2); 21089 (65.13.1a); 21090 (65.50.1a).

Fuchs & Collenette 21439 (56.2.1).

Furtado (He collected under Clemens numbers.)

Gibbs 3124 (54.1.1); 3125 (50.1.1a); 3956 (69.5.1); 3967 (47.3.6); 3968 (40.16.4); 3983 (40.7.7); 3996 (45.1.1); 4004 (47.1.3a); 4005 (65.26.1); 4007 (65.26.1); 4010 (39.14.6); 4011 (69.1.10); 4033 (41.1.2); 4037 (47.11.1); 4038 (47.19.4); 4041 (60.1.1); 4069 (65.23.4); 4070 (47.1.9b); 4091 (65.40.1); 4093 (47.1.3b); 4094 (47.2.1); 4095 (47.2.1); 4097 (47.1.9b); 4099 (39.6.3); 4100 (39.13.3); 4103 (45.1.1); 4113 (65.23.1); 4133 (65.32.1); 4157 (64.1.2); 4160 (47.7.2); 4185 (65.13.1a); 4188 (47.19.1); 4190 (47.20.1); 4191 (47.1.5); 4192 (65.13.1a); 4193 (47.1.5); 4194 (47.1.11); 4196 (47.1.2b); 4198 (47.19.1); 4207 (43.1.1); 4209 (43.1.1, 50.1.2); 4219 (40.16.4); 4220 (47.19.1); 4230 (47.1.5, 47.1.11); 4232 (65.40.1); 4234 (60.1.1); 4240 (47.1.9b); 4243 (47.1.5); 4270 (47.19.1); 4277 (47.20.1); 4278 (47.13.3); 4297 (48.1.6, 48.1.9); 4348 (40.3.8); 4352 (65.38.3, 65.38.4).

Good, J. B. & Minol 122489 (59.4.1).

Haviland 1125a (60.1.1); 1125b (60.1.1); 1125c (60.1.1); 1145 (66.1.3); 1153 (50.1.1a); 1153 bis (50.1.1a); 1156 (64.1.1); 1179 (54.1.1); 1204 (50.1.1a); 1226 (66.1.4); 1233 (69.1.7); 1259 (54.1.1); 1293 (66.1.6); 1329 (41.1.3); 1330 (44.6.1); 1390 (65.16.1); 1393 (47.1.2b); 1394 (47.9.1); 1395 (47.19.1); 1396 (47.19.1); 1397 (47.9.1); 1398 (47.20.1); 1399 (65.1.1a); 1400 (65.13.1a); 1401 (65.38.3); 1402 (47.1.5); 1403 (47.1.5); 1404 (47.1.9b); 1405 (47.13.3); 1406 (47.13.1); 1407 (47.19.4); 1408 (65.23.3); 1409 (65.32.1).

Hay 10046 (39.7.1).

Henry Tai 42527 (48.1.9).

Holttum 46 (65.51.1); 25113 (65.32.1); 25268 (65.29.1a); 25492 (47.20.1); 25590 (39.3.1).

Hotta 3812 (64.1.2); 20194 (40.3.22).

Hou 251 (60.1.1); 255 (43.1.1); 257 (65.12.1).

Jacobs 5704 (40.3.8); 5705 (40.1.2); 5705A (40.1.2); 5711 (40.3.8); 5719 (47.19.3); 5720 (47.20.1); 5722 (65.23.3); 5723 (65.23.4); 5756 (65.13.1a); 5759 (47.1.2a); 5760 (47.9.2); 5761 (65.38.4); 5791 (40.16.7).

Jamili Nais 4287 (39.14.1).

*Jatin Tungking 1 (40.3.18); 2 (40.16.4); 3 (40.7.1); 4 (40.2.1); 6 (40.7.6); 7 (40.7.6); 9 (40.2.4); 11 (40.3.20).

Jermy & Rankin 15082 (47.19.1); 15103 (65.12.1); 15114 (47.7.2); 15115 (47.1.9b); 15123 (47.20.1); 15203 (47.20.1); 15204 (47.7.1).

*Jibrin Sibil 1 (40.3.3); 6 (65.22.1); 10 (65.18.1); 59 (65.52.1); 61 (40.19.2); 72 (65.37.3); 73 (65.10.1); 76 (47.21.3a); 80 (65.7.1); 84 (69.1.1, 49.1.1); 86 (40.12.1); 96 (40.3.5); 97 (65.35.4); 100 (40.2.4); 101 (59.2.1); 102 (40.2.3); 103 (40.7.6); 104 (40.7.5); 115 (65.18.1); 118 (40.12.2); 142 (65.8.1); 150 (40.13.1); 152 (40.1.1); 155 (46.1.4); 161 (40.3.12b); 162 (40.11.6); 163 (40.7.11); 164 (40.7.8); 185 (65.18.1); 191 (65.52.1); 193 (40.4.1); 205 (65.18.1); 224 (65.22.1); 235 (65.8.1); 275 (46.1.4); 279

(65.20.1); 280 (47.21.3a); 283 (65.18.1); 294 (65.33.2); 296 (44.6.1); 297 (65.17.1); 303 (47.3.2); 319 (69.1.1, 49.1.1).

Julius K. 124867 (39.4.1).

*Jusimin Duaneh 82 (48.1.9); 92 (48.1.2); 101 (48.1.7); 102 (65.22.2); 112 (36.1.1); 113 (40.7.6); 114 (40.3.18); 118 (40.11.6); 119 (40.3.1); 120 (40.3.6); 121 (40.7.12); 123 (40.11.8); 124 (40.19.2); 125 (40.16.6); 126 (40.16.4); 129 (40.7.1); 136 (40.7.7); 160 (65.27.1); 169 (35.1.1); 177 (65.48.1); 181 (37.1.1); 182 (40.2.1); 183 (40.17.3); 231 (69.7.2); 248 (65.8.1); 252 (48.1.9); 255 (40.1.1); 271 (69.1.10); 274 (47.21.6); 300 (66.1.3); 342 (37.1.1); 385 (47.1.3a); 397 (40.16.4); 410 (66.1.4); 461 (39.10.1); 508 (64.1.1); 533 (69.7.5a); 544 (65.8.1).

Justin Jukian 196 (40.3.8).

Kadir 1674 (65.45.2); 1675 (65.43.3).

Kanis 51474 (69.4.1); 51478 (44.1.1); 53807 (65.40.2); 56149 (48.1.9).

Kanis & Kuripin 53954 (47.1.3a); 53968 (64.1.1).

Kanis & Sinanggul 50101 (66.1.2); 51490 (66.1.3).

*Kinsun Bakia 1 (40.7.6); 2 (40.2.3); 4 (59.1.1); 61 (44.1.4); 70 (69.7.2); 71 (44.6.1); 73 (59.3.4); 74 (59.4.1); 76 (69.1.9); 81 (39.13.2); 138 (40.12.1); 145 (65.11.1); 152 (69.1.1); 166 (57.1.1); 169 (65.33.1a); 173 (40.7.5); 175 (49.1.1); 178 (40.3.2); 192 (65.8.1); 200 (58.1.1); 206 (65.18.1); 210 (65.8.1); 212 (65.17.1); 213 (47.10.4); 232 (39.13.2); 233 (59.3.3); 237 (65.14.2a); 238 (65.7.1); 240 (40.7.11); 241 (40.11.2); 251 (65.11.1); 252 (65.17.1); 253 (40.12.2); 270 (53.1.3); 276 (65.46.3); 308 (65.9.1); 323 (65.8.1); 338 (40.4.1); 339 (40.11.9); 408 (48.1.13); 413 (65.10.2); 415 (47.21.3a); 416 (40.1.1); 424 (42.1.1); 435 (64.1.1); 444 (69.3.2); 446 (39.14.2); 450 (69.3.1); 451 (69.2.9); 470 (48.1.4); 518 (65.37.2); 519 (47.3.4); 520 (65.36.4).

Kokawa 6176 (69.1.7).

Kokawa & Hotta 2854 (45.1.1); 3089 (39.12.2); 3515 (47.1.5); 3995 (39.12.3); 4152 (39.14.1); 4217 (60.2.1); 4244 (66.1.4); 4481 (39.13.7); 4482 (69.1.10); 4506 (47.1.4); 4513 (39.13.7); 4843 (59.3.3); 4963 (59.3.3); 5177 (39.14.2); 5182 (66.1.7); 5247 (48.1.13); 5253 (66.1.1); 5300 (47.1.9b); 5667 (69.1.7); 5804 (66.1.4); 5854 (66.1.3).

Kuripin L. 28773 (48.1.8).

Lajangah 28772 (66.1.4); 33055 (49.1.2); 33113 (69.1.7); 44614 (66.1.4); 44765 (45.1.1).

Lassan 76697 (66.1.2).

Lau Choon Teng 8/LCT (39.14.1); 15/LCT (39.14.5).

*Lomudin Tadong 8 (59.1.1); 13 (65.35.4); 21 (53.1.2); 23 (47.17.1); 32 (39.13.7); 33 (44.5.1); 38 (65.18.1); 70 (47.5.7); 106 (44.3.1); 136 (69.7.2); 157 (40.7.11); 158 (40.7.6); 161 (65.36.1); 165 (47.5.3); 167 (65.7.1); 176 (47.3.4); 179 (40.11.4); 194 (53.1.2); 196 (49.1.1); 197 (40.3.2); 209 (66.1.1); 213 (47.5.5); 221 (44.1.4); 228 (39.13.2); 268 (44.5.1); 269 (47.6.1); 276 (65.16.1); 288 (47.3.7a); 298 (48.1.4); 299 (39.13.2); 305 (59.1.1); 307 (65.8.1); 312 (65.19.2); 313 (65.39.1); 337 (46.1.3); 356 (46.1.5); 422 (65.44.1); 487 (64.1.1); 516 (67.1.2); 518 (52.1.1).

*Lorence Lugas 24 (66.1.3); 46 (65.8.1); 49 (47.17.1); 58 (57.1.1); 65 (64.1.1); 68 (66.1.4); 69 (59.1.1); 80 (40.7.6); 82 (42.1.1); 89 (40.9.1); 95 (47.21.1); 103 (69.1.1); 105 (69.1.9); 144 (40.7.6); 154 (47.6.1); 160 (65.35.4); 161 (65.48.1); 174 (39.14.5); 176 (40.4.1); 180 (69.1.5); 229 (66.1.3); 270 (69.1.1); 272 (67.1.2); 289 (69.1.5); 292 (65.18.1); 300 (66.1.4); 307 (69.1.8); 317 (65.46.1); 320 (39.11.5); 327 (65.52.1); 386 (66.1.2); 409 (40.17.1a); 433 (65.7.1); 442 (40.1.1); 462 (61.1.1); 482

(65.29.1a); 483 (47.21.3a); 485 (47.5.7); 498 (59.1.1); 501 (65.46.3); 516 (40.6.1); 526 (65.22.1); 543 (65.36.2); 544 (65.36.3a); 549 (65.20.1); 565 (64.1.1); 571 (40.2.4); 572 (40.7.2); 577 (44.1.4); 578 (51.1.1); 581 (66.1.3); 582 (40.3.4); 586 (65.4.1); 592 (44.3.1); 594 (65.35.4); 595 (47.16.1); 599 (49.1.1); 604 (44.6.2); 609 (69.6.4); 610 (69.6.4); 620 (44.4.1); 622 (47.17.1); 624 (48.1.7); 636 (47.12.1); 638 (47.5.7); 640 (47.3.8); 641 (47.3.3); 668 (39.8.7); 678 (69.10.1); 729 (59.3.1); 730 (69.10.2); 731 (52.1.1); 775 (48.1.2); 783 (46.1.4); 793 (66.1.4); 799 (69.1.7); 802 (39.4.1); 818 (47.21.1); 822 (69.1.9); 824 (65.8.1); 845 (69.11.2); 860 (59.1.1); 869 (48.1.7); 880 (69.6.10); 893 (65.45.2); 915 (64.1.1); 931 (48.1.5); 973 (66.1.2); 995 (63.2.7); 1003 (59.1.1); 1026 (65.9.1); 1041 (47.21.3a); 1117 (66.1.1); 1124 (65.7.1); 1146 (47.12.1); 1148 (69.6.4); 1150 (47.17.1); 1188 (65.33.2); 1192 (44.1.1); 1193 (69.7.5a); 1198 (69.1.10); 1209 (69.10.4); 1215 (48.1.2); 1223 (65.22.1); 1232 (66.1.3); 1239 (69.1.5); 1249 (65.18.1); 1254 (65.52.1); 1258 (66.1.2); 1259 (48.1.9); 1278 (64.1.1); 1319 (66.1.3); 1330 (44.4.1); 1344 (65.8.1); 1356 (47.3.8); 1366 (47.6.1); 1368 (47.12.1); 1388 (59.1.1); 1423 (69.1.10); 1424 (65.49.1); 1425 (69.9.2); 1427 (69.1.5); 1436 (40.16.8); 1453 (39.5.1); 1513 (47.7.3); 1524 (47.21.1); 1536 (44.1.1); 1537 (55.1.1a); 1542 (40.16.3); 1548 (57.1.1); 1552 (47.8.1); 1558 (40.7.4); 1570 (40.17.3); 1571 (40.3.4); 1587 (51.1.1); 1588 (65.30.1); 1595 (40.9.1); 1600 (40.3.2); 1601 (40.2.4); 1605 (48.1.9); 1611 (39.3.3); 1625 (40.7.6); 1626 (40.3.7); 1627 (40.3.10); 1628 (40.7.2); 1642 (63.2.6); 1644 (40.14.1a); 1645 (40.7.6); 1646 (69.1.8); 1647 (65.36.1); 1658 (40.7.1); 1660 (65.48.1); 1666 (65.8.1); 1678 (40.7.12); 1679 (40.7.12); 1729 (46.1.4); 1750 (69.1.1); 1811 (48.1.13); 1837 (65.46.1); 1838 (65.49.1); 1841 (65.7.1); 1844 (47.5.7); 1847 (65.29.1a); 1851 (47.18.2); 1853 (66.1.1); 1854 (47.10.4); 1856 (65.36.3a); 1857 (65.19.1); 1861 (63.1.9); 1880 (65.45.1); 1895 (64.1.1); 1897 (69.7.3); 1903 (39.1.1); 1920 (47.10.3); 1921 (49.1.3); 1982 (47.13.4); 2075 (59.3.2); 2087 (69.9.4, 69.10.1); 2089 (63.2.10); 2090 (69.1.4); 2111 (69.1.7); 2130 (65.29.1a); 2189 (66.1.3); 2216 (69.1.1); 2293 (49.1.1); 2352 (65.46.3); 2364 (65.8.1); 2407 (63.2.9); 2425 (53.1.3); 2426 (47.21.1); 2462 (63.1.6); 2472 (63.2.8a); 2479 (65.48.1); 2521 (69.1.10); 2528 (63.2.6); 2544 (69.1.1); 2545 (63.2.9); 2558 (65.18.1); 2560 (47.3.8); 2590 (59.1.1); 2595 (65.29.1a); 2599 (39.4.1); 2603 (65.33.2); 2608 (39.3.3); 2622 (39.8.7); 2635 (51.1.1); 2637 (39.5.1); 2643 (47.3.9); 2644 (47.5.2); 2648 (65.24.2); 2651 (65.35.4); 2652 (40.13.1).

Madani 89500 (66.1.3); 89508 (44.5.1); 111616 (39.9.1).

Mason 2681 (69.8.1).

*Matamin Rumutom 3 (69.7.2); 11 (59.4.1); 12 (69.7.4); 25 (69.1.9); 26 (69.11.3); 49 (59.1.1); 54 (67.1.2); 71 (64.1.1); 77 (38.1.1); 128 (44.5.1); 135 (65.44.1); 183 (39.6.1); 202 (65.33.1a); 206 (51.1.1); 212 (47.14.3a); 236 (65.18.1); 247 (44.7.2); 267 (49.1.2); 269 (65.44.1); 281 (39.1.2); 287 (44.1.6); 290 (40.3.4); 304 (47.3.7a); 317 (66.1.4); 328 (47.21.3a); 333 (42.1.1); 335 (69.1.1); 417 (67.1.2).

Meijer 18823 (40.3.21); 20266 (65.40.3); 20268 (40.3.2); 20298 (69.1.10); 20360 (40.3.8); 20370 (47.1.11); 20391 (47.19.1, 47.19.4); 20391a (47.20.1); 20955 (40.16.6); 21039 (40.3.8); 21056 (65.12.1); 21064 (40.3.10); 21350 (40.3.10); 22017 (47.1.2a); 22020 (65.1.1b); 22035 (47.19.1); 22042 (47.1.5); 22050 (47.13.3); 22053 (47.20.1); 22053A (47.19.1); 22059 (47.9.2); 22080 (65.3.1a); 23451 (63.2.5); 24064 (40.16.4); 24085 (47.1.9a); 24098 (47.1.8); 24116 (47.1.9b); 24207 (65.13.1a); 24208 (65.50.1a); 24217 (65.1.1b); 26413 (40.3.1); 26433 (47.21.2); 28566 (47.15.1); 28567 (65.12.1); 28569 (50.1.2); 28570 (43.1.1); 28709 (40.16.7); 28810 (47.1.9b); 29015 (40.1.2); 29020 (39.14.6); 29114 (65.29.1a); 29276 (47.19.4); 34615 (40.3.8); 34646 (63.2.5); 38062 (69.2.5); 38078 (39.13.7); 38570 (47.19.4); 42470 (66.1.2); 42771 (66.1.4); 48092 (40.16.7); 48116 (40.3.10); 48117 (65.40.2); 54008 (47.14.5); 54278 (65.1.1b); 122414 (39.12.1); 122429 (39.12.1); 131924 (39.14.4).

Meijer & Abu Bakar 131913 (39.5.1).

Meijer et al. 122513 (39.14.2); 122520 (69.9.2).

*Meliden Giking 17 (40.4.1, 40.7.12); 20 (40.18.1); 21 (40.11.7); 22 (40.11.4); 23 (40.12.3); 24 (40.17.3); 25 (40.14.1a); 39 (65.16.1); 40 (40.3.18); 53 (47.21.1); 54 (65.20.1); 55 (65.43.1); 76 (59.1.1); 77 (69.6.3); 78 (65.5.1); 98 (57.1.1); 102 (65.5.1); 121 (40.3.17); 122 (40.7.9); 123 (40.6.1); 124 (40.2.4); 202 (65.18.1); 208 (65.43.2); 210 (65.10.1).

Mikil 28149 (66.1.4); 29245 (69.1.10); 37710 (49.1.2); 38475 (65.25.1); 38635 (63.1.8); 38691 (39.12.1); 38896 (40.3.10).

Molesworth-Allen 3261 (65.40.1); 3286 (65.38.4); 3287 (65.38.4); 3288 (65.1.1b); 3289 (65.13.1a); 3290 (65.38.2).

Newell 143 (55.1.1a); 144 (55.1.1a); 145 (55.1.1a); 149 (55.1.1a); 151 (55.1.1a).

Nooteboom & Aban 1508 (39.10.1); 1534 (66.1.6); 1544 (39.1.2); 1561 (65.21.1); 1571 (40.16.6); 1582 (39.14.7); 1583 (39.4.1); 1585 (69.1.10).

Ogata 11083 (39.9.1); 11097 (65.45.2); 11146 (66.1.4).

*PEK 1 (40.7.7); 2 (40.3.10); 3 (40.7.6); 4 (40.7.12); 5 (40.7.6); 6 (40.3.10); 7 (40.7.5); 8 (40.3.20); 9 (40.17.3); 10 (40.7.12); 11 (40.3.10); 12 (40.1.2); 14 (40.16.4); 15 (40.3.18); 16 (40.1.2); 17 (40.16.6); 20 (40.7.8); 23 (40.16.7); 45 (40.17.3); 46 (40.7.8).

Pereira 47214 (63.2.5).

Peter William 706 (63.2.5).

Phillipps, A. 190 (50.1.1a); 1012 (45.1.1).

Poore 124 (65.40.2); 131 (39.12.1); 199 (65.23.4); 242 (47.1.9b); 244 (47.20.1); 258 (65.40.2); 261 (69.6.9); 502 (47.7.2); 504 (47.19.4); 509 (47.1.5); 2133 (65.40.2); 2600 (65.40.2); 3587 (65.40.2); 3726 (65.40.2); 3863 (44.1.1, 44.1.6).

Price 118 (69.8.1); 124 (39.14.6); 129 (64.1.1); 132 (41.1.2); 147 (52.1.1); 155 (47.1.9b); 165 (39.10.1); 168 (39.14.6); 169 (45.2.1a); 179 (69.6.9); 195 (65.40.1); 196 (64.1.1); 201 (40.3.8); 203 (64.1.2); 208 (64.1.2); 222 (60.2.1); 250 (69.8.1).

Puasa 1542 (66.1.2); 1548 (64.1.1).

Rickards 160 (60.1.1); 163 (47.13.3).

Sadau 42888 (63.1.8).

Salick et al. 9023 (56.1.1).

*Sani Sambuling 1 (40.15.1); 2 (40.7.8); 3 (40.11.9); 8 (40.1.1); 9 (46.1.4); 10 (49.1.1); 12 (49.1.2); 13 (40.3.2); 17 (40.3.10); 21 (65.46.1); 22 (69.1.9); 23 (40.18.1); 24 (40.3.18); 25 (40.7.6); 27 (40.3.1); 29 (39.11.6); 33 (40.12.3); 41 (40.7.12); 42 (40.11.7); 43 (40.3.10); 44 (40.3.18); 45 (40.11.4); 46 (40.17.3); 50 (69.7.2); 51 (40.7.12, 59.1.1); 52 (65.43.5); 53 (40.12.3); 55 (40.11.4); 56 (40.3.1); 57 (40.7.12); 58 (40.3.2); 59 (40.3.1); 60 (40.7.5); 67 (40.7.12); 68 (40.3.2); 70 (69.1.1, 40.3.18); 71 (40.7.12); 72 (40.3.1); 73 (40.7.5); 74 (40.6.1); 75 (40.3.15); 76 (40.1.1); 77 (40.11.7, 66.1.3); 78 (40.7.12); 107 (40.7.11); 109 (40.11.5); 110 (40.3.10); 112 (40.5.1); 114 (40.11.5); 117 (40.11.5); 119 (40.3.1); 133 (40.11.9); 134 (40.7.12); 135 (40.3.18); 137 (40.3.1); 138 (40.7.5); 139 (40.18.2); 140 (40.3.10); 144 (40.11.4); 146 (40.3.2); 150 (40.3.3); 152 (59.1.1); 158 (40.7.12); 159 (40.3.3); 160 (40.4.1); 162 (40.3.18); 163 (40.3.2); 164 (40.11.9); 165 (40.11.8); 168 (40.3.1); 169 (40.11.4); 170 (40.7.12); 171 (40.3.1); 172 (40.3.2); 173 (40.3.2); 174 (40.3.10); 175 (40.12.3); 176 (40.3.10); 177 (40.12.4); 178 (40.3.2); 179 (40.3.1); 180 (40.3.2); 181 (40.12.3); 182 (40.3.10); 183 (40.3.1); 184 (40.3.10); 197 (40.14.1a); 198 (40.11.5); 201 (40.3.2); 202 (40.11.4); 203 (40.7.12); 204 (40.3.3); 205 (40.3.18); 217 (40.3.18); 218 (40.14.1a); 219 (40.3.3); 220 (40.3.10); 221 (40.7.12); 227 (40.14.1a); 228 (40.11.5); 229 (40.3.2); 230 (40.3.3); 231 (40.7.12); 232 (40.7.11); 237 (40.14.1a); 238 (40.7.12); 239 (40.3.2); 240 (40.7.11); 241 (40.3.1); 246 (40.14.1a); 247 (40.7.12); 248 (40.7.11); 249 (40.3.2); 250 (40.11.5); 251 (40.3.3); 252 (40.3.3); 253 (40.3.2); 255 (40.14.1a); 256 (40.3.3); 257 (40.3.10); 258 (40.3.2); 315 (40.3.3); 336 (69.7.2); 339 (44.6.1); 343 (39.14.2); 345 (69.11.3); 368 (69.1.5); 376 (46.1.4); 379 (69.1.1); 401 (47.21.1); 421 (39.11.6); 425 (65.20.1);

474 (66.1.3); 497 (65.8.1); 499 (44.6.1); 500 (69.7.2); 510 (59.1.1); 580 (40.7.12); 581 (40.3.18); 617 (69.10.1); 623 (69.7.3); 627 (40.12.3); 641 (66.1.3); 687 (40.7.12); 696 (65.43.5); 707 (65.24.1); 726 (59.1.1).

Sario 28509 (45.1.1).

Shea & Aban 76376 (63.2.2); 76727 (63.1.6); 76749 (65.44.1); 76759 (69.1.9); 76765 (39.11.6); 76770 (69.11.2); 76771 (49.1.1); 76834 (59.1.1); 76898 (65.49.1); 76938 (44.1.2); 76956 (39.5.1); 76969 (69.1.3); 77153 (40.2.4); 77195 (44.5.1); 77226 (65.33.1a); 77245 (44.8.1); 77260 (66.1.6); 77263 (39.4.1); 77311 (39.8.5); 77324 (44.2.1).

Sidek bin Kiah 52 (40.3.8).

Sigin *et al.* 110647 (45.1.1); 112257 (48.1.13); 112266 (65.8.1).

Simpson 184 (47.17.3); 185 (47.1.7); 186 (47.1.9b); 188 (47.5.2); 189 (47.12.1); 190 (47.10.2); 191 (47.10.1); 194 (47.21.4, 47.21.5); 195 (47.14.5); 197 (47.5.2); 198 (47.7.2); 199 (47.5.2); 200 (47.1.3a); 202 (47.14.1); 203 (47.3.8); 204 (47.3.6); 205 (47.3.8); 206 (47.16.2a); 207 (47.5.4); 208 (47.10.3); 209 (47.5.1); 210 (47.5.2); 212 (47.5.2); 213 (47.21.3a); 214 (47.3.5); 216 (47.21.2); 217 (47.1.9b); 218 (47.4.1); 219 (47.20.2); 220 (47.1.11).

Simpson & Casserly 196 (65.45.2).

Sinanggul 38368 (45.1.1).

Sinclair *et al.* 8959 (65.46.2); 9066 (65.23.4); 9083 (65.23.3); 9120 (47.20.1); 9122 (65.1.1b); 9124 (47.1.5); 9135 (47.19.1); 9136 (43.1.1); 9140 (47.13.3); 9142 (65.12.1); 9153 (65.13.1a); 9160 (65.38.4); 9162 (65.1.1a); 9167 (60.1.1); 9185 (65.50.1a); 9187 (47.1.5); 9188 (47.1.2b); 9194 (47.7.2); 9199 (47.20.1); 9204 (65.40.1); 9220 (47.1.7); 9230 (47.1.3a); 9237 (65.28.2); 9274 (65.16.1).

Singh 28335 (40.3.10).

Sleumer 4705 (50.1.2); 4706 (43.1.1); 4708 (65.38.4); 4714 (47.19.1); 4715 (47.20.1); 4716 (65.13.1a); 4717 (47.15.1); 4718 (43.1.1); 4719 (50.1.2); 4721 (65.12.1); 4722 (47.15.1); 4736 (47.19.4).

Smith, J. M. B. 456 (65.23.3); 458 (65.1.1b); 459 (65.50.1a); 460 (65.23.4); 461 (65.1.2a); 465 (47.1.2a); 466 (47.20.1); 467 (65.13.1a); 500 (65.38.1); 501 (47.1.5); 505 (65.38.3); 506 (65.38.4); 507 (65.1.1c); 510 (65.38.1); 511 (65.13.1a); 512 (65.50.1a); 513 (65.1.1b); 514 (65.3.1a); 517 (65.12.1); 521 (47.1.11); 522 (47.1.9b); 537 (47.1.1b); 540 (47.13.3); 554 (65.6.1); 556 (65.3.1a).

Smith, R. M. 11/86 (69.6.7); 12/86 (69.2.3); 15/86 (69.2.5); 16/86 (69.2.5); 21/86 (69.9.3); 28/86 (69.1.10); 31/86 (69.6.5); 33/86 (69.7.2); 34/86 (69.9.2); 35/86 (69.6.4); 40/86 (69.2.6); 48/86 (69.7.3).

Smith, R. M. & Phillipps 24 (69.9.1); 24a (69.7.4); 25 (69.2.3); 27 (69.2.5).

Smith, R. M. et al. 36/86 (69.9.1); 44/86 (69.2.7); 50/86 (69.2.1); 51/86 (69.2.7).

Smith & Everard 145 (40.2.4); 147 (65.40.2).

Smitinand 8182 (65.23.4).

Smythies 10609 (65.40.2).

Stein 50 (65.50.1a); 225 (65.1.1b); 227 (65.3.1a).

Stevens *et al.* 619 (40.1.2); 625 (66.1.3); 641 (69.6.7); 651 (66.1.3); 661 (69.9.1).

Stone 11325 (63.2.3); 11326 (63.2.3); 11327 (63.2.3); 11329 (65.1.1b); 11342 (65.50.1a); 11401 (63.1.8); 11412 (69.1.7); 11414 (39.14.1); 11429 (63.2.3); 11430 (63.1.5); 11442 (63.2.3); 12905 (63.2.4); 12906 (63.2.4); 12922 (63.1.7a); 12923 (63.2.3).

Stone & Littke 11443 (63.1.5); 11444 (63.2.5).

Stone *et al.* 11408 (63.2.5); 11425 (63.2.7); 11437 (63.2.9).

Tai 42538 (49.1.2).

Tamura & Hotta 446 (59.3.3).

Tikau 28920 (66.1.4).

Tiong 88036 (65.42.1, 65.43.1); 88612 (65.43.2); 88660 (65.47.1).

Tiong & George 88042 (65.5.1); 88045 (65.5.1).

Topping 1579 (65.43.2); 1596 (39.14.2); 1789 (66.1.7); 1885 (50.1.1a); 1887 (50.1.1a).

*Tungking Simbayan 5 (40.7.4); 8 (40.7.6); 10 (40.13.1); 23 (40.3.3); 24 (40.3.1); 25 (40.11.8); 27 (69.9.2); 29 (65.8.1); 30 (65.49.1).

Vermeulen & Chan 392 (69.10.5).

Wong 35154 (65.47.1); 35159 (65.40.2).

Wood 601 (39.6.2); 856 (39.6.1).

*Yalin Surunda 30 (65.18.1); 33 (65.8.1); 36 (39.12.3); 38 (69.2.10); 101 (44.1.6).

LITERATURE CITED

Beaman, J. H., & Beaman, R. S. (1990). Diversity and distribution patterns in the flora of Mount Kinabalu. Pp. 147–160 in Baas, P., Kalkman, K. & Geesink, R. (eds.). *The Plant Diversity of Malesia*. Kluwer Academic Publishers, Dordrecht/Boston/London.

Bentham, G., & Hooker, J. D. (1883). *Genera Plantarum*. Vol 3, part 2. L. Reeve, London.

Bouman, F. (1995). Seed structure and systematics in Dioscoreales. Pp. 139–156 in Rudall, P. J., Cribb, P. J., Cutler, D. F. & Humphries, C. J. (eds.). *Monocotyledons: Systematics and Evolution*. Royal Botanic Gardens, Kew.

Burtt, B. L. (1977). *Curcuma zedoaria. Gard. Bull. Singapore* 30: 59–62.

Chase, M. W., Stevenson, D. W., Wilkin, P. & Rudall, P. J. (1995). Monocot systematics: a combined analysis. Pp. 685–730 in Rudall, P. J., Cribb, P. J., Cutler, D. F. & Humphries, C. J. (eds.). *Monocotyledons: Systematics and Evolution*. Royal Botanic Gardens, Kew.

Corner, E. J. H., with minor revisions by Beaman, J. H. (1996). The plant life of Kinabalu—an introduction. Pp. 101–149 in Wong, K. M., & Phillipps, A. (eds.). *Kinabalu: Summit of Borneo, a revised and expanded edition*. The Sabah Society in association with Sabah Parks, Kota Kinabalu, Malaysia.

Cronquist, A. (1981). *An Integrated System of Classification of Flowering Plants*. Columbia Univ. Press, New York.

Dahlgren, R. M. T., & Clifford, H. T. (1982). *The Monocotyledons—a Comparative Study*. Academic Press, London.

Dahlgren, R. M. T., & Rasmussen, F. N. (1983). Monocotyledon evolution; characters and phylogenetic estimation. *Evol. Biol.* 16: 255–395.

Dahlgren, R. M. T., Clifford, H. T. and Yeo, P. F. (1985). *The Families of the Monocotyledons*. Springer-Verlag, Berlin.

Duvall, M. R., Learn, G. H., Eguiarte, L. E. and Clegg, M. T. (1993). Phylogenetic analysis of rbcL sequences identifies *Acorus calamus* as the primal extant monocotyledon. *Proc. Natl. Acad. Sci. U. S. A.* 90: 4641–4644.

French, J. C., Chung, M. G. & Yoon, K. H. (1995). Chloroplast DNA phylogeny of the Ariflorae. Pp. 255–275 in Rudall, P. J., Cribb, P. J., Cutler, D. F. & Humphries, C. J. (eds.). *Monocotyledons: Systematics and Evolution*. Royal Botanic Gardens, Kew.

Gibbs, L. S. (1914). A contribution to the flora and plant formations of Mount Kinabalu and the highlands of British North Borneo. *J. Linn. Soc., Bot.* 42: 1–240, 8 pl.

Hutchinson, J. (1934). *The Families of Flowering Plants. Vol. 2. Monocotyledons.* MacMillan, London.

Kellogg, E. A., & Linder, H. P. (1995). Phylogeny of Poales. Pp. 511–542 in Rudall, P. J., Cribb, P. J., Cutler, D. F. & Humphries, C. J. (eds.). *Monocotyledons: Systematics and Evolution.* Royal Botanic Gardens, Kew.

Kern, J. H. (1974). Cyperaceae. *Fl. Males.* I, 7: 435–753.

Kern, J. H., & Nooteboom, H. P. (1979). Cyperaceae–II. *Fl. Males.* I, 9: 107–187.

Kress, W. J. (1990). The phylogeny and classification of the Zingiberales. *Ann. Missouri Bot. Gard.* 77: 698–721.

Kress, W. J. (1995). Phylogeny of the Zingiberanae: morphology and molecules. Pp. 443–460 in Rudall, P. J., Cribb, P. J., Cutler, D. F. & Humphries, C. J. (eds.). *Monocotyledons: Systematics and Evolution.* Royal Botanic Gardens, Kew.

Les, D. H., & Haynes, R. R. (1995). Systematics of subclass Alismatidae: a synthesis of approaches. Pp. 353–377 in Rudall, P. J., Cribb, P. J., Cutler, D. F. & Humphries, C. J. (eds.). *Monocotyledons: Systematics and Evolution.* Royal Botanic Gardens, Kew.

Mayo, S. J., Bogner, J. & Boyce, P. (1995). The Arales. Pp. 277–286 in Rudall, P. J., Cribb, P. J., Cutler, D. F. & Humphries, C. J. (eds.). *Monocotyledons: Systematics and Evolution.* Royal Botanic Gardens, Kew.

Mayo, S. J., Bogner, J. & Boyce, P. C. (1997). *The Genera of Araceae.* Royal Botanic Gardens, Kew.

Parris, B. S., Beaman, R. S. & Beaman, J. H. (1992). *The Plants of Mount Kinabalu: 1. Ferns and Fern Allies.* Royal Botanic Gardens, Kew.

Putz, F. E., & Holbrook, N. M. (1986). Notes on the natural history of hemiepiphytes. *Selbyana* 9: 61–69.

Regis, P. (1996). The people and folklore of Kinabalu. Pp. 31–39 in Wong, K. M., & Phillipps, A., eds. *Kinabalu: Summit of Borneo, a revised and expanded edition.* The Sabah Society in association with Sabah Parks, Kota Kinabalu, Malaysia.

Rudall, P. J., Cribb, P. J., Cutler, D. F. & Humphries, C. J. (eds.) (1995). *Monocotyledons: Systematics and Evolution.* Royal Botanic Gardens, Kew. 750 pp.

Rudall, P. J., & Cutler, D. F. (1995). Asparagales: a reappraisal. Pp. 157–168 in Rudall, P. J., Cribb, P. J., Cutler, D. F. & Humphries, C. J. (eds.). *Monocotyledons: Systematics and Evolution*. Royal Botanic Gardens, Kew.

Rudall, P. J., Furness, C. A., Chase, M. W. & Fay, M. F. (1997). Microsporogenesis and pollen sulcus type in Asparagales (Lilianae). *Can. J. Bot.* 75: 408–430.

Simmonds, N. W. (1962). *The Evolution of the Bananas*. Longmans, Green and Co. Ltd. London.

Simpson, D. (1995). Relationships within Cyperales. Pp. 497–509 in Rudall, P. J., Cribb, P. J., Cutler, D. F. & Humphries, C. J. (eds.). *Monocotyledons: Systematics and Evolution*. Royal Botanic Gardens, Kew.

Stapf, O. (1894). On the flora of Mount Kinabalu, in North Borneo. *Trans. Linn. Soc. London, Bot.* 4: 69–263, pl. ll–20.

Steenis, C. G. G. J. (1954). Butomaceae. *Fl. Males.* I, 5: 118–120.

Steenis, C. G. G. J. (1964). Plant geography of the mountain flora of Mt Kinabalu. *Proc. Roy. Soc.*, Ser. B 161: 7–38.

Stevenson, D. W., & Loconte, H. (1995). Cladistic analysis of monocot families. Pp. 543–578 in Rudall, P. J., Cribb, P. J., Cutler, D. F. & Humphries, C. J. (eds.). *Monocotyledons: Systematics and Evolution*. Royal Botanic Gardens, Kew.

Takhtajan, A. (1980). Outline of the Classification of Flowering Plants (Magnoliophyta). *Bot. Rev.* 46: 225–359.

Thorne, R. F. (1976). A Phylogenetic Classification of the Angiospermae. *Evol. Biol.* 9: 35–106.

Tomlinson, P. B. (1995). Non-homology of vascular organisation in monocotyledons and dicotyledons. Pp. 589–622 in Rudall, P. J., Cribb, P. J., Cutler, D. F. & Humphries, C. J. (eds.). *Monocotyledons: Systematics and Evolution*. Royal Botanic Gardens, Kew.

Uhl, N. W., & Dransfield, J. (1987). *Genera Palmarum: a Classification of Palms Based on the Work of H. E. Moore, Jr.* International Palm Society and L. H. Bailey Hortorium, Lawrence, Kansas.

Uhl, N. W., Dransfield, J., Davis, J. I., Luckow, M. A., Hansen, K. S. & Doyle, J. J. (1995). Phylogenetic relationships among palms: cladistic analyses of morphological and chloroplast DNA restriction site variation. Pp. 623–661 in Rudall, P. J., Cribb, P. J., Cutler, D. F. & Humphries, C. J. (eds.). *Monocotyledons: Systematics and Evolution*. Royal Botanic Gardens, Kew.

Wilson, K. L. (1991). *Cyperus haspan*—spelling of the epithet. *Cyperaceae News-letter* 9: 8.

Wood, J. J., & Barkman, T. J. (In press). Notes on the orchid flora of Mount Kinabalu, Borneo. *Sandakania*.

Wood, J. J., Beaman, R. S. & Beaman, J. H. (1993). *The Plants of Mount Kinabalu: 2. Orchids*. Royal Botanic Gardens, Kew.

INDEX TO SCIENTIFIC NAMES

Accepted names are in roman type. Synonyms are in *italics*. Page numbers are indicated in **bold** where taxa are formally enumerated.

PLATE 1.

A. Mixed upper montane forest of *Dacrydium gibbsiae* (Podocarpaceae) and *Leptospermum recurvum* (Myrtaceae). *Dacrydium* is in the foreground, especially at left. Between Layang-layang and Paka-paka Cave on ultramafic substrate. Photo: A. Lamb.

B. *Agathis dammara* (Araucariaceae). Near summit of Hempuen Hill on ultramafic substrate. Photo: J. H. Beaman.

PLATE 2.

A. *Agathis kinabaluensis* (Araucariaceae). Upper end of Kiau View Trail. Photo: J. H. Beaman.

B. *Agathis kinabaluensis* (Araucariaceae). Trunk of young tree. Upper end of Kiau View Trail. Photo: J. H. Beaman.

C. *Agathis lenticula* (Araucariaceae). Park Headquarters. Photo: J. H. Beaman

PLATE 3.

A. *Agathis lenticula* (Araucariaceae). Female cones. Mount Kinabalu. Photo: A. Lamb.
B. *Phyllocladus hypophyllus* (Phyllocladaceae). The structures that look like leaves are technically cladophylls (modified stems). Park Headquarters. Photo: J. H. Beaman.
C. *Dacrycarpus imbricatus* var. *patulus* (Podocarpaceae). Park Headquarters. Photo: J. H. Beaman.

PLATE 4.

A. *Dacrycarpus kinabaluensis* (Podocarpacae). Between Layang-layang and Paka-paka Cave on ultramafic substrate. Photo: J. H. Beaman.

B. *Dacrydium gibbsiae* (Podocarpaceae). Between Layang-layang and Paka-paka Cave on ultramafic substrate. Photo: J. H. Beaman.

C. *Dacrydium gracilis* (Podocarpaceae). Park Headquarters. Photo: J. H. Beaman.

D. *Dacrydium pectinatum* (Podocarpaceae). Near summit of Hempuen Hill on ultramafic substrate. Photo: J. H. Beaman.

E. *Dacrydium xanthandrum* (Podocarpaceae). Summit Trail near Mempening Shelter. Photo: J. H. Beaman.

F. *Falcatifolium falciforme* (Podocarpaceae). Park Headquarters. Photo: J. H. Beaman.

PLATE 5.

A. *Podocarpus brevifolius* (Podocarpaceae). Between Layang-layang and Paka-paka Cave on ultramafic substrate. Photo: J. H. Beaman.

B. *Podocarpus neriifolius* (Podocarpaceae). Near summit of Hempuen Hill on ultramafic substrate. Photo: J. H. Beaman.

C. *Gnetum leptostachyum* var. *leptostachyum* (Gnetaceae). Hempuen Hill. Photo: A. Lamb.

D. *Aglaonema simplex* (Araceae). Maliau Basin. Photo: Au Kam Wah

E. *Alocasia cuprea* (Araceae). Tenom Orchid Centre. Photo: S. P. Lim.

F. *Alocasia cuprea* (Araceae). Poring. Photo: C. L. Chan.

PLATE 6.

A. *Alocasia longiloba* (Araceae). Sepilok Forest Reserve. Photo: C. L. Chan.
B. *Alocasia macrorrhizos* (Araceae). Tenom. Photo: C. L. Chan.
C. *Amorphophallus lambii* (Araceae). Tenom. Photo: A. Lamb.

PLATE 7.

A. *Amydrium medium* (Araceae). Brunei. Photo: A. Lamb.
B. *Arisaema filiforme* (Araceae). Liwagu River Trail. Photo: A. Lamb.
C. *Arisaema laminatum* (Araceae). Lugas Hill south of Bundu Tuhan. Photo: R. S. Beaman.
D. *Arisaema umbrinum* (Araceae). Melangkap Kapa. Photo: R. S. Beaman.

PLATE 8.

A. *Piptospatha elongata* (Araceae). Poring. Photo: A. Lamb.
B. *Rhaphidophora korthalsii* (Araceae). Juvenile foliage. Sepilok Forest Reserve. Photo: C. L. Chan.
C. *Rhaphidophora korthalsii* (Araceae). Mature foliage. Sepilok Forest Reserve. Photo: C. L. Chan.
D. *Schismatoglottis calyptrata* (Araceae). Mt. Matang, Sarawak. Photo: P. C. Boyce.

PLATE 9.

Scindapsus pictus (Araceae). Juvenile foliage. Sepilok Forest Reserve. Photo: C. L. Chan.
Areca catechu (Arecaceae). Poring. Photo: J. Dransfield.
Areca kinabaluensis (Arecaceae). Park Headquarters. Photo: J. Dransfield.
Arenga undulatifolia (Arecaceae). Poring. Photo: J. H. Beaman.

PLATE 10.

A. *Arenga undulatifolia* (Arecaceae). Poring. Photo: J. Dransfield.
B. *Calamus gibbsianus* (Arecaceae). Marai Parai. Photo: J. Dransfield.
C. *Calamus ornatus* (Arecaceae). Poring Canopy Walkway. Photo: J. H. Beaman.
D. *Caryota no* (Arecaceae). Poring. Photo: J. Dransfield.

PLATE 11.

A. *Daemonorops fissa* (Arecaceae). Poring. Photo: J. H. Beaman.
B. *Daemonorops longistipes* (Arecaceae). Male inflorescence. Park Headquarters. Photo: J. H. Beaman.
C. *Eugeissona utilis* (Arecaceae). Kampung Kilimu. Photo: J. H. Beaman.
D. *Korthalsia echinometra* (Arecaceae). Poring. Photo: J. H. Beaman.

PLATE 12.

A. *Korthalsia robusta* (Arecaceae). Poring. Photo: J. H. Beaman.
B. *Licuala valida* (Arecaceae) Kampung Kilimu. Photo: J. H. Beaman.
C. *Pinanga aristata* (Arecaceae). South Kalimantan. Photo: J. Dransfield.
D. *Pinanga capitata* (Arecaceae). Kemburongoh. Photo: J. Dransfield.
E. *Pinanga tenella* (Arecaceae). Belalong, Brunei. Photo: J. Dransfield.
F. *Salacca clemensiana* (Arecaceae). Crocker Range National Park. Photo: J. Dransfield.

PLATE 13.

A. *Salacca dolicholepis* (Arecaceae). Perhaps the one endemic palm on Mount Kinabalu. Park Headquarters. Photo: J. H. Beaman.
B. *Thismia episcopalis* (Burmanniaceae). Mt. Matang, Sarawak. Photo: J. Dransfield.
C. *Centrolepis philippinensis* (Centrolcpidaceae). Summit Area. Photo: J. Dransfield.
D. *Amischotolype mollissima* (Commelinaceae). Mt. Matang, Sarawak. Photo: J. Dransfield.
E. *Costus globosus* (Costaceae). Crocker Range National Park. Photo: A. Lamb.

PLATE 14.

A. *Carex filicina* (Cyperaceae). Panar Laban. Photo: C. L. Chan.
B. *Cyperus compactus* (Cyperaceae). Thailand. Photo: D. A. Simpson.

PLATE 15.

A. *Gahnia javanica* (Cyperaceae). Kiau View Trail. Photo: D. A. Simpson.
B. *Gahnia javanica* (Cyperaceae). Summit Trail. Photo: C. L. Chan.
C. *Hypolytrum nemorum* (Cyperaceae). Peninsular Malaysia. Photo: D. A. Simpson.
D. *Kyllinga nemoralis* (Cyperaceae). Thailand. Photo: D. A. Simpson.
E. *Machaerina falcata* (Cyperaceae). Between Layang-layang and Paka-paka Cave on ultramafic substrate. Photo: C. L. Chan.
F. *Machaerina glomerata* (Cyperaceae). Mt. Silam on ultramafic substrate. Photo: J. Dransfield.

PLATE 16.

A. *Mapania palustris* (Cyperaceae). Liwagu River Trail. Photo: D. A. Simpson.
B. *Mapania palustris* (Cyperaceae). Mountain Garden, Park Headquarters. Photo: C. L. Chan.
C. *Hanguana major* (Hanguanaceae). Mount Kinabalu. Photo: A. Lamb.
D. *Curculigo latifolia* (Hypoxidaceae). Mountain Garden, Park Headquarters. Photo: C. L. Chan.

PLATE 17.

A. *Patersonia lowii* (Iridaceae). Marai Parai. Photo: J. Dransfield.
B. *Joinvillea ascendens* subsp. *borneensis* (Joinvilleaceae). Crocker Range National Park. Photo: C. L. Chan.
C. *Aletris foliolosa* (Melanthiaceae). Mount Kinabalu. Photo: K. M. Wong.
D. *Petrosavia stellaris* (Melanthiaceae). Marai Parai. Photo: J. Dransfield.

PLATE 18.

A. *Musa* sp. (Musaceae). Not yet formally described. Mountain Garden, Park Headquarters. Photo: C L. Chan.

B. *Freycinetia javanica* (Pandanaceae). Lohan River. Photo: A. Lamb.

C. *Dianella ensifolia* (Phormiaceae). Park Headquarters. Photo: C. L. Chan.

D. *Freycinetia kinabaluana* (Pandanaceae). Ular Hill Trail. Photo: A. Phillipps.

PLATE 19.

A. *Dianella javanica* (Phormiaceae). Summit Trail. Photo: C. L. Chan.
B. *Dinochloa sublaevigata* (Poaceae). Poring. Photo: C. L. Chan.
C. *Gigantochloa levis* (Poaceae). Poring. Photo: A. Phillipps.
D. *Miscanthus floridulus* var. *malayanus* (Poaceae). Near the Power Station. Photo: C. L. Chan.

PLATE 20.

A. *Racemobambos gibbsiae* (Poaceae). Summit Trail near Mempening Shelter. Photo: J. Dransfield.
B. *Racemobambos rigidifolia* (Poaceae). Marai Parai. Photo: J. Dransfield.
C. *Racemobambos rigidifolia* (Poaceae). Marai Parai. Photo: J. Dransfield.

PLATE 21.

A. *Smilax* sp. (Smilacaceae). The plant cannot be identified to species without leaves, but this is an excellent illustration of an umbel of female flowers of the genus. Upper Silau-silau Trail near Park Headquarters. Photo: C. L. Chan.

B. *Trisetum spicatum* subsp. *kinabaluense* (Poaceae). Panar Laban. Photo: C. L. Chan.

C. *Tacca integrifolia* (Taccaceae). Mt. Matang, Sarawak. Photo: P. C. Boyce.

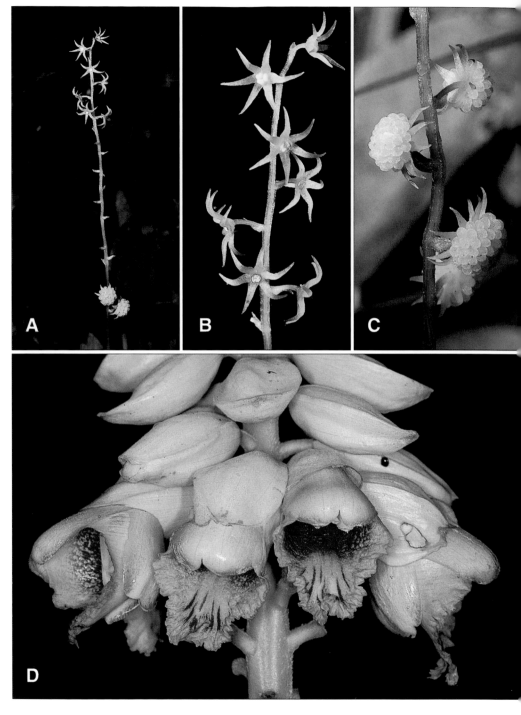

PLATE 22.

A. *Sciaphila secundiflora* (Triuridaceae). Plant habit. Bako National Park, Sarawak. Photo: J Dransfield.

B. *Sciaphila secundiflora* (Triuridaceae). Male flowers. Bako National Park, Sarawak. Photo: J Dransfield.

C. *Sciaphila secundiflora* (Triuridaceae). Female flowers. Bako National Park, Sarawak. Photo: J Dransfield.

D. *Alpinia latilabris* (Zingiberaceae). Plant grown in garden from Crocker Range National Park. Photo A. Lamb.

PLATE 23.

A. *Alpinia havilandii* (Zingiberaceae). Poring. Photo: A. Lamb.
B. *Amomum oliganthum* (Zingiberaceae). Tenom Orchid Centre. Photo: A. Lamb.
C. *Amomum sceletescens* (Zingiberaceae). Crocker Range National Park. Photo: A. Lamb.
D. *Etlingera cf. metriocheilos* (Zingiberaceae). Lugas Hill south of Bundu Tuhan. Photo: R. S. Beaman.
E. *Globba tricolor* var. *gibbsiae* (Zingiberaceae). Mt. Trus Madi. Photo: A. Lamb.

PLATE 24.

A. *Etlingera fimbriobracteata* (Zingiberaceae). Tenom. Photo: A. Lamb.
B. *Etlingera punicea* (Zingiberaceae). Between Kiau and the Haye-Haye River. Photo: A. Lamb.

PLATE 25.

A. *Hedychium cylindricum* (Zingiberaceae). Park Headquarters. Photo: C. L. Chan.
B. *Hornstedtia gracilis* (Zingiberaceae). Park Headquarters. Photo: A. Lamb.
C. *Hornstedtia incana* (Zingiberaceae). Mesilau River. Photo: A. Lamb.
D. *Zingiber coloratum* (Zingiberaceae). Tenom Orchid Centre. Photo: A. Lamb.